AF527768

Microbes and Sustainable Agriculture

Microbes and Sustainable Agriculture

Editor
Ram Prasad
Assistant Professor
Amity Institute of Microbial Technology
Amity University
Noida (Uttar Pradesh)

Co-editor
Narendra Kumar
Assistant Professor
Department of Biotechnology
IMSEC, Ghaziabad (U.P.)

I.K. International Publishing House Pvt. Ltd.
NEW DELHI

Published by
I.K. International Publishing House Pvt. Ltd.
S-25, Green Park Extension
Uphaar Cinema Market
New Delhi–110 016 (india)
E-mail: info@ikinternational.com
Website: www.ikbooks.com

ISBN: 978-93-85909-48-1

Published by Krishan Makhijani for I.K. International Publishing House Pvt. Ltd., S-25, Green Park Extension, Uphaar Cinema Market, New Delhi–110 016 and Printed by Rekha Printers Pvt. Ltd., Okhla industrial Area, Phase II, New Delhi–110 020.

Preface

Microbes are everywhere, all over the world and in every imaginable environment. For example, just one gram of soil contains billions of microorganisms and all their associated activities. This book covers recent outbreaks, highlighting the major role of microbes in agriculture sustainability. It also examines the role of bacteria, cyanobacteria and fungi in diverse facets of agriculture to preserve the sustainability including arbuscular mycorrhizal association, microbial biopesticides, heavy metal resistant bacterial strains, herbicides for weed management, biochar and its role in soil fertility management and plant disease management. The authors explore the relationship between microbial use in different agricultural practices like use of biofertilizers, pesticides, for the reclamation & restoration of soil and plant disease management.

This book is designed for undergraduate, postgraduate and research students working or interested in agriculture microbiology and to serve as a reference for scientists, professionals and engineers in this field. The overall objectives of the text are to define the important role of microbes in agricultural sustainability. This book represents a joint effort led by the editors and authors who have diverse yet complementary backgrounds in agricultural microbiology. The authors are actively involved in agricultural microbiology or related areas. They have worked together extensively on a variety of practical problems using advanced, interdisciplinary approaches. We thank all the contributors who spared their precious time, thoughts and expertise in the writing of the chapters and hope that their contribution will stimulate further high-quality teaching and research. As a result, we have particularly enjoyed editing the book, as the opportunity to do so has provided us stimulating cooperation with an in-depth opportunity to reflect on the important contributions. We are grateful to all the contributors for their work, which so vividly shows the new advances that can occur in what many might consider an established field, unlikely to offer new horizons.

Ram Prasad
Narendra Kumar

Contents

1

Arbuscular Mycorrhizal Association Improves Secondary Metabolite Production in Medicinal and Aromatic Plants: A Sustainable Approach

Bhawna Saxena and Bhoopander Giri[1*]

Department of Botany, Swami Shraddhanand College
University of Delhi, Delhi

[1]Present address: Department of Plant Pathology
North Carolina State University, Raleigh NC
United States *Email: *bhoopg@yahoo.com*

1. INTRODUCTION

Metabolites in plants can be grouped under two broad classes, primary and secondary. Primary metabolites have a role in the basic processes of life, such as photosynthesis and respiration while secondary metabolites provide phytochemicals, defense and specialized compounds required for interaction of plant with its environment. These chemical compounds have important role in defense, attraction and protection of plants from both biotic and abiotic stresses (Kliebenstein, 2012).

Secondary metabolite refers to those compounds, which are produced by the plant, and are not essential for its growth and development. Earlier they were thought to be wasteful products, but since the time their importance has been realized in medicines, pesticides, dyes, cosmetics, neutraceuticals and other commercial products (Bourgaud et al., 2001), plants are considered chemical factories for their production (Zhi-lin et al., 2007). However, their production thought to be costly, consuming 10 percent of the total energy of the plant, and thereby, affecting the plant growth and biomass production.

Arbuscular mycorrhizal fungi (AMF) establish symbiotic association with more than 80 percent of the vascular plant species (Smith et al., 2010; Zolfaghari et al., 2013). This 450 million years old symbiotic relationship is of great ecological importance, as it increases the plant productivity by extending the root-fungal penetration zone in the soil and increasing nutrient transfer. Medicinal and aromatic plants are the source of various phytochemicals having high healing properties with fewer side effects, therefore, they have gained importance in treating several mammalian diseases. According to a survey conducted by WHO, near about 80 percent populations of developing countries rely on the traditional medicines (Soetan and Ayelaagbe, 1998).

Research studies indicate that more than 19 families with 50 species of medicinal plants have been benefitted by Arbuscular Mycorrhizal (AM) association. This association not only enhances the production, but also improves the quality of herbal material (Zeng et al., 2013). Researchers have highlighted the importance of the AM association for the production of important secondary metabolites which are the organic compounds like terpenoids (Akiyama and Hayashi 2002; Gupta et al., 2002), phenolics (Jurkiewicz et al., 2010; Rosa-Mera et al., 2011) and alkaloids (Zhou and Fan 2007; Yu et al., 2010; Rosa-Mera et al., 2011). In this article, we attempt to summarize the relationship between arbuscular mycorrhizal association and secondary metabolites accumulation in medicinal and aromatic plants, and also consider that this knowledge will improve the quality of useful ingredients of medicinal herbs for the sustainable use of mankind.

2. PLANT SECONDARY METABOLITES

Plants produce a diverse array of secondary metabolites. Till now 1,20,000 of them had been identified as well. Though, they have no direct role in growth and development related processes of plants, but appear to be multifunctional in relation to their adaptation to various biotic and abiotic factors. They have protective role against predators and pathogens, repellence to herbivores, resistance to abiotic stresses like UV damage, osmotic stress, attraction to pollinators, and also have role in horizontal communication with other soil microbes.

In addition to their physiological role in plants, these compounds have a severe impact on human survival, and appear to be an excellent reservoir having medicinal and pharmacological properties. Secondary metabolites with antioxidant, anticancer, antifungal, antibacterial and laxative properties have been studied from the members of plant families like Lamiaceae (Gupta et al., 2002; Sailo and Bagyaraj, 2005; Copetta et al., 2006; Lee and Seagel, 2009), Solanaceae (Wei and Wang, 1989), Liliaceae (Gong et al., 2002; Pan et al., 2008; Zhou et al., 2009; 2010), Asteraceae (Lu and He, 2005; Lu et al., 2008a; 2008b; Araim et al., 2009; Zubek et al., 2010), Apocynaceae (Rosa-Mera et al., 2011), Apiaceae (Umbelliferae) (Kapoor et al., 2002a, b, 2004; Zhao et al., 2009) and many more. Due to restricted distribution of these compounds to certain families and species, they can also be used in the taxonomic classification (Bennett and Wallsgrove, 1994).

Up till now, merely 30 percent of the plants have been analyzed for their presence, and reports suggest that each plant produces more than 20,000 different secondary metabolites in its lifespan. Beside this, some are expected to be present at very low-level amount, beyond the point of their detection and identification. These low molecular weight compounds are often accumulated at a concentration of 1–3 percent of the dry biomass. Once they are isolated by using a suitable solvent (n-hexane, chloroform, ethyl acetate, methanol, ethanol), their structure can be analyzed through X-ray diffraction,

NMR or mass spectroscopy techniques, like EI-MS, CI-MS, ESI-LC-MS (Eisenreich and Bacher, 2007; Marston, 2007). However, about 30 percent of the photosynthate is directed towards the biosynthesis of secondary metabolites. The precursors are derived from the glycolysis, Kreb cycle or shikimate pathway.

The distribution pattern of secondary metabolites is also not uniform and it may vary with the stage of growth and plant tissues. Some secondary metabolites are abundant in young tissues (which need more protection from the environment) and decreases with age, while others are restricted to a specific organ of plant. Besides, the co-evolution of plant-pathogen/herbivore has also led to the compartmentalization of these secondary metabolites to specialized tissues such as lacticifers, glandular trichomes, resin ducts, or cuticular waxes for lipophilic compounds and vacuole for hydrophilic compounds (Gershenson, 1994). This sequestration might decrease the development of resistance against these phytochemicals in the herbivores and pathogens (Shelton, 2004, 2005).

3. CLASSIFICATION OF PLANT SECONDARY METABOLITES

Plant secondary metabolites are usually classified according to their biosynthetic pathways into three chemically distinct groups like **phenolics**, **terpenes** and **alkaloids**.

3.1 Phenolic Compounds

These compounds had played an important role in their evolutionary adaptation to land. Besides, being an important constituent of cell wall, they also have important role in defense against harmful microbes and herbivores, impart color and odour for successful pollination and provide flavor to various plant products. They are characterized by the presence of hydroxyl (-OH) to benzene ring or to other aromatic rings. On the basis of their structure, they can be classified into two broad classes, i.e., (a) non-flavanoid containing phenolics and (b) flavanoid containing phenolics.

3.1.1 Non- Flavanoid Containing Phenolics

Simple phenols (C_6-C_1) mainly include hydroxyl-benzoic acid, vanillic acid and protocatechuic acid. In grains, they are present in bound form, while fruits and vegetables contain phenolics in free form. Gallic acid is the main component of simple phenolics and is formed through shikimate pathway. It further gets converted into ellagic acid, which upon modifications with sugar components yield a range of hydroziable tannins and lignins. Tannins have the ability to precipitate animal protein and thus are used to convert rawhides to leather. Tannins are also known to bind the salivary protein, thus providing the astringency taste. They had also been reported to have anti-oxidant and anti-inflammatory effect on cancerous cells (Liu, 2004; Fresco et al., 2004).

Hydroxycinnamate is generated by the deamination of amino acid phenylalanine, which through further hydroxylation produces coumaric acid. It mainly includes ferrulic acid, caffeic acid and sinapic acid. Coumarins are often associated with anti-oxidant properties and thus have anti-cancer effects on mammalian cells (Liu, 2004; Fresco et al., 2004). Curcumin present in *curcuma longa* is effective against all the three stages of carcinogenesis (Thangapazham et al., 2006). Polyphenolics (C6-C2-C6), namely, stilbene is mainly the phytoalexins produced in response to bacterial, fungal and viral attack. Reversetol, the main phenolic in red wine, is responsible for its cardio-protective effect.

3.1.2 *Flavanoid Containing Phenolics*

Flavanoids are polyphenolic compounds with two aromatic rings joined by a 3 carbon bridge. They are found in high concentration in leaf epidermis and fruit skin. They play a role in pigmentation, pathogen resistance and UV protection. The basic flavanoid molecule may have different substitution on it, such as of hydroxyl group, sugars, methylated, farnesylated and acetylated compounds. Depending on the modifications of central carbon ring, they can be divided into different structural classes including flavanols, flavanones, isoflavanoids and anthocyanidins. Those with sugar conjugates are hydrophilic in nature, while methylation confers lipophilicity to them. Their structure and glycosylation pattern determine antioxidant property. The isoflavones present in leguminous plants reduce blood cholesterol level, blood pressure, oxidative stress and also prevent osteoporosis and post-menopausal disorders (Anderson et al., 1999). Table 1.1 shows the medicinal plants and applications of phenolic compounds.

Table 1.1: Medicinal applications of phenolic compounds

Phenolics	Role in Medicine	Family	Source Plant	Reference
Apiole	Diuretic	Apiaceae	*Petroselinum crispum*	Louli et al., 2004
Galangin	Antibacterial activity	Zingiberaeceae	*Alpina officinarum*	Cushnie et al., 2003
Coumestrol	Phytoestrogen activity	Fabaceae	*Medicago sativa*	Cornwell et al., 2004
Silybin	Liver disorder	Asteraceae	*Silybum marianum*	Sridar et al., 2004
Genistein	Anticancer properties	Fabaceae	*Glycine max*	Kaufman et al., 1997
Gossypol	Contraceptive activity	Malvaceae	*Gossypium* sp.	Han et al., 2007
Proanthocyanidins	Antioxidant activity, anticancerous	Vitaceae	*Vitis* sp.	Huh et al., 2004
Curcumin	Anti-inflammatory, anti-ulcer, hepato-protective properties	Zingiberaceae	*Curcuma longa*	Ammom and Wahl, 1990
Benzoin	Stimulant, expectorant	Styraceae	*Styrax* spp.	Daniel, 2006

Coumarin	Astringent, improves digestion	Rutaceae	*Aegle marmelos*	Daniel, 2006
Apocynin	Antiasthamatic	Scrophulariaceae	*Picrorhiza kurroa*	Dorsch et al., 1994
Chicoric acid	Inhibit HIV integrase, antioxidant property	Asteraceae	*Echinacea purpurea*	Charvat et al., 2006

3.2 Terpenoids (Terpenes)

They are derived from the fusion of five-carbon (C_5) isoprene units based on iso-pentane skeleton, and are commonly known as isoprenoids. Terpenoids constitute the largest group of bioactive phytochemicals, accounting for 40,000 primary and secondary metabolites (Table 1.2). Their nature is generally lipophilic and is found mainly in the glandular trichomes of leaf, resins and bud exudates. In plants, terpenoids are generated upon the condensation of two interconvertible precursors, isopentenyl pyrophosphate (IPP) and dimethyl allyl pyrophosphate (DMAPP), through two different pathways; (a) mevelonate or MEV pathway, which occurs in cytosol and (b) methyl-erythritol phosphate pathway or MEP, which is a plastid bound pathway. On the basis of number of isoprene units present, they are grouped into:

1. Monoterpene, with two isoprene units commonly includes menthol, camphor, lavandulol, geraniol and is commercially important in perfume industry as well as in aromatherapies. *Eucalyptus* oil has been reported to be effective in reducing inflammation and pain.
2. Sesquiterpene, with three isoprene units includes farnesol, farnesene, humulene. A sesquiterpene lactone, artemisinin obtained from *Artemisia annua*, is currently used as an antimalarial drug. Its various derivatives are under clinical trial.
3. Diterpenoids, with four isoprene units are mainly derived from geranyl geranyl pyrophosphate. It includes mainly resins and are mostly toxic in nature, but some such as taxol and ginkgoloides, are used in cancer and memory loss treatment (Heinrich et al., 2004; Gurib-Fakim et al., 2005). Diterpenes are also involved in the synthesis of biologically active compounds, retinol, retinal and phytol.

Table 1.2: Medicinal applications of terpenoids

Terpenoid	Role in Medicine	Family	Source Plant	Reference
Carvone	Antifungal	Apiaceae	*Carum carvi*	McGeady et al., 2002
Eucalyptol	Leukemia cell death	Myrtaceae	*Eucalyptus globulus*	Moteki et al., 2002
Helenalin	Anti trypanosomal activity	Asteraceae	*Arnica montana*	Schmidt et al., 2002
Clerodane	Antifeedant activity	Lamiaceae	*Salvia* sp.	Kumari et al., 2003
Glycyrrhizic acid	Expectorant	Fabaceae	*Glycyrrhiza glabra*	Fenwick et al., 1990

4. Triterpenes, with six isoprene units are the most abundant among the plant terpenes and mainly include steroids.
5. Tetraterpenoids, with eight isoprene units include carotenoid (yellow-orange plant pigment), lycopene (red pigment of tomato).
6. Polyterpenes with many isoprene units also occur in nature with double bond in cis-conformation as in natural rubber or with double bonds in trans-conformation as in gutta-percha.

3.3 Alkaloids

Alkaloids constitute a class of naturally occurring cyclic and nitrogen containing compounds. Till date, about twelve thousand of them have been isolated from 150 families of land plants. The basic nature of compounds with one or more hetrocyclic nitrogen atoms are derived either from amino acids (such as tryptophan, tyrosine, phenylalanine, lysine and arginine) alone or in combination with other compounds (such as terpenoids, purine, polyketides). They generally occur as salts of organic acids (e.g., atropine, cocaine), while a few with weak base (e.g., nicotine) are found freely in nature.

They have been utilized by humans for centuries and are important even today. Morphine was the first crude alkaloid isolated from *Papaver somniferum* latex and used till date as an analgesic in pain. *Catharanthus rosesus* (family apocynaceae) has been reported to produce anticancer vinblastine, vincristine, vinorelbine and vinflunine. The alkaloids are used as poisons, narcotic, stimulant and in other medicines since 400 B.C. Due to their complex structure, alkaloids are difficult to synthesize, thus they are usually extracted from the plant tissues. Plants often produce alkaloids to defend themselves against pests, diseases, and other external biological stimuli (Zeng et al., 2013). Some of the commonly used alkaloids in medicines are listed in Table 1.3. Most of them are infamous due to their toxic effects on animal system. Saponins make the cell membrane leaky and phalloidin obtained from *Amanita phalloides,* has been reported to be poisonous to humans. Berberine acts as a potent mutagen by intercalating between the stacks of nucleotides in DNA helix. Some alkaloids like emetine interact with nucleic acids and inhibit the ribosomal protein biosynthesis. These cytotoxic effects are not only limited to animals, but are also effective on microbial system, thus defending the plants against its enemies. However, they can also become toxic to plants, thus plants have evolved the mechanism of compartmentalization and target site inactivation as a possible mechanism to defend them.

Instead plant kingdom, alkaloids are produced by several species of amphibians, arthropods, mammals, bacteria, fungi and few marine animals, while most of the insects use plant alkaloids as a source of attractant and pheromones. However, some amphibians, such as *Bufo* toad have been reported to secrete a variety of toxic alkaloids for protection from their enemies (Daly and Spande, 1986).

Table 1.3: Medicinal applications of alkaloids

Alkaloid	Role in Medicine	Source Plant	Family	Reference
Atropine	Antidote to nerve gas poisoning	*Atropa belladonna*	Solanaceae	Icon health Publisher, 2004
Morphine	Pain killer	*Papaver somniferum*	Papavereaceae	Bercovitch et al., 1999
Caffeine	Stimulant for CNS	Tea	Theaceae	Weinberg and Bealer, 2002
Quinine	Antimalarial drug	*Chinchona officinalis*	Rubiaceae	Honigsbaum, 2002
Isovoacangine	Apoptosis of human leukemia cell line THP-1	*Tabernaemontana salzmannii*	Apocynaceae	Figueiredo et al., 2010
Vincristine and Vinblastine	Antitumor	*Catharanthus roseus*	Apocynaceae	Salmon and Sartorelli, 1989
Reserpine	Antihypertensive	*Rauwofia serpentine*	Apocynaceae	Duke, 1992a,b
Colchicine	Anticancerous	*Colchicum autumnale*	Liliaceae	Jordon, 2002
Berberine	Antibacterial	*Berberis*	Berberidaceae	Birdsall and Kelly, 1997
Hyoscine	Travel sickness	*Atropa*	Solanaceae	Golding and Stott, 1997
Papaverine	Anti-impotence/ vasodialator	*Papaver*	Papaveraceae	Kalsi et al., 2002
Calystegine	Diabetes	*Morus*	Moraceae	Chen et al., 1995
Cocaine	Local anesthesia	*Coca*	Erythroxylaceae	Ruetsch et al., 2001

4. HOW DOES SECONDARY METABOLITE SYNTHESIS OCCUR IN PLANT?

The building blocks for a large number of secondary metabolites are derived from three basic processes in plants, i.e., photosynthesis, glycolysis and krebs cycle (Fig. 1.1). The precursor molecules for secondary metabolite biosynthesis are acetyl CoA, mevalonate, and shikimate. Acetyl CoA is formed from the oxidative decarboxylation of pyruvate, the end product of glycolysis. It is involved in the biosynthesis of a variety of phenolic compounds in plants. Three molecules of acetyl CoA join to form one molecule of mevalonate. Mevalonate along with deoxyxylulose pathway generates array of terpenoids in plants. The phosphoenol pyruvate (PEP), a glycolytic intermediate combines with erythrose 4-phosphate to form shikimic acid. Shikimic acid acts as a precursor for the biosynthesis of a variety of alkaloids and phenolic metabolite. The precursors undergo a variety of alkylation, decarboxylation, Mannich reaction, oxidation-reduction, glycosylation and transamination reaction to generate a diverse array of secondary metabolites (Dewick, 2002).

Their biosynthesis processes are either compartmentalized to cytosol (hydrophilic compounds), chloroplast (caffeine), mitochondria (amines, alkaloids), vacuole or endoplasmic reticulum (lipophilic compounds). Once they are synthesized they are sequestered either in the vacuole (hydrophilic compounds) or in lacticifers, glandular hairs and resin ducts (lipophillic compound).

4.1 Biosynthesis of Phenolic Compounds

Phenolic compounds alone account for approximately 40 percent of the organic molecules in the biosphere. In plants, they are derived mainly through phenyl-propanoid and phenyl-propanoid acetate pathway. While a few phenolic compounds, such as hydrolysable tannins are the condensation products of carbohydrate with gallic acid (shikimate pathway derivative).

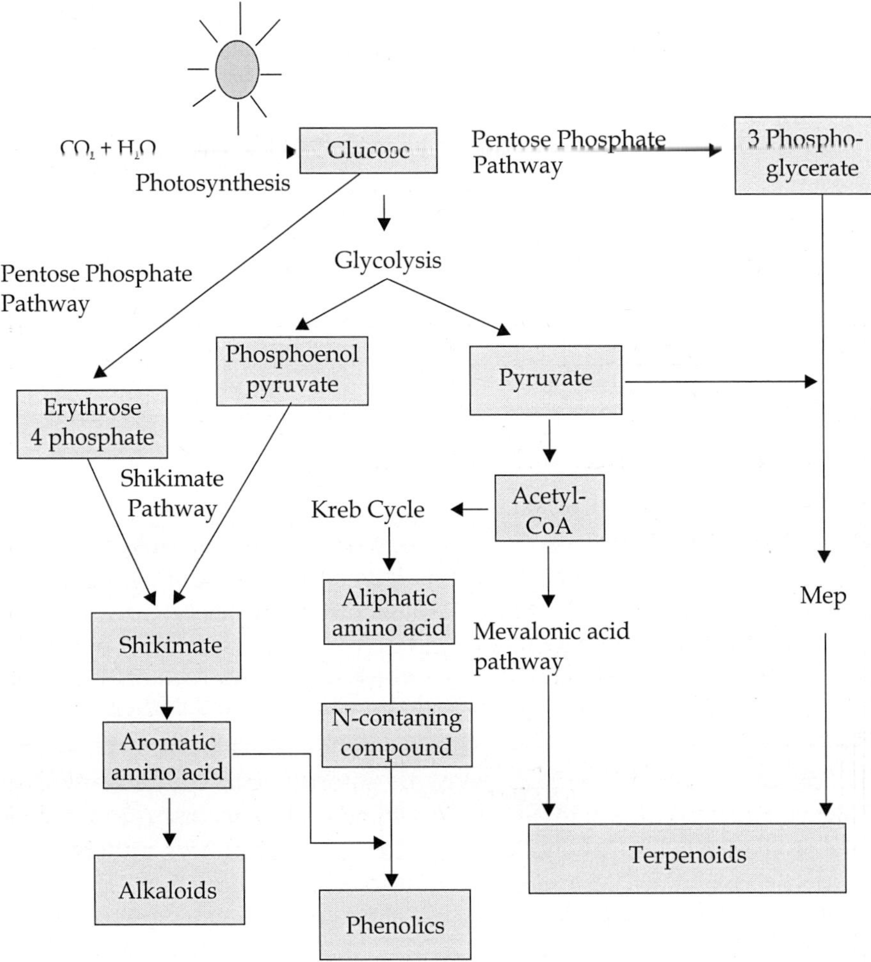

Fig. 1.1 Biosynthesis of all the three secondary metabolites

The biosynthesis pathway begins with the conversion of phenylalanine to cinnamic acid by the elimination of an ammonia molecule in the presence of enzyme phenylalanine ammonia lyase (PAL). Phenylalanine is derived from the shikimate (Fig. 1.2). Shikimate upon phosphorylation and condensation with phosphoenolpyruvate (PEP) yields 5-enolpyruvate shikimate 3-phosphate (EPSP), which upon dephosphorylation yields chorsimate, the phenylalanine precursor molecule.

Phenylalanine ammonia lyase (PAL), a homotetrameric protein, catalyses the removal of ammonia and adds a trans double bond between C7 and C8 position (Calabrese et al., 2004). PAL is the most widely studied enzyme of secondary metabolism present in a number of dicots while in monocots tyrosine ammonia lyase (TAL) predominates. The presence of PAL has also been reported in some aquatic plants, where they were thought to be involved in flavanoid biosynthesis. PAL does not require any cofactor for its activity, while pathogen attack, low temperature, light and hormone influence the activity of this enzyme (Hahlbrock and Grisebach, 1979).

Cinnamate once formed, undergoes a series of hydroxylation and methylation steps to generate four different hydroxy cinnamates, i.e., 4-coumarate, caffeate, ferulate and sinapate. Activation of these hydroxy cinnamate is carried out by the enzyme hydroxycinnamate-CoA ligase. Medicinally important coumarin glycosides (dicumarol or warfarin) are basically synthesized from O-hydroxylation of cinnamic acid or p-coumaric acid.

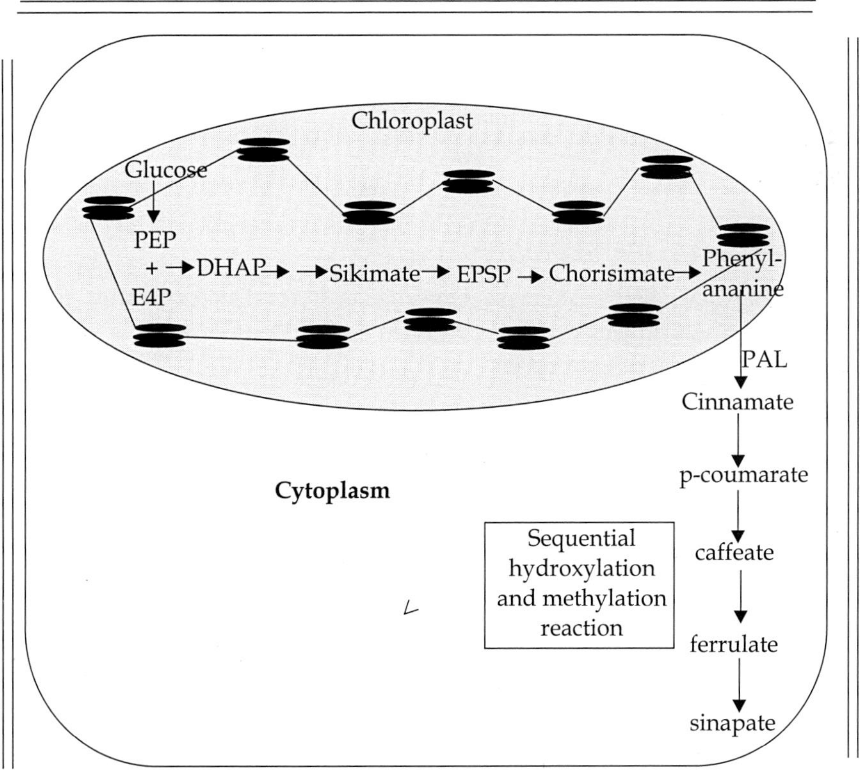

Fig. 1.2 Biosynthesis of hydroxycinnamate derivates through phenylpropanoid pathway

4.2 Biosynthesis of Terpenoids

As mentioned above in this chapter, terpenoids are derived from the head to tail condensation of a C5 compound—isoprene. Isoprene is generated through phosphorylation of mevalonic acid. The whole terpenoid biosynthesis (Fig. 1.3) pathway can be divided into four stages.

4.2.1 *Isopentenyl Pyro Phosphate (IPP) Biosynthesis*

There are altogether two pathways that lead to IPP biosynthesis in plants: (i) mevalonate pathway, and (ii) non-mevalonate pathway. Mevalonate pathway involves condensation of three acetyl-CoA molecules to generate 3-hydroxy-3-methyl glutaryl coenzyme A (HMG-CoA), which upon reduction yields mevalonate. The reduction of HMG-CoA to mevalonate is the rate limiting step of IPP biosynthesis and is catalyzed by HMG-CoA reductase (HMGR). HMGR has thus become a priority target of gene manipulation in plants. Overexpression of the truncated or full length HMGR from hetrologous sources has been reported to enhance the terpenoid biosynthesis in plants (Chappell et al., 1995; Harker et al., 2003). Mevalonate once formed, undergoes phosphorylation and decarboxylation step to generate IPP.

In non-mevalonate pathway, commonly known as MEP or 2-methyl 3-erythritol pathway, the first step is the transketolose condensation between pyruvate and glyceraldehyde 3-phosphate. This condensation results in the generation of deoxy-xylulose 5-phosphate in plants (DXP). Deoxy-xylulose 5-phosphate reductoisomerase (DXR) converts DXP to MEP. The overexpression of DXR in tobacco plants from the chloroplast genome has elevated levels of sterols. Then through series of phosphorylation and rearrangement steps MEP results in the generation of IPP.

4.2.2 *Isomerization of IPP to DMAPP*

Once IPP is generated, IPP isomerase isomerizes it to dimethyl allyl pyrophosphate (DMAPP). This isomerization step involves the addition of proton to sp^2 carbon of IPP molecule leading to DMAPP synthesis.

4.2.3 *Condensation of Isoprene to Yield Prenyl Diphosphate*

During this stage, IPP reacts with DMAPP to form geranyl pyrophosphate (GPP). The enzyme prenylphosphate transferase carries out the addition of IPP to DMAPP, GPP and FPP (farnesyl pyrophosphate) resulting in the formation of GPP, FPP and GGPP (geranyl geranyl pyrophosphate), respectively. GPP acts as a precursor for monoterpene, FPP as a precursor for sesquiterpenoid and sterol, and GGPP as carotenoid precursor.

4.2.4 *Conversion of Prenyl Diphosphate to Terpene Backbone*

This step is catalysed by the enzyme terpene cyclase, which uses GPP, FPP and GGPP as substrate for mono-, sesqui- and diterpenoid biosynthesis. The mode of action of this enzyme is similar to prenylphosphate transferase. It also carries out H-atom transfer and Wagner-Meerwein rearrangement. The enzyme requires Mg^{2+} or Mn^{2+} as cofactor.

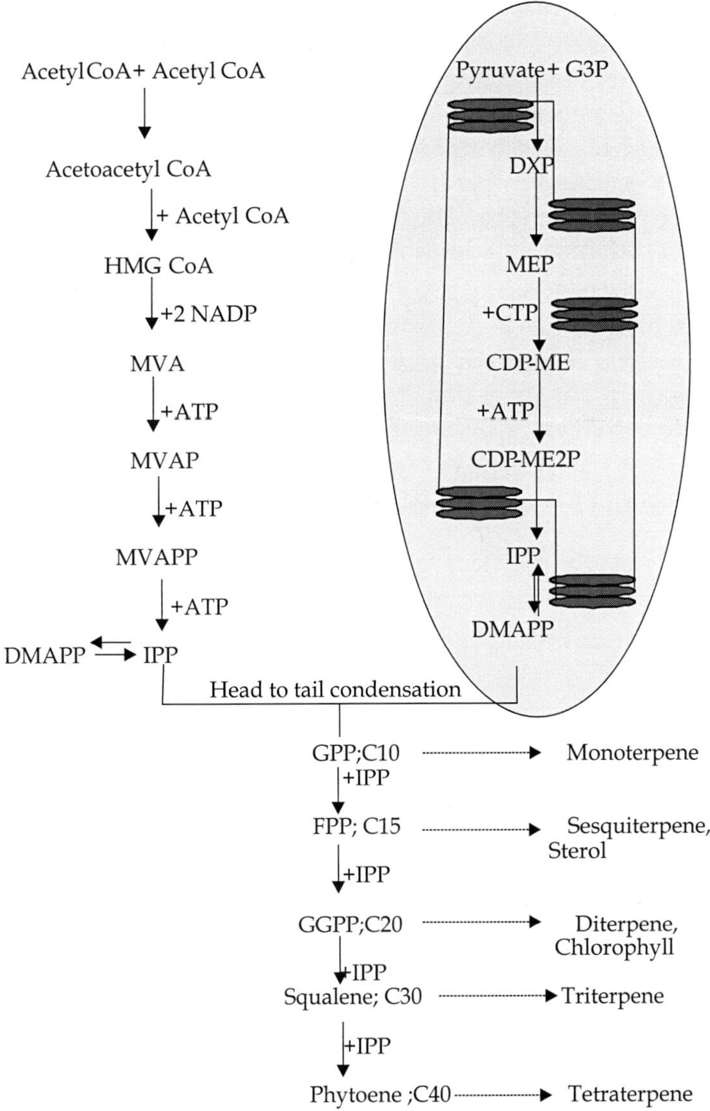

Fig. 1.3 Biosynthesis of major classes of terpenoids from five-carbon precursor IPP and DMAPP

4.3 Biosynthesis of Alkaloids

Alkaloids are mainly derived from amino acids such as tyrosine, tryptophan, phenylalanine, arginine, lysine, histidine and nicotinic acid, with majority of the insect and microorganism alkaloid appeared to be either polyketide or terpenoid in origin. Though different alkaloids are synthesized through different biosynthetic pathways. This chapter mainly focuses on the biosynthetic pathways for monoterpenoid indole alkaloid (MIA) as the most of the medicinally important alkaloids (vinblastine, vincristine, quinine, ajmaline) are MIA in nature. MIA is mainly the condensation product of nitrogen containing indole moiety with a monoterpenoid.

Ajmalicine was the first MIA for which enzyme leveled biosynthetic pathway was given (Fig. 1.4). It is formed from the tryptamine. Tryptamine is a decarboxylation product of L-tryptophan, which upon condensation with secologanin in the presence of the enzyme strictosidine synthase, yield strictosidine. Strictosidine through multiple enzymatic steps form ajmalicine. The strictosidine in species-specific manner is also capable of forming different MIA in different plants, such as vinblastine and vincristine in *Vinca rosea*, quinine in *Chinchona* sp., ajmaline in *Rauwolfia serpentine*.

Sequestration of alkaloids is necessary otherwise they will not be able to function in dose dependent manner. Being hydrophilic in nature, they are generally stored in vacuole. Epidermal cells of the plant are generally rich in such vacuole, as the epidermal cells are the sites where the pathogen or herbivore first comes in contact. Even light microscopic studies have also showed the presence of *Beta vulgaris* pigment (betalin) in the vacuole.

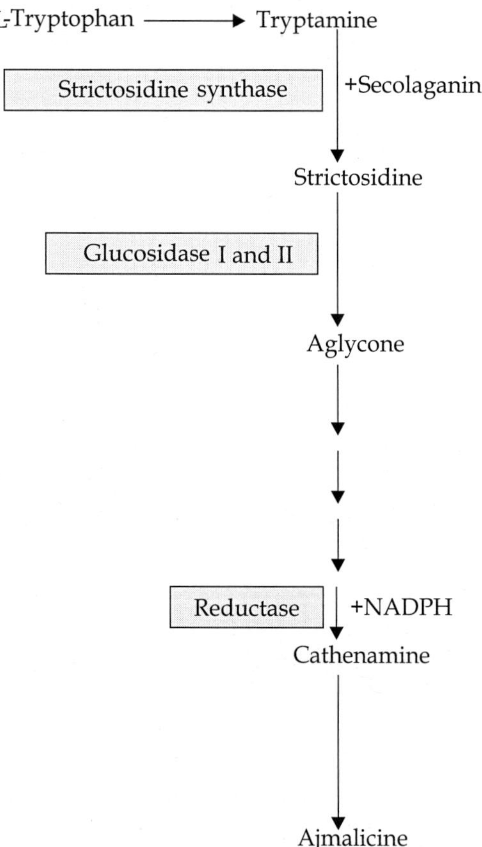

Fig. 1.4 Biosynthetic pathway of alkaloid Ajmalicine

5. HOW TO ENHANCE SECONDARY METABOLITE PRODUCTION IN PLANTS?

It is estimated that around 75 percent of the world population depends on the traditional medicinal herbs. Even in a developed nation like United States, pharmaceutical industry depends on plant-derived chemicals (Payne et al., 1991; Farnsworth, 1985). Due to their complex chemical structures, some drugs could not be synthesized in the laboratory. Besides, most of the secondary metabolites are synthesized by plants only, and could not be produced by microorganisms (Rao and Ravishankar, 2002).

One of the major issues with the plant-derived medicines is the requirement of large arable land, for their bulk production. In this way, they would compete with the food grain production. Being secondary metabolites in nature, their production is also very low in plants. Keeping all these constraints in mind, we have to enhance their production to meet the demand in near future without compromising with the production of foodgrains. To accomplish the goal, various biotechnological approaches have been followed and some are discussed below for sustainable production of these medicinal and aromatic plants.

5.1 Tissue Culture

Plant tissue culture technique along with the bioreactor is the most commonly used way to enhance the secondary metabolite production for the pharmaceutical industries. Using tissue culture techniques thousands of plants can be grown in a limited area under the controlled environmental conditions. Beside this, plants remain protected from the various insects and diseases. Once the high yielding cultivars are selected under the laboratory conditions, their production can be further enhanced through genetic transformation. The use of these high yielding cultivars reduces the cost of the secondary metabolites in the long run.

However, the yield of plant through tissue culture technique depends on various factors, such as selecting high yielding clones, type of media used, different types of precursors added to media, optimization of culture condition and elicitation (Mulabagal and Tsay, 2004). The concentration of hormones has been found to be an indispensable factor in the success of *in vitro* establishment of medicinal plant (Benjamin et al., 1987; Lal and Ahuja, 1996; Borthakur and Singh, 2002; Rai, 2002). The callus and suspension cultures as well as somatic embryogenesis have been used to produce capsaicin, vinblastine, vincristine, pyrethrin, diosgenin, morphine, codeine, and reserpine.

5.2 Genetic Transformation

With the development of rDNA technology, the secondary metabolite production is increasing rapidly. Though a number of transformation techniques like, gene gun, electroporation, microinjection, liposome mediated transfer are available, still *Agrobacterium* mediated transformation is considered the most preferred method for gene transfer. It is a natural mode of gene transfer and is capable of infecting the intact tissue. The transformation of *Atropa belladonna* using *Agrobacterium tumefaciens* has increased alkaloid content in this plant (Yun et al., 1992). Two- to threefold higher content of artemisinin in *Artemisia annua* was reported upon the transfer of chimeric farensyl diphosphate synthase gene (Chen et al., 2000). Besides increasing secondary metabolite production, another strategy can be developing transgenic medicinal plants for various agronomic traits, as various biotic and abiotic stresses also effect the yield of these metabolites.

5.3 Hairy Root Culture

Hairy root culture is the most commonly used approach to boost the production of secondary metabolite synthesized in the roots of the medicinal plant. T-DNA of *Agrobacterium rhizogens* Ri plasmid results in multibranched hairy roots at the site of infection. These hairy roots mimic the mother plant root in metabolite production. To scale up the production of secondary metabolites, this approach is often combined with bioreactor, some hairy root culture work well under submerged conditions, while other requires support. Thus, different kinds if bioreactors are available depending on the type of mother plant. Hairy root of *Beta vulgaris* produces more betaine under submerged condition as compared to stirred tank (Taya et al., 1999), while chichory requires air sparging system for efficient production of esculin production (Bais et al., 1999). Further, biotransformation of hairy roots can also be used to enhance the production of secondary metabolites. Many plants have been transformed with recombinant Ri plasmid including *Hyoscyamus niger, Chinchona ledgeriana, Atropa belladonna, Nicotiana* spp. (Subroto et al., 1996; Walton and Belshaw, 1988).

5.4 Metabolic Engineering

For enhancing the secondary metabolite production, a holistic approach is required to increase the anabolism of the desired metabolite and reduce its catabolism. In certain medicinal plants like the rate limiting steps should be checked and competitive pathways have to be blocked, so that all the carbon flux can be increased towards the desired compound. For example, Artemisinin is an anticancer drug obtained from *Artemisia annua*. The drug is produced at a very low level in the plant. Chen and his co-workers (2000) transferred the recombinant and chimeric gene respectively for the enzyme farnesyl diphosphate, carrying out the rate-limiting step in artemisin biosynthesis. This resulted in threefold increase in the metabolite production in plant.

The overexpression of branching point, enzyme putresine N-methyltransferase (PMT) in *Nicotiana sylvestris* enhanced the nicotine content in plant by 40 percent, while the same gene in *Atropa belladona* resulted in unmodified alkaloid profile in transformants (Sato et al., 2001). They also transformed *Coptis* with scoulerine 9-O methyltransferase (SMT), converting scoulerine into successive compounds leading to berberine. The scoulerine, may also get converted into sanguinarine by a parallel pathway. The overexpression of SMT resulted in 20 percent increase in berberine content in the transformant. Thus, showing that overexpression of branch point enzymes in the pathway can be used as a strategy to enhance secondary metabolite production. Some secondary metabolites, if produced at higher level, lead to slow growth of the plant cell. For this microbial system appears to be the best alternative. If whole metabolic pathway is known, then the genes required for biosynthesis can be introduced into it. However, this approach is feasible only if the pathway is short or additional genes required are less in number.

All the above-mentioned methods have the drawback of being expensive and low in field practicability. Further, the use of transgene leads to ethical issues, public acceptance, patenting, and screening for the risk associated with the use of transgenic. Therefore, a better and sustainable alternative for secondary metabolite production could be the use of arbuscular mycorrhizal fungi.

6. ARBUSCULAR MYCORRHIZA ASSOCIATION

Arbuscular mycorrhiza is a symbiotic association between the terrestrial plants and the fungi belonging to the phylum Glomeromycota (Morton and Benny, 1990). Their terrestrial partners range from bryophytes to angiosperms. Nearly 80 percent of the plant species were reported to be colonized by arbuscular mycorrhizal fungi (AMF) (Smith and Read, 2008). During AM establishment fungal hyphae penetrate the cortical cells, become either balloon like or branch dichotomously and form a tuft-like structure called arbuscule. AMF have possibly been asexual for millions of years and contain many genetically different nuclei, a phenomenon called heterokaryosis. The hyphae of AMF produce a glycoprotein called glomalin, which may be one of the major storages of carbon in the soil. Glomalin accumulates in the soil for longer duration and maintains soil aggregation, holds soil moisture and sequester soil organic carbon thus playing an important role in the terrestrial ecosystems.

In the last few decades, extensive research on mycorrhizal associations has emphasized that the symbiosis offers several aids to the crop plants like (i) increases the overall rate of photosynthesis; (ii) increases nutrient uptake and cycling of N, P, C, Zn, Cu; (iii) provides tolerance to high soil temperature, salinity and water stress; (iv) emerged as a potential biofertilizer; (v) initiates host defense reaction in the root by producing phenols, peroxidase, phytoalexins; (vi) improves the soil structure; (vii) improves rooting and plant development (Smith and Read, 2008). Thus, its presence in soil is beneficial for sustainable agriculture (Gianinazzi et al., 2010). The establishment of this beneficiary association involves a cross-talk between the plant and fungal partner. Phytohormones and secondary metabolites are also involved in this symbiotic interaction. The flavanoids produced by the plant roots increase the rate of spore germination of AMF (Gianinazzi-Pearson et al., 1989; Tsai and Phillips, 1991). The occurrence of a yellow colored compound mycorradicin, in the roots of most of the AMF associated plants and the carotenoid nature of mycorradicin substantiates the role of secondary metabolites in the establishment process of mycorrhizal symbiosis (Klingner et al., 1995a,b).

6.1 AMF-mediated Production of Secondary Metabolites

6.1.1 *Phenolics Production*

Mycorrhizal symbiosis positively influences plant secondary metabolism in both underground and aerial parts of medicinal and aromatic plants (Larose et al., 2002; Araim et al., 2009). Indeed, accumulation of β-carotene in the tuber of *Ipomoea batatas* (Farmer et al., 2007), glycyrrhizin in roots of *Glycyrrhiza uralensis* (Liu et al., 2007) and a diterpene forskol in the roots of *Coleus forskohlii* was reported on mycorrhization (Salio and Bagyaraj 2005). Instead, mycorrhization also improves the concentration of essential oil in the leaves of *Foeniculum vulgare* (Fennel), *Coriandrium sativum* (coriander), *Anethum graveolens* (Dill), *Trachyspermum ammi* (Carum), and *Origanum vulgare* (Oregano) (Kapoor et al., 2002a,b; 2004; Khaossad et al., 2006) and in the trichomes of the *Ocimum basilicum* var. genovese (sweet basil) (Copetta et al., 2006), the antioxidant production in the leaves of sweet basil and enhanced total phenol and antioxidant activity in the leaves and flower head extracts of *Cynara cardunculus* L. var. *scalymus* (Toussaint et al., 2007; Ceccarelli et al., 2010), increase in the ajmalacine production in *Catharanthus roseus* (Karthikeyan et al., 2008) and increase in the phenols, flavanoids and tannins in cebil (*Anadenanthera colubrina*) (Pedone-Bonfirm et al., 2013). Recently, a 21 percent increase in the gallic acid content has been reported on AMF inoculation of *Libidibia ferrea* (Silva et al., 2014). The highest level of antioxidant secondary plant metabolites rosmarinic acid and diterpene carnosic acid in *Salvia officinalis* (garden sage) leaves was observed under optimal P fertilization and inoculation with AMF (Nell et al., 2009).

Phenolic compounds protect plants from pathogens and oxidative damage. Several studies have shown that AMF colonization increased the concentration of phenolic compounds in herbal medicinal plants, however the exact mechanism involved is still unknown (Rosa-Mera et al., 2011). The defence-related phenolic compounds, such as flavonoids and coumarin are the active ingredients in many herbal medicinal plants. AM symbiosis has been found to positively influence the production of secondary plant metabolites, coumarin and imperatorin in *Angelica dahurica* (Zhao and He, 2011), flavonoids in *Bupleurum chinense, S. miltiorrhiza, Ginkgo biloba* and *Astragalus membranaceus* (Meng and He, 2011; Zeng et al., 2013), phenolic acid in *Arnica montana* and *Catharanthus roseus* (Jurkiewicz et al., 2010; Rosa-Mera et al., 2011), sesquiterpenic acid in *Valeriana officinalis* (Nell et al., 2010), and a thymol derivative in *Inula ensifolia* (Zubek et al., 2010).

Sweet basil is considered a natural producer of up to 20 different phenolic compounds (Copetta et al., 2006) that impelled it to traditionally use for the treatment of many ailments, such as headache, cough and diarrhea. It is a rich source of phenolic antioxidants and flavonoids. The rosmarinic acid is probably the major compound that imparts antioxidant activity in sweet basil (Toussaint et al., 2007). Among three different species of AMF studied by Copetta and his colleagues (2006), inoculation with *Gigaspora rosae,* significantly increased essential oil content in basil. This increase in essential oil may be attributed to the increased biomass production by AMF inoculation. Increased production of rosmarinic acid and caffeic acid in the shoots of sweet basil was also observed on inoculation of AMF, *Glomus caledonium* and *Glomus mosseae*. In fact, this increase in the accumulation of rosmarinic acid and caffeic acid in sweet basil could mediate by direct effects of the AMF (Toussaint et al., 2007). Toussaint et al., (2007) concluded that AMF potentially represents an alternative way of promoting growth and secondary metabolites of sweet basil. Gupta et al., (2002) also reported an increase in essential oil content in *Mentha arvensis* L. as compared to their non-mycorrhizal counterpart.

However, some researchers did not show a positive control of AM symbiosis on the composition of phenolic ingredients. Lee and Seagel (2009) did not find a notable increase in total polyphenol synthesis in leaves and stems of basil inoculated with *G. intraradices*. Moreover, AM symbiosis failed to impact total concentration of phenolic and rosmarinic acid in common sage (Nell et al., 2009). A decrease was also noticed in total phenols and flavonoids contents of mycorrhizal common sage plants (Geneva et al., 2010).

6.1.2 Terpenoids Production

Essential oils comprise some low-molecular weight terpenoids, which are active ingredients of herbal medicines. In the last two decades, research studies carried out by a number of researchers to understand arbuscular mycorrhiza-medicinal and aromatic plants interaction have shown that mycorrhizal fungi improve and enhance the essential oil content of various medicinal and aromatic plants like *Mentha arvensis* (Gupta et al., 2002), *Foeniculum vulgare* (Kapoor et al., 2004), *Origanum vulgare* (Khaosaad et al., 2006; Karagiannidis et al., 2011), *Coriandrum sativum* (Kapoor, 2002, Farahani et al., 2008), *Ocimum basilicum* (Copetta et al., 2006). There was severalfold increase in glycyrrhizic acid content (a triterpenoid saponin) on inoculation of liquorice with AMF as compared with control plants (Liu et al., 2007). Morone Fortunato and Avato (2008) applied AMF to micropropagated oregano seedlings and observed a significant increase in the content of essential oils.

6.1.3 Alkaloids Production

Alkaloids, a diverse class of plant secondary metabolite, are present in members of the family papaveraceae, apocynaceae, liliaceae, solanaceae. Increased alkaloids content upon mycorrhization has been reported in several plants (Rosa-Merra et al., 2011; Andrade et al., 2013).

AM symbiosis has been found to increase the accumulation of berberine, jatrorrhizine, and palmatine in *Phellodendron amurense,* hyoscyamine and scopolamine in *Datura stramonium*, camptothecin in *Camptotheca acuminata* and vinca in *Catharanthus roseus* (show anticancer properties), castanospermine in *Castanospermum australe* (active against AIDS virus), and berberine in *Phellodendron chinense,* forskolin in *Coleus forskohlii* (Hadwiger et al., 1986; Wei and Wang, 1989; Sailo and Bagyaraj, 2005; Fan et al., 2006; Zhou and Fan, 2007; Yu et al., 2010; Rosa-Mera et al., 2011). Coleus acquires forskolin in its root which is useful for the treatment of glaucoma, heart diseases, asthma, and some cancer cases (Shah et al., 1980). Mycorrhizal inoculation of *Allium sativum* (garlic), a very popular cooking preservative, increased accumulation of an alkaloid allicin, which contains antioxidative and anticancerogenic properties and could improve the functions of the cardiovascular and immune systems (Harris et al., 2001; Borde et al., 2009) Recently, increased production of nicotine was reported on mycorrhization of *Nicotiana tobaccum* by controlling the gene expression pattern for the same (Andrade et al., 2013).

6.2 Mechanism by which AMF Promotes Secondary Metabolites Production

AM-induced increase in the concentrations of several groups of plant secondary metabolites has been considered a defense response to fungal colonization (Zeng et al., 2013). Indeed, some researchers suggest that increased secondary metabolites accumulation in mycorrhizal plants may be due to mycorrhiza-assisted improved P nutrition (Kapoor et al 2002 a,b; 2004). However, till now, it is not clear that how mycorrhizal fungi trigger the accumulation of secondary metabolites in plant tissue. In this article, some AMF-mediated strategies, which could help in boosting secondary metabolite accumulation in medicinal and aromatic plants are discussed.

6.2.1 *Improved Nutrients Supply*

Phosphorus is an important constituent of ATP and NADPH, nucleic acid, phospholipids (in biomembrane), and is required for the biosynthesis of a large number of plant metabolites. There is a direct correlation between the amount of phosphorus absorbed by the plant and amount of essential oil formed. Phosphorus is generally absorbed by the plant in trivalent ionic state, which is insoluble in water. The mycorrhizal extraradical hyphae explore soil and go beyond the zone of root, and root hair to absorb phosphorus. The mycorrhizal hyphae also have higher affinity for phosphate ions and a lower threshold concentration for absorption compared to plant roots. The ability of the extraradical hyphae to store larger amounts of absorbed phosphorus than the plant roots also facilitates the continued movement of phosphorus into the hyphae. The improved AMF-mediated phosphorus nutrition could have a positive influence on the biosynthesis of secondary compounds (Liu and Zhong, 1998). Phosphorus attributes to secondary metabolite biosynthesis either by increasing the biomass of the plant or the concentration of the pyrophosphate compounds (IPP and DMAPP), which are precursors for the formation of secondary metabolites (Kapoor et al., 2004; Zubek et al., 2010). Moreover, the synthesis of secondary metabolites also requires acetyl-CoA, ATP and NADPH, in which P forms a major structural component (Kapoor et al., 2002b; 2004).

Nitrogen is an important constituent of several molecules including amino acids, proteins, nucleic acids and chlorophyll, which are necessary for the basic structure and functioning of cell. N is not only required for plant growth and development it is also a major constituent of several precursors of secondary metabolic pathways (Ram et al., 2006). Several studies have shown that an increase in the uptake of N leads to increased secondary metabolite production in medicinal plants (Sreevalli et al., 2004; Ram et al., 2006). AMF activates a plant ammonium transporter thus influences the uptake of N in medicinal plants (Guether et al., 2009; Rasouli-Sadaghiani et al., 2010; Miransari, 2011). AM-induced increased uptake of N has been shown to influence secondary metabolite accumulation in medicinal plants. In an experiment, improved content of phenolic compounds was found due to higher nitrogen assimilation in AM plants (Toussaint et al., 2007). Toussaint et al., (2007) suggested that the higher concentration of N in mycorrhizal plants may contribute to the synthesis of amino acids that serve as precursors for secondary metabolite biosynthesis. However, only a few studies have reported a correlation between AM induced N-nutrition and secondary metabolite biosynthesis, hence further investigations are required to establish such a correlation.

Besides N and P, micronutrients, such as Cu, Zn, Fe and Mn also play a crucial role in the growth and development of plant thereby on the production of secondary metabolites. The deficiency of these micronutrients greatly disturbs various physiological and metabolic processes of plant. They are the integral components of various enzymes and act as functional, structural and regulatory cofactors of some enzymes. They are also associated with protein and photosynthesis in plants. Mycorrhizal inoculation of plants improves the uptake of these micronutrients (Smith and Read, 2008) thus helps in increased production of secondary metabolites. Maudinas et al. (1997) demonstrated that Mn stimulates the production of phytoene synthase in plant, which otherwise involved in carotenoids biosynthesis. Mg accelerates activity of ent-kaurene synthase, which is involved in the synthesis of phytohormone gibberellins. Similarly, the activity of mevalonic kinase depends on the Mn and carbonic anhydrase on Zn.

6.2.2 *Increased Biomass and Trichomes Density*

Biomass is a sum total of growth, development and resistance to various biotic and abiotic stresses in plant. Most of the secondary metabolites are associated with providing resistance to biotic and abiotic stresses. After forming a symbiotic relation with medicinal plant, mycorrhiza improves the microenvironment of the plant and increases resistance to pathogens and disease, drought, salt and heavy metal stresses, thereby improving overall growth and development and secondary metabolite production.

Mycorrhizal inoculation has shown increased biomass production in *Mentha arvensis, Mentha piperata* (Gupta et al., 2002; Cabello et al., 2005), red clover (Khaosaad et al., 2008) and sweet basil (Rasouli-Sadaghiani et al., 2010; Zolfaghari et al., 2013), increased shoot length and number of nodes in basil (Copetta et al., 2006). An increase in stevol glycoside content in *Stevia rebaudiana* was reported due to increased biomass content in mycorrhizal plant (Mandal et al., 2013). Mandal et al. (2013) have demonstrated that the increased biomass in stevia was due to increased nutrient uptake and improved photosynthetic efficiency of mycorrhizal plants. Indeed, increased yield of the secondary metabolites was associated with the larger number of glandular trichomes on leaf surface of mycorrhizal compared to nonmycorrhizal plants.

Glandular trichomes are one of the most common secretory structures that produce and store essential oils in plants (Cabello et al., 2005). They are the sites of secondary metabolites synthesis and their accumulation (Wang et al., 2009). Therefore, increased trichomes density has often been credited with increased secondary metabolite production. A strong correlation was found in the number of trichomes formed and secondary metabolite content of peppermint (Ringer et al., 2005), lima bean (Bartram et al., 2006; Pinto et al., 2007) lavender (Behnam et al., 2006; Munoz-Bertomeu et al., 2006), sweet basil and *Artemisia* (Coppetta et al., 2006; kapoor et al., 2007). Recently, Liu et al., (2014), found a correlation between roots and shoot biomass as well as P uptake and secondary metabolite accumulation in *Glycyrrhiza glabra*. Besides increasing the production, mycorrhization also improved the pharmaceutical quality of glycyrrhizin. However, a negative influence of AMF on shoot biomass was reported by Zubek and coworkers (2015) in *Viola tricolor*. The possible reason for this decrease could be the increased carbon allocation from the plant to the fungal partner. However, decreased shoot biomass had no effect on secondary metabolite production.

6.2.3 *Enhanced Photosynthetic Rate*

AM-mediated improved plant growth is mainly attributed to improved uptake of nutrients from soil solution. Mycorrhizal colonization increases the surface area of absorption of nutrients through the development of a network of extra radical hyphae, which complements the root system. By enhancing the uptake of mineral nutrients, the overall rate of photosynthesis in mycorrhizal plant is increased, which subsequently enhances the biomass, bringing about an increase in secondary metabolite production. *Hypericum perforatum*, an important medicinal plant, produces the vital active ingredients like hypericin and pseudohypericin, polycyclic aromatic diones with anti-viral and anti-AIDS activities. The increased production of hypericin and pseudohypericin in *Hypericum perforatum* due to increased photosynthetic performance index in AMF treated plants was reported by Zubek and coworker (2012). Similar results were also reported in case of *Inula ensifolia*. An increase in thymol content in the roots of *Rhizophagus clarus* was also due to the increased photosynthetic performance index after AMF inoculation. However, at the same time, a decrease in photosynthetic performance index was also recorded in *R. intraradices* (Zubek et al., 2010).

6.2.4 *Modulating Level of Plant Growth Regulators*

Plant growth regulators play a crucial role in the establishment of AM symbiosis as well as AM-mediated secondary metabolites production (Hause et al., 2002; Shaul-Keinan et al., 2002; Copetta et al., 2006; Kapoor et al., 2007; Toussaint 2007; Toussaint et al., 2007). It has been demonstrated by a number of researchers that phytohormones like jasmonic acid (JA), gibberellic acid (GA_3), cytokinins and brassinosteroids alter epidermal differentiation programmes and result in the increased glandular trichome densities, by this means, accelerate plant secondary metabolites production in medicinal and aromatic plants (Perazza et al., 1998; Traw and Bergelson, 2003; Li et al., 2004; Maes et al., 2008; Maes and Gossens, 2010). While working on *Artemisia annua*, Maes and his colleagues (2011) further investigated the role of JA, GA_3 and cytokinins and observed that GA_3 and

JA encourage hairy trichome formation and development. Instead, jasmonic acid also promotes cytokinins and gland formation, and coordinates transcriptional activation of biosynthetic gene expression, thereby improves the production of a secondary metabolite, sesquiterpenoid (Dermastia et al., 1994; Maes et al., 2011).

7. FUTURE PERSPECTIVE

This review paper displays a comprehensive review of information available about the research on arbuscular mycorrhizal fungi and secondary metabolites of medicinal and aromatic plants, in order to set a basis for future work to investigate the exact mechanism of AM-induced increased yield of secondary metabolites. The mechanisms discussed in this paper are at preliminary stage, which are yet to be established therefore, necessitate further investigations at physiological, biological and molecular levels.

References

Akiyama K, and Hayashi H (2002). Arbuscular mycorrhizal fungus promoted accumulation of two new triterpenoids in cucumber roots. Biosci Biotech and Biochem 66(4):762–769.

Andrade SAL, Malik S, Sawaya ACHF, Bottcher A, and Mazzafera P (2013). Association with arbuscular mycorrhizal fungi influences alkaloid synthesis and accumulation in *Catharanthus roseus* and *Nicotiana tabacum* plants. Acta Physiol Plant 35(3): 867–880.

Anderson JJ, Anthony MS, Cline JM, Washburn SA, and Garner SC (1999). Health potential of soy isoflavones for menopausal women. Public Health Nutr 2:489–504.

Araim G, Saleem A, Arnason JT, and Charest AC (2009). Root colonization by an arbuscular mycorrhizal (AM) fungus increases growth and secondary metabolism of purple coneflower, *Echinacea purpurea* (L.) Moench. J Agric Food Chem 57:2255–2258.

Bais HP, George J, and Ravishankar GA (1998). Production of esculin by hairy root cultures of *Cichorium intybus* L. cv. Lucknow Local. Indian J Expt Biol 37:269–273.

Bartram S, Jux A, Gleixner G, and Boland W (2006). Dynamic pathway allocation in early terpenoid biosynthesis of stress-induced lima bean leaves. Phytochem 67(15): 1661–1672.

Behnam S, Farzaneh M, Ahmadzadeh M, and Tehrani AS (2006). Composition and antifungal activity of essential oils of Mentha piperita and *Lavendula angustifolia* on post-harvest phytopathogens. Commun Agric Appl Biol Sci 71(3): 1321–1326.

Benjamin BD, Roja P, Heble MR, and Chadha MS (1987). Multiple shoot cultures of *Atropa belladona*: effect of physicochemical factors on growth and alkaloid formation. J Plant Nutr 129: 129–135.

Bennett RN, and Wallsgrove RM (1994). Secondary metabolites in plant defence mechanisms. New Phytol 127: 617–633.

Bercovitch M, Waller A, and Adunsky A (1999). High dose morphine use in hospice setting: A database survey of patient characteristics and effect on life expectancy. Cancer 86: 871–877.

Birdsall T, and Kelly G (1997). Berberine: therapeutic potential of an alkaloid in several medicinal plants. Altern Med Rev 2: 94–103.

Borde M, Dudhane M, and Jite PK (2009). Role bioinoculant (AM fungi) increasing in growth, flavor content and yield in *Allium sativum* L. under field condition. Not Bot Hort Agrobot Cluj 37(2):124–128.

Borthakur M, and Singh RS (2002). Direct plantlet regeneration from male inflorescences of medicinal yam (*Dioscorea floribunda* Mart. & Gal.). *In vitro* Cell Dev Biol-Plant 38:183–185.

Bourgaud F, Gravot A, Milesi S, and Gontier E (2001). Production of plant secondary metabolites: a historical perspective. Plant Sci 161:839–851.

Cabello M, Irrazabal G, Bucsinszky AM, Saparrat M, Schalamuk S (2005). Effect of an arbuscular mycorrhizal fungus, *Glomus mosseae*, and a rock-phosphate-solubilizing fungus, *Penicillium thomii* on *Mentha piperita* growth in a soilless medium. J Basic Microbiol 45: 182–189.

Calabrese JC, Jordan DB, Boodhoo A, Sariaslani S, and Vannelli T (2004). Crystal structure of phenylalanine ammonia lyase: multiple helix dipoles implicated in catalysis. Biochem 43(36): 11403–11416.

Ceccarelli N, Curadi M, Martelloni L, Sbrana C, Picciarelli P, and Giovannetti M (2010). Mycorrhizal colonization impacts on phenolic content and antioxidant properties of artichoke leaves and flower heads two years after field transplant. Plant Soil 335:311–323.

Chappell J, Wolf F, Proulx J, Cuellar R, and Saunders C (1995). Is the reaction catalyzed by 3-hydroxy-3-methylglutaryl coenzyme-a reductase a rate-limiting step for isoprenoid biosynthesis in plants. Plant Physiol 109:1337–1343.

Charvat TT, Lee DJ, Robinson WE, and Chamberlin AR (2006). Design, synthesis and biological valuation of chicoric acid analogs as inhibitors of HIVa integrase. Bioorg Med Chem 14: 4552–4567.

Chen DH, Ye HC, and Li HF (2000). Expression of chimeric farnesyl diphosphate synthase gene in *Artemisia annua* using *Agrobacterium tumefaciens*-mediated transformation. Plant Sci 155:179–185.

Chen F, Nakashima N, and Kimura M (1995). Hypoglycemic activity and mechanism of extracts from mulberry leaves and cortex mori radices in streptozotocin induced diabetic mice. Yakugaku Zasshi 115: 476–482.

Copetta A, Lingua G, and Berta G (2006). Effects of three AM fungi on growth, distribution of glandular hairs, and essential oil production in *Ocimum basilicum* L. var. Genovese. Mycorrhiza 16:485–494.

Cornwell T, Cohick W, and Raskin I (2004). Dietary phytoestrogens and health. Phytochem 65:995–1016.

Cushnie TP, Hamilton VE, and Lamb AJ (2003). Assessment of the antibacterial activity of selected flavonoids and consideration of discrepancies between previous reports. Microbiol Res 158: 281–289.

Daly JW, and Spande TF (1986). Amphibian alkaloids: Chemistry, pharmacology, and biology. In: Alkaloids: chemical and physiological perspectives. SW Pelletier (ed.) J. Wiley and Sons, New York. pp. 1–274.

Daniel M (2006). Medicinal Plants: Chemistry and Properties. Taylor and Fransis, US. p. 230.

Dermastia M, Ravnikar M, Vilhar B, and Kovac M (1994). Increased levels of cytokinin riboside in jasmonic acid-treated potato (*Solanum tuberosum*) stem node cultures. Physiol Plant 92: 241–246.

Dewick PM (2002). Secondary metabolism: the building blocks and construction mechanism. In: Medicinal Natural Products. John Wiley & Sons, Ltd.

Dorsch W, Müller A, Christoffel V, Stuppner H, Antus S, Gottsegen A, and Wagner H (1994). Antiasthmatic *acetophenones* – an *in vivo* study on structure activity relationship. Phytomed 1:47–54.

Duke JA (1992a). Handbook of Biologically Active Phytochemicals and Their Activities. CRC Press, Boca Raton, Florida.

Duke JA (1992b). Handbook of Phytochemical Constituents of GRAS Herbs and Other Economic Plants. CRC Press, Boca Raton, Florida.

Eisenreich W, and Bacher A (2007). Advances of high-resolution NMR techniques in the structural and metabolic analysis of plant biochemistry. Phytochem 68: 2799–2815.

Fan JH, Yang GT, Mu LQ, and Zhou JH (2006). Effect of AM fungi on the content of berberine, jatrorrhizine and palmatine of *Phellodendron amurense* seedings. Prot Forest Sci Technol 5:24–26.

Farahani HA, Lebaschi MH, and Hamidi A (2008). Effects of arbuscular mycorrhizal fungi, phosphorus and water stress on quantity and quality characteristics of coriander. J Adv Nat Appl Sci 2(2): 55–59.

Farmer MJ, Li X, Feng G, Zhao B, Chatagnier O, Gianinazzi S, Gianinazzi Pearson V, and Van Tuinen D (2007). Molecular monitoring of field-inoculated AMF to evaluate persistence in sweet potato crops in China. Appl Soil Ecol 35: 599–609.

Farnsworth NR (1985). The role of medicinal plants in drug development. In: Kroogsgard-Larsen P, Brogger Christenses S (eds) Natural products and drug development. Denmark: Munsgaard Copenhagen, pp. 17–30.

Fenwick GR, Lutomski J, and Nieman C (1990). Liquorice, *Glycyrrhiza glabra* L.- Composition, uses and analysis. Food Chem 38: 119–143.

Figueiredo ER, Vieira IJC, Souza J, Braz-Filho R, Mathias L, Kanashiro MM, and Côrtes FH (2010). Isolamento, identificação e avaliação da atividade antileucêmica de alcaloides indólicos monoterpênicos de *Tabernaemontana salzmannii* (A. DC.), Apocynaceae. Brazilian Journal of Pharmacognosy 20:675–681.

Fresco P, Borges F, Diniz C, and Marques MPM (2004). New insights into the anticancer properties of polyphenols. Med Res Rev 26: 747–766.

Geneva MP, Stancheva IV, Boychinova MM, Mincheva NH, and Yonova PA (2010). Effects of foliar fertilization and arbuscular mycorrhizal colonization on *Salvia officinalis* L. growth, antioxidant capacity, and essential oil composition. J Sci Food Agric 90: 696–702.

Gershenson J (1994). The cost of plant chemical defense against herbivory: a biochemical perspective. In: Insect–plant Interactions, Vol. 5. Edited by Barnays EA. CRC Press, pp105–173.

Gianinazzi -Pearson V, Branzanti, B, and Gianinazzi S (1989). *In vitro* enhancement of spore germination and early hyphal growth of a vesicular arbuscular mycorrhizal fungus by host root exudates and plant flavonoids. Symbiosis 7: 243–255.

Gianinazzi S, Gollotte A, Binet MN, van Tuinen D, Redecker D, Wipf D (2010). Agroecology: the key role of arbuscular mycorrhizas in ecosystem services. Mycorrhiza 20: 519–530.

Gupta ML, Prasad A, Ram M, and Kumar S (2002). Effect of the vesicular arbuscular mycorrhizal (VAM) fungus *Glomus fasciculatum* on the essential oil yield related characters and nutrient acquisition in the crops of different cultivars of menthol mint (*Mentha arvensis*) under field conditions. Bioresour Technol 81(1): 7 7–79.

Gurib-Fakim A, Subratty H, Narod F, Govinden-Soulange J, and Mahomoodally F (2005). Biological activity from indigenous medicinal plants of Mauritius. Pure and Appl Chem 77: 41–51.

Golding JF, and Stott JRR (1997). Comparison of the effects of a selective muscarinic receptor antagonist and hyoscine (scopolamine) on motion sickness, skin conductance and heart rate. Br J Clin Pharmacol 43: 633–637.

Gong MQ, Wang FZ, and Chen Y (2002). Study on application of arbuscular-mycorrhizas in growing seedling of *Aloe vera*. J Chin Med Mater 25(1): 1–3.

Guether M, Neuhäuser B, Balestrini R, Dynowski M, Ludewig U, and Bonfante P (2009). A mycorrhizal-specific ammonium transporter from *Lotus japonicus* acquires nitrogen released by arbuscular mycorrhizal fungi. Plant Physiol 150(1): 73–83.

Hadwiger A, Neimann H, Kaebisch A, Bauer H, and Tamura T (1986). Appropriate glucosylation of the FMS gene product is a prerequisite for its transforming potency. EMBO J 5: 689–694.

Hahlbrock K, and Grisebach H (1979). Enzymatic controls in the biosynthesis of lignin and flavonoids. Ann Rev Plant Physiol 30(1): 105–130.

Han XZ, Shen T, and Lou HX (2007). Dietary polyphenols and their biological significance. Int J Mol Sci 8: 950–988.

Harker M, Holmberg N, Clayton JC, Gibbard CL, Wallace AD, Rawlins S, Hellyer SA, Lanot A, and Safford R (2003). Enhancement of seed phytosterol levels by expression of an N-terminal truncated *Hevea brasiliensis* (rubber tree) 3-hydroxy-3-methylglutaryl-CoA reductase. Plant Biotech J 1: 113–121.

Harris JC, Cottrell S, Plummer S, and Lloyd D (2001). Antimicrobial properties of *Allium sativum* (garlic). Appl Microbiol Biotechnol 57: 282–286.

Hause B, Maier W, Miersch O, Kramell R, and Strack D (2002). Induction of jasmonate biosynthesis in arbuscular mycorrhizal barley roots. Plant Physiol 130: 1213–1220.

Heinrich M, Barnes J, Gibbons S, and Williamson EM (2004). Fundamentals of Pharmacognosy and Phytotherapy. Churchill Livingstone, Elsevier Science Ltd., UK.

Honigsbaum M (2002). The Fever Trail: In Search of the Cure for Malaria. New York: Farrar, Straus, & Giroux.

Huh YS, Hong TH, and Hong WH (2004). Effective extraction of oligomeric proanthocyanidin (OPC) from wild grape seeds. Biotechnol Bioprocess E 9: 471–475.

Icon health (2004). Atropine: A medical dictionary, bibliography, and annonated research guide to internet references. Icon health Publishers, San Diego, California.

Jurkiewicz A, Ryszka P, Anielska T, Waligórski P, Białońska D, Góralska K, Michael MT, and Turnau K (2010). Optimization of culture conditions of *Arnica montana* L.: effects of mycorrhizal fungi and competing plants. Mycorrhiza 20: 293–306.

Kalsi J, Cellek S, Muneer A, Kell P, Ralph D, and Minhas S (2002). Current oral treatments for erectile dysfunction. Expert Opin Pharmacother 3: 1613–1629.

Kapoor R, Giri B, and Mukerji KG (2002a). *Glomus macrocarpum*: a potential bioinoculant to improve essential oil quality and concentration in Dill (*Anethum graveolens* L.) and Carum (*Trachyspermum ammi* (Linn.) Sprague). World J Microb Biot 18: 459–463.

Kapoor R, Giri B, and Mukerji KG (2002b). Mycorrhization of coriander (*Coriandrum sativum* L.) to enhance the concentration and quality of essential oil. J Sci Food Agr 82: 339–342.

Kapoor R, Giri B, and Mukerji KG (2004). Improved growth and essential oil yield and quality in *Foeniculum vulgare* mill on mycorrhizal inoculation supplemented with P-fertilizer. Bioresour Technol 93: 307–311.

Kapoor R, Chaudhary V, and Bhatnagar AK (2007). Effects of arbuscular mycorrhiza and phosphorus application on artemisinin concentration in *Artemisia annua* L. Mycorrihiza 17: 581-587.

Karagiannidis N, Thomidisa T, Lazari D, Filotheou EP, and Karagiannidou C (2011). Effect of three Greek arbuscular mycorrhizal fungi in improving the growth, nutrient concentration, and production of essential oils of oregano and mint plants. Sci Hortic 129: 329–334.

Karthikeyan B, Jaleel CA, Changxing Z, Joe MM, Srimannarayan J, and Deiveekasundaram M (2008). The effect of AM fungi and phosphorous level on the biomass yield and ajmalicine production in *Catharanthus roseus*. EurAsia J Bio Sci 2(3): 26–33.

Kaufman PB, Duke JA, Brielmann H, Boik J, and Hoyt JE (1997). A comparative survey of leguminous plants as sources of the isoflavones, genistein and daidzein: implications for human nutrition and health. J Altern Complement Med 3: 7–12.

Khaosaad T, Krenn L, Medjakovic S, Ranner A, Lossl A, Nell M (2008). Effect of mycorrhization on the isoflavone content and the phytoestrogen activity of red clover. Journal of Plant Physiology 165: 1161-1167.

Khaosaad T, Vierheilig H, Nell M, Zitterl-Eglseer K, and Novak J (2006). Arbuscular mycorrhiza alter the concentration of essential oils in oregano (*Origanum* sp., Lamiaceae). Mycorrhiza 16: 443–446.

Kliebenstein DJ (2012). Plant Defense Compounds: Systems Approaches to Metabolic Analysis. Phytopath 50: 155–173.

Klingner A, Bothe H, Wray V, and Marner FJ (1995a). Identification of a yellow pigment formed in maize roots upon mycorrhizal colonization. Phytochem 38: 53–55.

Klingner A, Hundershagen B, Kernebeck H, and Bothe H (1995b). Localization of the yellow pigment formed in roots of gramineous plants colonized by arbuscular fungi. Protoplasma 185: 50–57.

Lal N, and Ahuja PS (1996). Plantlet regeneration from callus in *Picrorhiza kurroa* Royle ex Benth -An endangered medicinal plant. Plant Tissue Cult 6: 127–134.

Larose G, Chenevert R, Moutoglis P, Gagne S, Piche Y, and Vierheilig H (2002). Flavonoid levels in roots of *Medicago sativa* are modulated by the developmental stage of the symbiosis and the root colonizing arbuscular mycorrhizal fungus. J Plant Physiol 159: 1329–1339.

Lee J, Scagel CF (2009). Chicoric acid found in basil (*Ocimum basilicum* L.) leaves. Food Chem 115(2): 650–656.

Li L, Zhao Y, McCaig BC, Wingerd BA, Wang J, Whalon ME, Pichersky E, and Howe GA (2004). The tomato homolog of CORONATINE-INSENSITIVE1 is required for the maternal control of seed maturation, jasmonate-signaled defense responses, and glandular trichome development. Plant Cell 16: 126–143.

Liu HL, Tan Y, Nell M, Zitter-Eglseer, Wswscrah C, Kopp B, Wang SM, and Novak J (2014). Arbuscular mycorrhizal fungal colonization of *Glycyrrhiza glabra* roots enhances plant biomass, phosphorus uptake and concentration of root secondary metabolites. J Arid Land 6(2): 186–194.

Liu I, Wu L, Wei S, Xiao X, Su C, Jiang P, Song Z, Wang T, and Yu Z (2007). Effects of arbuscular mycorrhizal fungi on the growth, nutrient uptake and glycyrrhizin production of licorice (Glycyrrhiza uralensis Fisch). Plant Growth Regul 52(1): 29–39.

Liu RH (2004). Potential synergism of phytochemicals in cancer prevention: mechanisms of action. J Nutr 134: 3479–3485.

Liu S, and Zhong JJ (1998). Phosphate effect on production of ginseng saponin and polysaccharide by cell suspension cultures of *Panax ginseng* and *Panax quinquefolium*. Process Biochem 33(1): 69–67.

Louli V, Folas G,Voutsas E, and Magoulas K (2004). Extraction of parsley seed oil by supercritical CO_2. J Supercrit Fluids 30: 163–174.

Lu YQ, He XL (2005). Effects of AM fungi on the chemical composition and growth amount of Atractylodes macrocephala koidz seedling on diffetent N levels. J Hebei Univers 25(6): 650–653.

Lu YQ, Cui Y, and He XL (2008a). Effects of AM fungi on biomass and nitrogen content of Atractylodes macrocephala under different nitrogen levels. J Henan Agr Sci 4: 94–96.

Lu YQ, He XL, and Li LZ (2008b). Effects of AM fungi on leaf protective enzymes of Atractrlodes macrocephala under different nitrogen levels. Hubei Agr Sci 47(6): 659–660.

Maes L, and Goossens A (2010). Hormone-mediated promotion of trichome initiation in plants is conserved but utilizes species and trichome-specific regulatory mechanisms. Plant Signal Behav 5: 205–207.

Maes L, Inzé D, and Goossens A (2008). Functional specialization of the TRANSPARENT TESTA GLABRA1 network allows differential hormonal control of laminal and marginal trichome initiation in *Arabidopsis* rosette leaves. Plant Physiol 148: 1453–1464.

Maes L, Van Nieuwerburgh FCW, Zhang Y, Reed DW, Pollier J, Vande Casteele SRF, Inzé D, Covello PS, Deforce DLD, and Goossens A (2011). Dissection of the phytohormonal regulation of trichome formation and biosynthesis of the anti-malarial compound artemisinin in *Artemisia annua* plants. New Phytol 189: 176–189.

Mandal S, Evelin H, Giri B, Singh VP, and Kapoor R (2013). Arbuscular mycorrhiza enhances the production of stevioside and rebaudioside-A in *Stevia rebaudiana* via nutritional and non-nutritional mechanisms. Appl Soil Eco 72: 187–194.

Marston A (2007). Roles of advances in chromatographic techniques in phytochemistry. Phytochem 68: 2786–2798.

Maudinas B, Bucholtz ML, Papastephanou C, Katiyar SS, Briedis AV, and Porter JW (1997). The partial purification and properties of a phytoene synthesizing enzyme system. Arch Biochem Biophys 180: 354–362.

McGeady P, Wansley DL, and Logan DA (2002). Carvone and perillaldehyde interfere with the seruminduced formation of filamentous structures in *Candida albicans* at substantially lower concentrations than those causing significant inhibition of growth. J Nat Prod 65: 953–955.

Meng JJ, He XL (2011). Effects of AM fungi on growth and nutritional contents of *Salvia miltiorrhiza* Bge. under drought stress. Journal of Agricultural University of Hebei 1: 012.

Miransari M (2011). Arbuscular mycorrhizal fungi and nitrogen uptake. Review article. Arch Microbiol 193: 77–81.

Morone-Fortunato I, and Avato P (2008). Plant development and synthesis of essential oils in micropropagated and mycorrhiza inoculated plants of *Origanum vulgare* L. spp. hirtum (Link) Ietswaart. Plant cell tissue organ cult 93(2): 139–149.

Morton, and Benny GL (1990). Revised classification of arbuscular mycorrhizal fungi (Zygomycetes): a new order Glomales and Gigasporineae and two new families Acaulosporaceae and Gigasporaceae with an emendation of Glomaceae. Mycotaxon 37: 471–491.

Moteki H, Hibasami H, Yamada Y, Katsuzaki H, Imai K, and Komiya T (2002). Specific induction of apoptosis by 1,8-cineole in two human leukemia cell lines, but not in a human stomach cancer cell line. Oncol Rep 9: 757–760.

Mulabagal V, Tsay HS (2004). Plant Cell Cultures – An Alternative and Efficient Source for the Production of Biologically Important Secondary Metabolites. Int J Appl Sci Engin 2(1): 29–48.

Muñoz-Bertomeu J, Arrillaga I, Ros R, and Segura J (2006). Up-regulation of 1-deoxy-D-xylulose-5-phosphate synthase enhances production of essential oils in transgenic spike lavender. Plant Physiol 142(3): 890–900.

Nell M, Vötsch M, Vierheilig H, Steinkellner S, Zitterl-Eglseer K,Franz C, and Novak J (2009). Effect of phosphorus uptake on growth and secondary metabolites of garden (*Salvia officinalis* L.). J Sci Food Agric 89:1090–1096.

Nell M, Wawrosch C, Steinkellner S, Vierheilig H, Kopp B, Lössl A, Franz C, Novak J, Zitterl-Eglseer K (2010). Root colonization by symbiotic arbuscular mycorrhizal fungi increases sesquiterpenic acid concentrations in *Valeriana officinalis* L. Planta Medica 76(4):393–398.

Pan PL, Chen DQ, Chen YT, and Zhou FM (2008). The research on the sift and germinate of AM fungi spore of *Ophiopogon japonicas*. Mod Chin Med 10(10):13–14.

Payne GF, Bringi V, Prince C, and Shuler ML (1991). Plant cell and tissue culture in liquid systems. Munich: Hanser Publ. pp. 1–10.

Pedone-Bonfim MV, Lins MA, Coelho IR, Santana AS, Silva FS, and Maia LC (2013). Mycorrhizal technology and phosphorus in the production of primary and secondary metabolites in cebil (*Anadenanthera colubrina* (Vell.) Brenan) seedlings. J Sci Food Agric 93: 1479–1484.

Perazza D, Vachon G, and Herzog M (1998). Gibberellins promote trichome formation by up-regulating *GLABROUS1* in Arabidopsis. Plant Physiol 117: 375–383.

Pinto DM, Nerg AM, and Holopainen JK (2007). The role of ozone-reactive compounds, terpenes, and green leaf volatiles (glvs), in the orientation of *Cotesia plutellae*. J Chem Ecol 33: 2218–2228.

Rai VR (2002). Rapid clonal propagation of Nothapodytes foetida (wight) sleumer - a threatened medicinal tree. In-vitro Cell Dev Biol Plant 38: 347–51.

Ram D, Ram M, and Singh R (2006). Optimization of water and nitrogen application to menthol mint (*Mentha arvensis* L.) through sugarcane trash mulch in a sandy loam soil of semi-arid subtropical climate. Biores tech 97(7): 886–893.

Rao SR, and Ravishankar GA (2002). Plant cell cultures: Chemical factories of secondary metabolites. Biotechnol Adv 20: 101–153.

Rasouli-Sadaghiani MH, Hassani A, Barin M, Rezaee YD, and Sefidkon Y (2010). Effects of arbuscular mycorrhizal (AM) fungi on growth, essential oil production and nutrients uptake in basil. J Med Plants Res 4(21): 2222–2228.

Ringer KL, Davis EM, and Croteau R (2005). Monoterpene metabolism. cloning, expression and characterization of (−)-isopiperitenol/(−)-carveol dehydrogease of peppermint and spearmint. Plant Physiol 137(3): 863–872.

Rosa-Mera CJD, Ferrera-Cerrato R, Alarcón A, Sánchez-Colín MJ, and David OD (2011). Arbuscular mycorrhizal fungi and potassium bicarbonate enhance the foliar content of the vinblastine alkaloid in *Catharanthus roseus*. Plant Soil 349: 367–376.

Ruetsch YA, Böni T, and Borgeat A (2001). From cocaine to ropivacaine: the history of local anesthetic drugs. Curr Top Med Chem 1(3): 175–82.

Salio GL, and Bagyaraj DJ (2005). Influence of different AM-fungi on the growth, nutrition and forskolin content of *Coleus forskohlii*. Mycol Res 109: 795–798.

Salmon SE, and Sartorelli AC (1989). Cancer chemotherapy. In: Basic and Clinical Pharmacology. B.G. Katzung (Ed). Appleton-Lange, Norwalk, Connecticut, p. 694.

Sato F, Hashimoto T, Hachiya A, Tamura K, Choi K-B, Morishige T, Fujimoto H, Yamada Y (2001). Metabolic engineering of plant alkaloid biosynthesis. PNAS 98: 367-372.

Schmidt TJ, Brun R, Willuhn G, Khalid SA (2002). Anti-trypanosomal activity of helenalin and somestructurally related sesquiterpene lactones. Planta Medica 68: 750–751.

Shah V, Bhat SV, Bajwa BS, Dornaeur H, and DeSouza NJ (1980). The occurrence of forskolin in Labiatae. Planta Medica 39: 183–185.

Sreevalli Y, Kulkarni RN, Baskaran K, and Chandrashekara RS (2004). Increasing the content of leaf and root alkaloids of high alkaloid content mutants of periwinkle through nitrogen fertilization. Ind Crops Prod 19(2): 191–195.

Shaul-Keinan O, Gadkar V, Ginzberg I, Grünzweig JM, Chet I, Elad Y, Wininger S, Belausov E, Eshed Y, Atzmon N, Ben-Tal Y, Kapulnik Y (2002). Hormone concentrations in tobacco roots change during arbuscular mycorrhizal colonization with *Glomus intraradices*. New Phytol 154: 501–507.

Shelton AL (2004). Variation in chemical defences of plants may improve the effectiveness of defence. Evol Ecol Res 6 (23): 709–726.

Shelton AL (2005). Within-plant variation in glucosinolate concentrations of *Raphanus sativus* across multiple scales. J Chem Ecol 31: 1711–1732.

Silva FA, Ferreira MR, Soares LA, Sampaio EV, Maia LC (2014). Arbuscular mycorrhizal fungi increase gallic acid production in leaves of field grown *Libidibia ferrea* (Mart. ex Tul.) LP Queiroz. Journal of Medicinal Plants Research 8(36): 1110-1115.

Smith EE, Facelli E, Pope S, and Smith FA (2010). Plant performance in stressful environments. Interpreting new and established knowledge of the roles of arbuscular mycorrhizas. Plant and Soil 326: 3–20.

Smith SE, and Read DJ (2008). Mycorrhizal symbiosis. Acad. Press, London, UK.

Soetan KO, Aiyelaagbe OO (1998). The need for bioactivity-safety evaluation and conservation of medicinal plants - A review. J Med Plants Res 3(5): 324–328.

Sridar C, Goosen TC, Kent UM, Williams JA, and Hollenberg PF (2004). Silybin inactivates cytochromes P450 3A4 and 2C9 and inhibits major hepatic glucuronosyltransferases. Drug Metab Dispos 32: 587–594.

Subroto MA, Kwok K, Hamill JD, and Doran PM (1996). Coculture of genetically transformed roots and shoots for synthesis, translocation and biotransformation of secondary metabolites. Biotechnol Bioeng 49: 481–494.

Taya M, Yoyama A, Kando O, Kobayashi T, and Matsui C (1989). Growth and characteristics of plant hairy roots and their cultures in bioreactors. J Chem Eng Jpn 22: 74–89.

Thangapazham RL, Sharma A, and Maheshwari RK (2006). Multiple molecular targets in cancer chemoprevention by curcumin. APPSJ 8: 443–449.

Traw MB, and Bergelson J (2003). Interactive effects of jasmonic acid, salicylic acid, and gibberellin on induction of trichomes in *Arabidopsis*. Plant Physiol 133: 1367–1375.

Toussaint JP (2007). Investigating physiological changes in the aerial parts of AM plants: what do we know and where should we be heading? Mycorrhiza 17:349–353.

Toussaint JP, Smith FA, and Smith SE (2007). Arbuscular mycorrhizal fungi can induce the production of phytochemicals in sweet basil irrespective of phosphorus nutrition. Mycorrhiza 17: 291–297.

Tsai SM, and Phillips DA (1991). Flavonoids released naturally from alfalfa promote development of symbiotic *Glomus* spores *in vitro*. Appl Environ Microbiol 57: 1485–1488.

Wang W, Wang Y, Zhang Q, Qi Y, and Gou D (2009). Global characterization of Artemisia annua glandular trichome transcriptome using 454 pyrosequencing. BMC Genom 10:465–475.

Walton NJ, Belshaw NJ (1988). The effect of cadaverine on the formation of anabasine from lysine in hairy root cultures of *Nicotiana hesperis*. Plant Cell Rep 7: 115–118.

Wei GT, and Wang HG (1989). Effects of VA mycorrhizal fungi on growth, nutrient uptake and effective compounds in Chinese medicinal herb *Datura stramonium* L. Sci Agr Sin 22(5): 56–61.

Weinberg BA, and Bealer BA (2002). The world of caffeine: The science and culture of the world's most popular drug. Bennett Alan Weinberg, Bonnie K. B: 384.

Yun DJ, Hashimoto T, and Yamada Y (1992). Metabolic engineering of medicinal plants: transgenic *Atropa belladonna* with an improved alkaloid composition. Proc Nat Acad Sci 89: 11799–11803.

Yu Y, Yu T, Wang Y, and Yan XF (2010). Effect of inoculation time on camptothecin content in arbuscular mycorrhizal *Camptotheca acuminate* seedlings. Chin J Plant Ecol 34(6): 687–694.

Zeng Y, Guo LP, Chen BD, Huo ZP, Wang JY, Huang LQ, Yang G,Cui XM, Yang L, Wu ZX, Chen ML, and Zeng Y (2013). Arbuscular mycorrhizal symbiosis and active ingredients of medicinal plants: current research status and prospective. Mycorrhiza 23(4): 253–265.

Zhang J, Liu DH, Guo LP, Jin H, Zhou J, and Yang G (2010). Effects of four AM fungi on growth and essential oil composition in rhizome of *Atractylodes lancea*. World Sci Technol/Moder TCM Mater Med 12(5): 779–782.

Zhao JL, and He XL (2011). Effects of AM fungi on drought resistance and content of chemical components in *Angelica dahurica*. Acta Agr Bor Occi Sin 20(3):184–189.

Zhi-lin Y, Chuan-chao D, and Lian-qing C (2007). Regulation and accumulation of secondary metabolites in plant-fungus symbiotic system. Afr J Biotech 6(11): 1266–1271.

Zhou JH, and Fan JH (2007). Effects of AM fungi on the berberine content in *Phellodendron chinense* seedings. North Hortic 12: 25–27.

Zhou N, Xia CL, Jiang B, Bai ZC, Liu GN, and Ma XK (2009). Arbuscular mycorrhiza in *Paris polyphylla* var. *yunnanensis*. Chin J Chin Med 34(14): 1768–1772.

Zhou N, Zou L, Wang GZ, and Jiang B (2010). Primary explore to relation of arbuscular mycorrhizae and its secondary metabolite steroidal saponin in *Paris polyphylla*. Chin J Exper Trad Med Formulae16(16): 85–88.

Zolfaghari M, Nazeri V, Sefidkon F, and Rejali F (2013). Effects Effect of arbuscular mycorrhizal fungi on plant growth and essential oil content and composition of *Ocimum basilicum* L. Iranian J Plant Physiol 3 (2): 643–650.

Zubek S, Mielcarek S, and Turnau K (2012). Hypericin and pseudohypericin concentrations of a valuable medicinal plant *Hypericum perforatum* L. are enhanced by arbuscular mycorrhizal fungi. Mycorrhiza 22(2): 149–156.

Zubek S, Rola K, Szewczyk A, Marta LM, and Turnau K (2015). Enhanced concentrations of elements and secondary metabolites in *Viola tricolor* L. induced by arbuscular mycorrhizal fungi. Plant and Soil 1: 1–14.

Zubek S, Stojakowska A, Anielska T, and Turnau K (2010). Arbuscular mycorrhizal fungi alter thymol derivative contents of *Inula ensifolia* L. Mycorrhiza 20: 497–504.

2

Microbial Biopesticides with a Focus on *Bacillus thuringiensis* and Baculoviruses

ARUN KUMAR SHARMA AND VINAY SHARMA*
Department of Bioscience and Biotechnology
Banasthali University, Rajasthan
*Corresponding author email: *vinaysharma30@yahoo.co.uk*

1. INTRODUCTION

Agriculture has been facing the destructive activities of numerous pests since long, leading to radical decrease in yields. Chemical pesticides for plant protection occupy the leading place as regards their total volume of application in integrated pest management. But these chemical pesticides cause toxicity to humans and warm-blooded animals as well as pollute our environment.

Despite many years of effective control by conventional chemical insecticides, continued use of chemical pesticides and some other factors are threatening the effectiveness of these agents. This leads to development of pesticide resistance and de-registration of some insecticides due to human health and environmental concerns. Therefore, an alternative is needed which must be ecofriendly. Higher quality and greater quantity of agricultural products can be generated by improvement in pest control strategies. Therefore, there is a need to develop biopesticides which are effective, biodegradable and ecofriendly (Mazid et al., 2011).

Biopesticides are natural substances that control pests (insects). Biopesticides are living organisms (bacteria, virus, fungi, etc.) or their products (phytochemicals, microbial products) or byproducts (semiochemicals), which can be used for the management of pests that are injurious to crops. Biopesticides have an important role in crop protection. Biopesticides based on pathogenic microorganisms specific to a target pest offer an ecologically sound and effective solution to pest problems. They are not harmful to the environment and human beings. The most commonly used biopesticides are living organisms, which are pathogenic for the pest of interest. These include biofungicides (*Trichoderma*), bioherbicides (*Phytopthora*), bioinsecticides (*Bacillus thuringiensis*) and baculoviruses (Mazid et al., 2011). The biopesticides offer the following advantages:

(i) inherently less harmful and less environmental load, (ii) designed to affect only one specific pest or, in some cases, a few target organisms, but not harmful for beneficial insects, (iii) usually effective in very small quantities and often decompose quickly, thereby resulting in lower exposures and largely avoiding the pollution problems and (iv) biopesticides can contribute greatly when used as a component of integrated pest management program (Gupta and Dikshit, 2010).

2. MAJOR CATEGORIES OF BIOPESTICIDES

2.1 Microbial Pesticides

They contain a microorganism (virus, fungus, bacteria, protozoan or alga) as the active ingredient. Microbial pesticides can control many different kinds of pests. Each separate microbial pesticide is specific for its target pest. The most widely known microbial pesticides are different varieties of the bacterium *Bacillus thuringiensis*, or Bt, which can control certain insects in cabbage, potatoes, and other crop plants. Bt produces a crystal protein that is harmful to specific insect pests.

2.2 Plant Pesticides

They are pesticidal substances, produced by foreign gene, which is inserted into the genome of a plant by recombinant DNA technology. For example, scientists can take the gene for the insecticidal crystal protein from *Bacillus thuringiensis* genome and introduce the gene into the crop plant's own genetic material by recombinant DNA technology. Then the genetically modified plant, instead of the Bt bacterium manufactures the substance (insecticidal crystal protein) that destroys the pest.

2.3 Biochemical Pesticides

Conventional pesticides, by contrast, are synthetic materials that usually kill or inactivate the pest, but biochemical pesticides are naturally occurring substances that control pests by non-toxic mechanisms. Biochemical pesticides include substances that interfere with growth or mating, such as plant growth regulators, or substances that attract or repel pests, such as pheromones (Gupta and Dikshit, 2010).

3. BIOPESTICIDES IN INDIA

Biopesticides represent only 2.89 percent (as on 2005) of the overall pesticide market in India and it is expected to exhibit an annual growth rate of about 2.3 percent in the coming years (Thakore, 2006). In India, only 12 types of biopesticides have been registered under the Insecticide Act, 1968. *Bacillus thuringensis*, *Trichoderma*, NPV and Neem based pesticides are the major biopesticides manufactured and used in India. Whereas more than 190 synthetic compounds are registered for use as chemical pesticides. Most of the biopesticides are used for public health, except a few that are used in agriculture.

Table 2.1: List of registered biopesticides under Insecticides Act, 1968 (Gupta and Dikshit, 2010)

S. No.	Name of the Biopesticide
1.	NPV of *Helicoverpa armigera*
2.	NPV of *Spodoptera litura*
3.	*Bacillus thuringiensis* var. *galleriae*
4.	*Bacillus thuringiensis* var. *kurstaki*
5.	*Bacillus thuringiensis* var. *israelensis*
6.	*Bacillus sphaericus*
7.	*Trichoderma harzianum*
8.	*Trichoderma viride*
9.	*Beauveria bassiana*
10.	*Pseudomonas fluorescens*
11.	Cymbopogan
12.	Neem based pesticides

4. BACTERIAL BIOPESTICIDES

Bacterial biopesticides are the most widely used and are cheaper than the other modes of pest bioregulation. Many species of bacteria can infect different varities of insects, but those bacteria belonging to the genus *Bacillus* are most widely used as pesticides. One of the *Bacillus* species, *Bacillus thuringiensis,* has developed many molecular mechanisms to produce insecticidal toxins; most of toxins are coded for by several Cry genes (Schnepf et al., 1998).

Since its discovery in 1901 as a microbial insecticide, *Bacillus thuringiensis* has been widely used to control insect pests important in forestry, agriculture, and medicine. Its principal characteristic is that during sporulation it synthesizes a crystalline inclusion containing proteins known as δ-endotoxins or Cry proteins, which have insecticidal properties. To date, more than one hundred *B. thuringiensis*-based bioinsecticides have been developed, which are mostly used against lepidopteran, coleopteran and dipteran larvae. In addition, the genes that code for the insecticidal crystal proteins have been successfully transferred into different crop plants (corn, brinjal, cotton, tomato, etc.), which has led to significant economic benefits (no need of spraying chemical pesticides and crop yield was increased). *B. thuringiensis* and Cry proteins are efficient, safe and sustainable alternatives to chemical pesticides for the control of insect pests because of their high specificity and their safety in the environment, (Roh et al., 2007; Kumar et al., 2008). The toxicity of the Cry proteins has traditionally been explained by the formation of transmembrane pores or ion channels that lead to efflux of ATP and finally osmotic cell lysis (Roh et al., 2007; Kumar et al., 2008).

4.1 *Bacillus thuringiensis*

Bacillus thuringiensis is capable of synthesizing during sporulation a protein crystal known as δ-endotoxins and it has insecticidal activity. This crystalline inclusion may contribute about 25 percent of the dry weight of the bacterium. Bt bacterium was first isolated in 1901, by the Japanese bacteriologist S. Ishiwata from infected silk worms, *Bombyx mori* (Ishiwata 1901). It was subsequently rediscovered by the German biologist Berliner in 1911, who isolated it from infected chrysalids of the Mediterranean flour moth, *Ephestia kuehniella*, collected from a mill in the province of Thuringe (Berliner, 1915). He called this bacterium *Bacillus thuringiensis*. Agronomists soon became interested in the insecticidal properties of Bt, because small amounts of preparations of this bacterium were sufficient to kill many insect larvae rapidly. The first formulation based on Bt was developed under the name "Sporéine" in France in 1938, but the first well-documented industrial procedure for producing a Bt-based product dates from 1959, with the manufacture of "Bactospéine". It was the first French patent for a biopesticide formulation. Commercial formulations of Bt composed of spore/crystal preparations obtained by culturing the bacterium in fermentors; the preparations are purified, dried and used in a granulated or wettable powder formulation for use as a spray. δ-endotoxins are highly diverse, resulting in a generally restricted activity spectrum for each individual toxin, and are innocuous to plants, animals and almost all non-target insects (ladybirds, bees and other auxiliary biological control agents) (Marvier et al., 2007). The industrial-scale production of Bt is now well controlled. It is relatively simple and competitive in terms of cost, and this obviously contributes to its success.

4.1.1 *Processing of Parasporal Crystal Protein in Insect Midgut after Ingestion*

The insecticidal activity (toxin) of *B. thuringiensis* subsp. *kurstaki* (first discovered in 1911) and other strains is contained within a very large structure called a parasporal crystal, which is synthesized during bacterial sporulation. Although no significant role on behalf of the bacterium has been attributed to the parasporal crystal structure, by synthesizing the crystal, the bacterium is "providing for its future" in that a dead insect provides sufficient nutrients to allow germination of the dormant spore. The parasporal crystal contains approximately 20 to 30 percent of the dry weight of a endospore culture and usually composed of mainly protein (~95%) and a small amount of carbohydrate (~5%). About 150 different parasporal crystal proteins (Cry proteins) are known. The parasporal crystal is an aggregate of individual crystal protein that can generally be dissociated by mild alkali treatment into individual subunits (Aronson and Shai, 2001). The individual subunits can be further dissociated *in vitro* by treatment with β-mercaptoethanol, which reduces disulfide linkages (Fig. 2.1).

The parasporal crystal does not usually contain the active form of the insecticide. Rather, once the crystal has been solubilized, the protein that is released is generally a protoxin, a precursor of the active toxin. The protoxin of many of the Cry toxins that are directed against lepidoptera have a molecular mass of approximately 130 kilodaltons (kDa) (Fig. 2.1). When a parasporal crystal is ingested by a target insect, it is processed by the combination of gut alkaline pH (7.5 to 8.0) and specific digestive proteases to produce an active toxin with a molecular mass of approximately 68 kDa (Fig. 2.1).

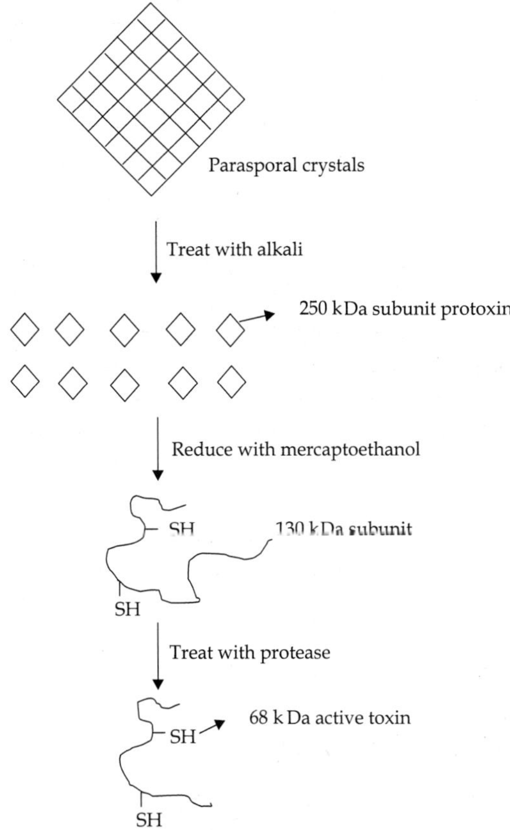

Fig. 2.1 Schematic representation of a *B. thuringiensis* parasporal crystal composed of Cry1 protoxin protein. Each 250 kDa protein subunit of the parasporal crystal contains two 130 kDa polypeptides (molecular masses determined by polyacrylamide gel electrophoresis are approximations and do not always provide exact multiples). 130 kDa protoxin can be converted into an active toxin (68 kDa) by the combination of a slightly alkaline pH (7.5 to 8) and the action of specific proteases, both of which are found in the insect midgut. The activated toxin peptide fragment binds to protein receptors on the surface of the gut epithelial cell membrane (Aronson and Shai, 2001).

4.1.2 *Mode of Action of Active Toxin Fragment (Killing Mechanism)*

In its active form, the toxic protein inserts itself into the membranes of the midgut epithelial cells of the insect and creates an ion channel. This results in the excessive loss of cellular ATP molecules (Fig. 2.2). About 15 minutes after this ion channel forms, cellular metabolism ceases; the insect stops feeding within a few hours, becomes dehydrated, and eventually dies (in about 2 to 5 days). Since the conversion of the protoxin to the active toxin requires both alkaline pH and the presence of specific proteases, it is extremely unlikely that nontarget species, such as humans and farm animals, will be affected.

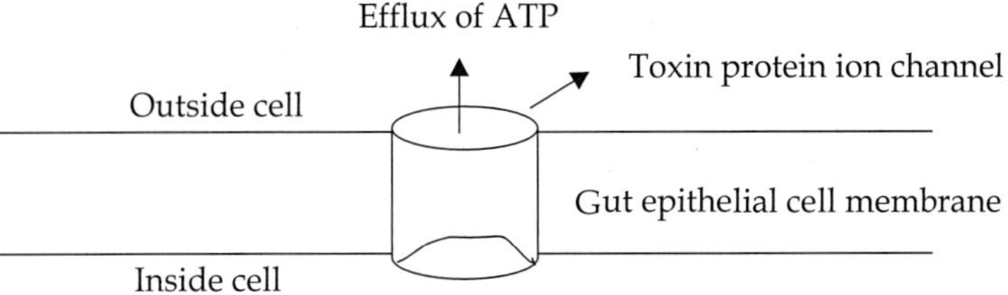

Fig. 2.2 Insertion of the Bt toxin into the membrane of an insect midgut epithelial cell. The Bt toxin creates an ion channel between the cell cytoplasm and the exterior environment.

Activated Bt toxins bind to specific proteins (cadherins) which are present on the surfaces of the microvilli of the insect midgut epithelial cells. Binding of active toxin monomers to cadherins, which are transmembrane glycoproteins containing 12 cadherin repeating domains and one membrane-proximal extracellular domain (Fig. 2.3), facilitates the development of a multimeric form of the toxin monomers and formation of a pore in the membrane. Loss of cadherin protein or mutation in cadherin genes is generally associated with resistance of insects to *B. thuringiensis* (Glick et al., 2010).

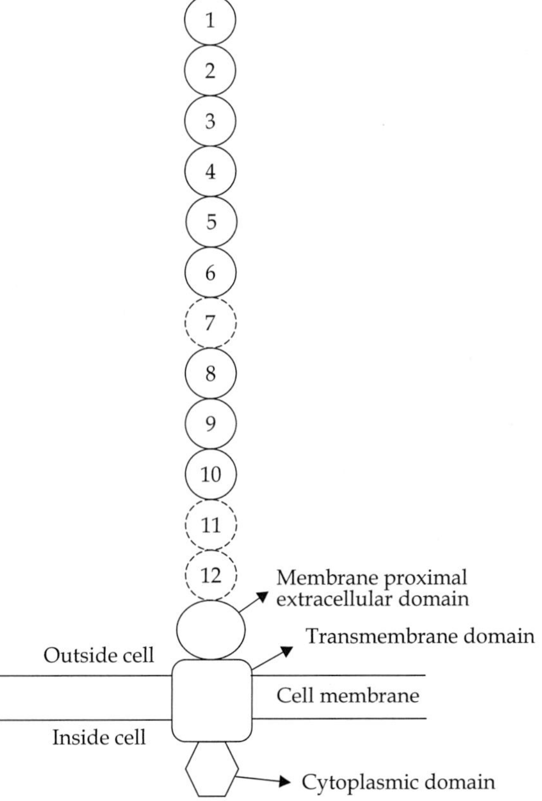

Fig. 2.3 Schematic representation of a cadherin protein molecule embedded in an insect midgut epithelial membrane. Twelve cadherin domains are labeled. Domains 7, 11, and 12, which are highlighted as dashed circles, have been implicated as binding sites for Cry proteins.

4.1.3 Properties of the Insecticidal Toxin Protein from Different Strains of B. thuringiensis

This bacterium comprises a large number of strains and subspecies, each of which produces a different crystal toxin that can kill specific insects only. There are more than 150 different subspecies of *B. thuringiensis* (Table 2.2).

Table 2.2: Some important properties of the insecticidal toxins from different strains of *B. thuringiensis* (Lereclus et al., 1993)

B. thuringiensis strain or subspecies	Protoxin size (kDa)	Target insects	Serotype
Berliner	130–140	Lepidoptera	1
kurstaki KTO, HD-1	130–140	Lepidoptera	3
entomocidus 6.01	130–140	Lepidoptera	6
aizawai 7.29	130–140	Lepidoptera	7
aizawai IC 1	135	Lepidoptera, Diptera	7
kurstaki HD-1	71	Lepidoptera, Diptera	3
tenebrionis (*san diego*)	66-73	Coleoptera	8
morrisoni PG14	125-145	Diptera	8
israelensis	68	Diptera	14

For example, *B. thuringiensis* subsp. *israelensis* kills diptera, such as blackflies and mosquitoes. *B. thuringiensis* subsp. *tenebrionis* (also known as *B. thuringiensis* subsp. *san diego*) kills coleoptera (beetles), such as the boll weevil and potato beetle. *B. thuringiensis* subsp. *kurstaki* is toxic to lepidopteran larvae, including those of butterflies, moths and skippers; cabbage worms; and spruce budworms. In addition, some subspecies of *B. thuringiensis* produce insecticidal toxins that are effective against orthoptera (grasshoppers, crickets, and locusts), hymenoptera (sawflies, wasps, bees, and ants) and mallophaga (lice).

4.1.4 Classification of B. thuringiensis Toxin Protein

The insecticidal toxins from the *B. thuringiensis* strains were previously grouped into four major classes, i.e., CryI, CryII, CryIII, and CryIV, based on the insecticidal activity of the toxin. These proteins were further organized into subclasses (A, B, C, etc.) and subgroups (a, b, c, etc.). In the past few years, as increasing numbers of *B. thuringiensis* strains were isolated and their genes were characterized, it became clear that the original classification was unable to accommodate many of the newly discovered *B. thuringiensis* toxin genes. Therefore, a new system of *B. thuringiensis* gene classification was introduced.

In the current classification scheme (established in 1998), *B. thuringiensis* insecticidal (Cry) proteins are assigned designations based on their degree of evolutionary divergence and this parameter was estimated by certain mathematical algorithms. In this scheme, a phylogenetic tree was constructed based on the amino acid sequences of *B. thuringiensis* toxin proteins, i.e., Cry proteins (Fig. 2.4). Basically, the amino acid sequences of the proteins are compared, and if the proteins are identical, then they are 100 percent homologous. If only 50 percent of the amino acids are the same, then the proteins have 50 percent identity. The relationship among a set of protein sequences can be deduced and represented as a branched tree. The nodes (branch points) of the tree represent points of divergence. For the classification of the *B. thuringiensis* Cry proteins, a four-part naming system was devised. Demarcations, set at 95, 78, and 45 percent homology, show the boundaries that define the different nomenclature ranks. The name that is assigned to a particular cry toxin basically depends on the location of the node where the toxin protein sequence enters the tree relative to these set boundaries. A toxin that joins the tree to the left boundary is assigned a new primary rank (an Arabic numeral), one that joins the tree between the central and left boundaries is assigned a new secondary rank (an uppercase letter), one that joins the tree between the right and the central boundaries is assigned a new tertiary rank (a lowercase letter), and one that joins to the right of the rightmost boundary is assigned a new quaternary rank (an Arabic numeral). For example, Cry proteins that are less than 45 percent homologous are given a number (e.g., Cry1 and Cry7) and are assigned to the primary rank. Cry proteins that are 45 to 78 percent identical to proteins of the primary rank are further designated with an uppercase letter (e.g., Cry1A and Cry1F). The complete Cry protein tree consists of the positions of all Cry proteins. This classification system is utilized throughout this chapter, even when referring to work that was published prior to the development of this system.

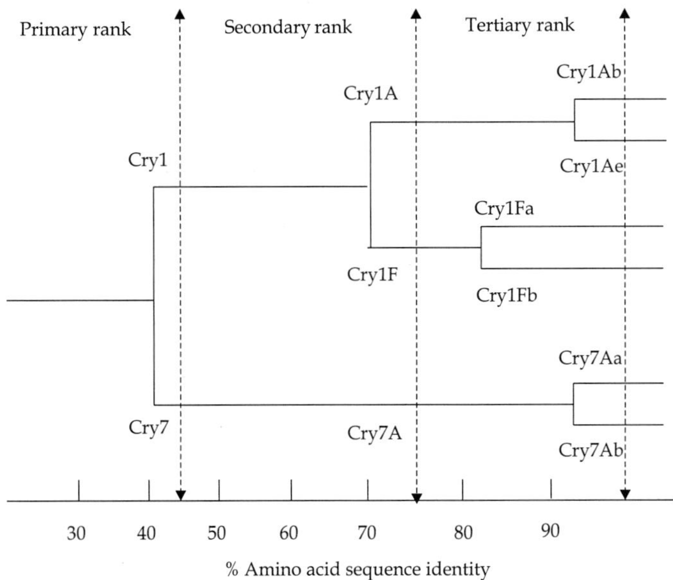

Fig. 2.4 Schematic representation of a portion of the phylogenetic tree of *B. thuringiensis* insecticidal (Cry) proteins. Different levels of nomenclature ranks, i.e., primary, secondary and tertiary are represented here. Cry1 and Cry7 share less than 45% identity, Cry1A and Cry1F are between 45% and 78% identical, and Cry1Ab and Cry1Ae are between 78% and 95% identical (Schnepf et al., 1998).

4.1.5 Limitations of Bt Toxin

The mode of action of *B. thuringiensis* toxins imposes certain constraints on its application. To kill an insect pest, *B. thuringiensis* parasporal crystals must be ingested. Contact of the bacterium or the insecticidal toxin with the surface of the target organism has no effect on it. The requirement that the insecticide be ingested, in part, limits the susceptibility of nontarget insects and other animals to the insecticide. *B. thuringiensis* is generally applied by spraying, so it is usually formulated with insect attractants to increase the probability that the target insect will ingest the toxin. However, insects that bore into plants or attack plant roots are less likely to ingest a *B. thuringiensis* toxin that has been sprayed on a host plant, so other strategies have been devised to control such pests. One approach is to create transgenic plants that carry and express a *B. thuringiensis* toxin gene so that they are protected from infestation throughout the growing season.

A limiting feature of the action of the *B. thuringiensis* toxin is that it can kill a susceptible insect only during a specific developmental stage. Therefore, the toxin must be applied when the pest population is at a particular stage in its life cycle (generally, the larval stage). The other major impediment to more widespread application of *B. thuringiensis* subsp. *kurstaki* is that it costs from 1.5 to 3 times as much as chemical insecticides. Some subspecies of *B. thuringiensis* that have been approved for use in the field and have rapidly gained widespread. These are *B. thuringiensis* subspecies *aizawai, kurstaki, israelensis, tenebrionis* (Glick et al., 2010).

4.1.5.1 *Bacillus thuringiensis subsp. kurstaki*

It was first discovered in 1901, although its commercial potential was ignored until 1951. Within recent decades, however, *B. thuringiensis* subsp. *kurstaki* has become the major means of controlling the spruce budworm in Canada. In 1979, approximately 1 percent of the forest area in Canada that was treated with an insecticide to combat the spruce budworm (about 2 million hectares, or 8000 square miles) was sprayed with *B. thuringiensis* subsp. *kurstaki*. The remainder of the treated forests were sprayed with chemical insecticides. By 1986, the use of *B. thuringiensis* subsp. *kurstaki* had increased dramatically. It was used to treat approximately 74 percent of the forests sprayed in that year for spruce budworm.

In other countries, *B. thuringiensis* subsp. *kurstaki* has been used against tent caterpillars, gypsy moths, cabbage worms, cabbage loopers, and tobacco hornworms. For the biological control (biocontrol) of insect pests, *B. thuringiensis* subsp. *kurstaki* is typically applied by spraying approximately 1.3×10^8 to 2.6×10^8 spores per square foot (1 square foot is equivalent to 0.093 m^2) of the target area. Administration of the spores is timed to coincide with the peak of the larval population of the target organism, because the parasporal crystals, being sensitive to sunlight, are short-lived in the environment. Under simulated conditions, sunlight degrades over 60 percent of the tryptophan residues of the parasporal crystal within a 24-hour period, thereby rendering the protein inactive. Depending on the amount of sunlight present, parasporal crystals may persist in the environment for as little as a day or as long as a month. The lack of persistence of the insecticidal protoxin in the natural environment means that natural selection of resistant insects is highly unlikely (Glick et al., 2010).

5. VIRUSES AS BIOPESTICIDES

Biopesticides mean the use of bacteria, fungi, protozoa, viruses, and nematodes for the biological control of insect pest, diseases, nematodes and weeds in agriculture, veterinary, medicinal, horticultural and forestry ecosystems. At present, the global market for microbial pesticides is more than US $ 125 million per annum which is still less than 1 percent of the total global market for chemical pesticide crop protection of $ 20–25 billion. The market is dominated by *Bacillus thuringiensis* (80%) followed by nematodes (13.3%) and others (6.67%). Nearly Rs. 14, 926 million have been spent during last decade by government of India for biocontrol programme in different crops. In India, some major states, adapting IPM strategy are Madhya Pradesh, Maharashtra, Uttar Pradesh, Andhra Pradesh and Karnataka. It is, therefore, clear that proper encouragement is needed for biological suppression of crop pests. Insect viruses are obligate, intracellular and pathogenic parasites. The viruses belonging to 11 families which are pathogenic to insects. Baculoviruses are associated with the orders of Hymenoptera, Lepidoptera, Coleoptera, Diptera, Trichoptera, Crustacea, Neuroptera, and mites. The virions are rod-shaped, 40–70 × 250–400 nm, comprising an envelope around a protein capsid containing DNA-protein core. The core (DNA) and protein capsid is known as the nucleo-capsid (Steinhaus, 1949).

These viruses are specific and highly virulent to their hosts. They are restricted in their virulency to the class insecta; they are often genus or species specific. About 60 percent of the total known insect viruses (1200) belong to the family Baculoviridae and it is estimated that such viruses can be used against nearly 30 percent of all the major pests of food and fibre crops.

Nucleopolyhedrosis virus (NPV) is known for high epizootic levels and is naturally occurring, self-perpetuating, safe to natural enemies due to host specificity and eco-friendly. Since, NPV is an obligate parasite, it multiplies only in living insect larvae. So, mass production of NPV is a tough job and requires skilled workmen (Mclntosh et al., 1987).

Nuclear polyhedrosis viruses recorded in India include *Helicoverpa armigera, S. exigua, S. litura, Agrotis ipsilon, A. segetum, Amsacta moorei, Anadividia peponis, Trichoplusiani, Thysanoplusia orichalcea, Plutella xylostella, Corcyra cephalonica, Adisura atkinsoni, Phthorimaea operculella* and *Mythimna separata* (Erayya et al., 2013).

5.1 Baculoviruses

Baculoviruses are one of the largest and most diverse groups of double-stranded DNA viruses (almost 1000 species have been described) pathogenic only for insects mostly of the orders Lepidoptera (butterflies, moths), Hymenoptera (sawflies) Coleoptera (beetles) and Diptera. The size of viral genome ranges from 80 to 200 kb. Individual baculovirus usually have a narrow host range restricted to a few closely related species. *Autographa californica* nucleopolyhedrovirus (*Ac*NPV) is the most widely studied baculovirus. Baculoviruses are arthropod viruses and they are well known due to their potential as agents of biological control of pests in agriculture, forestry and horticulture. They are also widely used as expression vectors in biotechnology. The family Baculoviridae contains diverse members and in the past, the classification was based on morphology of virus. It was divided into two genera: the Granulovirus (GVs) and Nucleopolyhedrovirus (NPVs). Recently, this division was challenged by Jehle et al. (2006). He compared 29 fully sequenced baculoviral genomes and indicated that virus phylogeny followed more closely the classification of the hosts than the virion morphological traits, but the traditional division into two genera is still widely used.

Baculoviruses infect and multiply in arthropods (invertebrates). They do not replicate in vertebrates, plants and microorganisms. Although they do not replicate, they may, under special conditions, enter animal cells. This unexpected property made them a important tool for studies of transient expression of foreign genes under vertebrate promoters introduced into baculovirus genome (Boyce and Bucher, 1996; Kost et al., 2005).

5.1.1 *Structure and Replication of Baculovirus (Killing Mechanism)*

The circular, covalently closed double-stranded DNA genome of *Ac*NPV is surrounded by a small basic protein which neutralizes the negative charge of the DNA. This structure is further enclosed by proteins forming a nucleocapsid. Mature virion consist of one or more nucleocapsids embedded in a membranous envelope. Two viral forms are produced at different times post-infection which are morphologically distinct, but genetically identical.

- Budded virus particles (BV) which are necessary for the transmission of the virus to other healthy tissues of the caterpillar body (Fig. 2.5).
- Occlusion bodies (OB) which are responsible for the survival of the virus in the environment and the spread of the virus from insect to insect (Fig. 2.5).

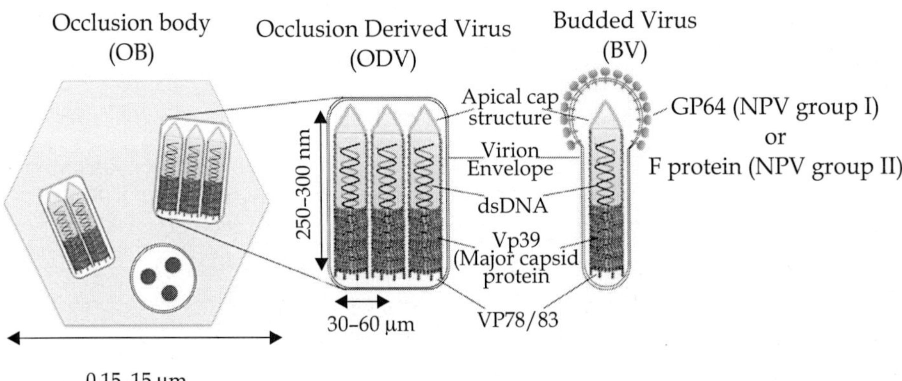

Fig. 2.5 Schematic diagrams of the structure of baculovirus occlusion bodies (OB), occlusion-derived virion (ODV) and budded virion (BV). ODVs are embedded in a crystalline matrix of polyhedron protein to form OBs. Shown here is the OB of nucleopolyhedroviruses (NPVs) (the vesicle or envelope containing three round structures is a cross section of an ODV). The ODV and BV envelope and their nucleocapsid contain numerous proteins. VP39 is the major nucleocapsid protein which constitutes the barrel of the nucleocapsid and is present in the whole nucleocapsid. VP78/83 is located at one end of the nucleocapsid, presumably at the blunt end. The fusion protein GP64 (NPV group I) or F protein (NPV group II) is found throughout the envelope of BV, but forms peplomers on one end, presumably the conical end (Shelly et al., 2013).

The occlusion bodies (polyhedra) of nucleopolyhedrovirus contain many occlusion-derived virions (ODV) surrounded by a matrix of polyhedrin protein, a major structural protein (Braunagel et al., 2003). Polyhedrin is produced in large quantities (nearly 30% of total protein mass at the time of host death), but it is not needed for the transmission of the virus from infected cell to healthy cell. Polyhedra are relatively stable and the protected virions can survive in the environment for more than twenty years under favourable conditions. Polyhedra resemble clear, irregular crystals of salt under 1000× magnification so they are big enough to be seen in a light microscope.

In nature, baculoviruses are found embedded within proteinaceous matrix known as 'polyhedra' on plant foliage, plant debris and soil. Insect larvae acquire infection when they feed on plant foliage, plant debris. Baculoviruses primarily infect insect larvae and it is uninfectious to the adult insects. The insect larvae, while feeding on the plant foliage, plant debrics accidentally feed upon the polyhedral. Upon ingestion by an insect, the polyhedra move to the midgut, where the alkaline environment facilitates the dissolution of the polyhedrin protein coat, releasing infectious nucleocapsids. The nucleocapsids are taken up by the insect midgut cells and then migrate through the cytoplasm to the nucleus, where the nucleocapsid is removed. After viral replication, which takes place in the nucleus, and nucleocapsid assembly, some nucleocapsids are released by budding through the plasma membranes of infected midgut epithelial cells into the circulatory system of the insect or get embedded in the polyhedral late in the infection process. Consequently, the infection spreads to other cells throughout the insect (Figs. 2.6 and 2.7). It usually takes about 10 rounds of viral replication, or about 5 to 9 days, for the insect to die. At that stage, about 25 percent of the dry weight of the insect consists of polyhedron/ occlusion bodies. Masses of these polyhedral are released in the soil environment upon death (after tissue liquefaction and then rupturing of midgut epithelial cells) of the infected larvae. From here, they are again ready to be ingested and infect their healthy caterpillar hosts (Huh and Weaver, 1990).

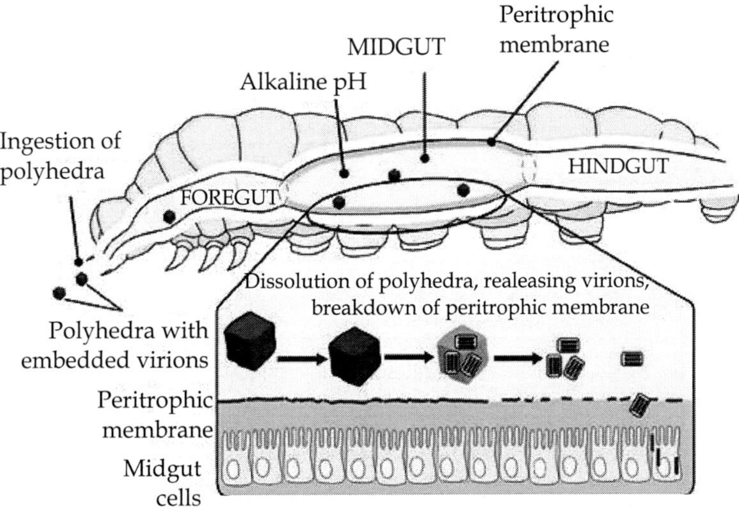

Fig. 2.6 Infection of an insect caterpillar larva by a nucleopolyhedrovirus

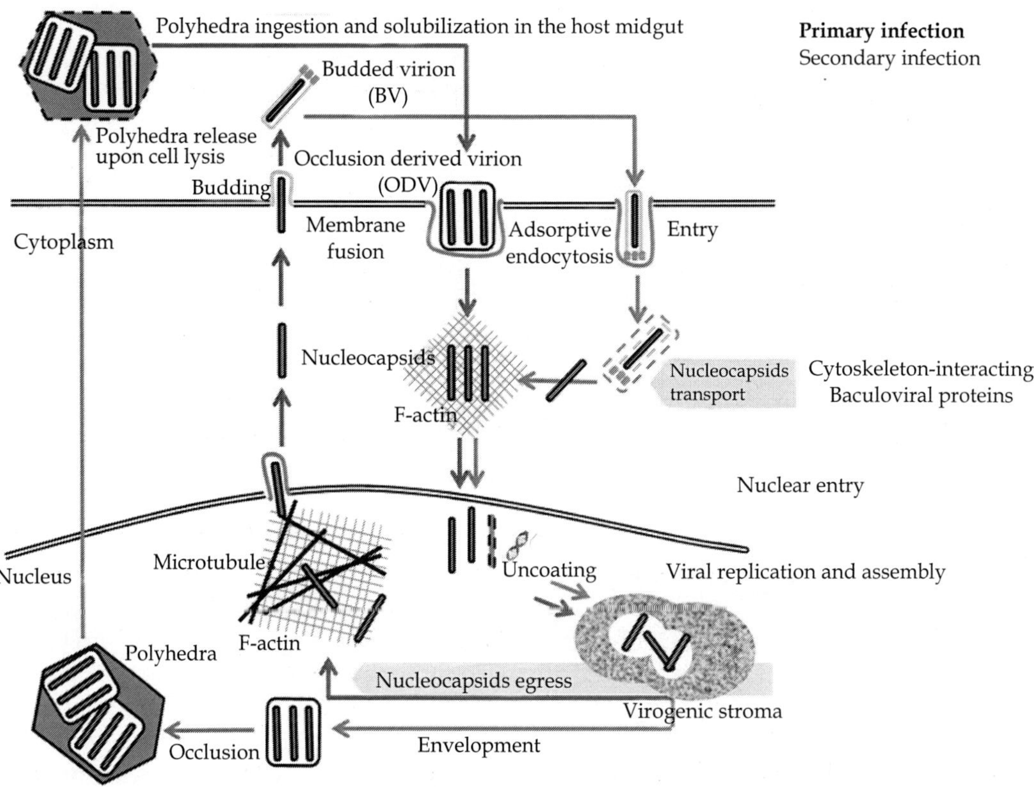

Fig. 2.7 Diagram representing different stages of life cycle of a typical baculovirus and formation of polyhedral, occlusion-derived virion (ODV) and budded virion (BV) (Monteiro et al., 2012).

Table 2.3: Some insect pests that are currently controlled with baculoviruses (Glick et al., 2010)

Pest	Common name	Crop
Anticarsia gemmatalis	Velvetbean caterpillar	Soybean
Chrysomela scripta	Cottonwood leaf beetle	Trees
Chrysomela scripta	Codling moth	Apple, walnut
Heliothis sp.	Cotton bollworm	Cotton, sorghum
Lymantria dispar	Gypsy moth	Deciduous trees
Mamestra brassicae	Cabbage moth	Vegetables
Neodiprion sertifer	European pine sawfly	Pine
Oryctes rhinoceros	Rhinoceros beetle	Coconut
Spodoptera littoralis	Egyptian cotton leaf worm	Cotton
Spodoptera exigua	Beet armyworm	Vegetables, flowers
Trichoplusia ni	Cabbage looper	Brassicas

Table 2.4: Baculovirus preparations registered for pest control in various parts of the world (Erayya et al., 2013)

Insect species	Trade name of the product	Country	Year of registration
Nuclear polyhedrosis viruses			
Mamestra brassicae	Mamestrin	Finland	1988
Orgyia pseudo seudotsugata	Biocontrol-1 and Virtuss	USA	1976
Lymantria dispar	Gypchek	USA	1978
Neodiprion lecontei	Lecont-virus	USA/Canada	1982/1983
Heliothis spp.	Elcar	USA	1975
Heliothis spp.	Biotrol VHZ and Virion H	USA	1973
Granulosis viruses			
Cydia pomonella	Madex	Czechoslovakia	1987
Agotis segetum	Agrovir Germany 1990	Germany	1990

5.1.2 *Baculovirus Production Technology*

At present, commercial production of baculovirus is carried out only by *in vivo* either by applying virus against the host insects in the field and then diseased or dead larvae are collected (production in the field condition) or target insect is produced in the laboratory on an artificial diet and then it is allowed to infect with baculovirus (production under laboratory condition). The latter is the most commonly used method for commercial production of baculoviruses in many countries, but for commercial production of the *Anticarsia gemmatalis* baculovirus (*Ag*NPV) in Brazil both methods have been successfully used (Moscardi, 1999, 2007). Their use has been promoted in some countries, e.g., China, Brazil, and the former Soviet Union. For some insects, artificial diet is not available and, therefore, commercial production of baculoviruses of these insects is generally too difficult or impossible under laboratory conditions. In such cases, field production of baculovirus stocks may be a method of choice, also from financial point of view (Moscardi, 1999). However, this approach is difficult when liquefaction of the insect body is very intense, as, for instance, in larvae infected by *Spodoptera* spp. Baculoviruses, making it too difficult to collect dead larvae. In this case, live larvae must be collected close to death when the body has not yet ruptured.

Production of baculovirus in insect cell cultures offers advantages over *in vivo* multiplication for being a sterile, controllable, highly pure product yield process. Production of baculovirus for agricultural pest control by *in vitro* process needs to be efficient, with competitive costs, leading to a final product which is highly pathogenic to the target pest. There is a strong limitation for *in vitro* production, however, genetic alteration leading to loss of virulence may appear during successive passages of the virus in cell culture (Krell, 1996; Rhodes, 1996). In laboratory culture, production of occlusion derived virions (ODV) is not necessary for survival of the virus. The budded virus (BV) particle is the form used for cell-to-cell transmission in cell culture. The main protein of the BV particle, essential for virus budding and necessary for entrance of the virus into the next host cell is the GP64 (Blissard 1996). Infection of lepidopteran cells by baculovirus is influenced by various culture conditions including dissolved oxygen concentration, pH, temperature, osmolarity and nutrient composition of the culture medium. For process optimization of *in vitro* baculovirus production, main requirements are the investigation of factors associated with loss of genetic stability and the use of new strategies, such as isolation of more stable variants, as well as the reduction in the costs of components of cell culture medium.

Some other challenges for *in vitro* production of baculoviruses are the requirements of productive insect cell lines (Jem et al., 1997) and highly productive culture media (Chakraborty et al., 1999). At present, many cell lines are available for production purposes and are derived from various sources, thus exhibiting a wide variety of growth and production characteristics. Screening or formulation of media must be performed carefully for a particular virus isolate cell line combination, as different media can greatly affect yields of polyhedra (Pedrini et al., 2006). Recently, a new strategy for *in vitro* baculovirus production was proposed based on many polyhedra (MP) variants. These are clones selected after several passages of the virus in cell culture using the plaque assay technique. MPs maintain the wild-type features such as formation of many polyhedra in the infected cell nucleus and high titer of budded virus (Slavicek et al., 2001; Pedrini et al., 2005) which allow them, in principle, to compete with the population of few polyhedra mutants accumulated in cell culture.

5.1.3 *Advantages of Baculovirus Based Biopesticides*

A positive feature of using baculoviruses as biocontrol agents is that they generally have limited host ranges and do not affect nontarget organisms. However, this means that any particular baculovirus can be used to control only a limited number of insect pests. Since baculoviruses coevolved with their insect hosts over thousands of years, they are well adapted to avoid the insect's defence mechanisms, and resistance to these viruses develops only rarely, and much less frequently than resistance to *B. thuringiensis* (Glick et al., 2010).

Control of the European spruce sawfly *(Gilipinia hercyniae)* population in eastern Canada is the best example of insect control by a baculovirus. Population of European sawfly were reduced to below economic threshold levels by 1943 and remain under control today. Ironically, the reason why some baculoviruses are not used commercially is related to the effectiveness of the virus. If a virus is effective for preventing proliferation of a particular insect species, the virus has to be applied only once every year or so, making it difficult for the industry to justify the high registration costs. Farmers and growers prefer to use a single insecticidal agent that can control many different insect pests rather than a number of different insecticides, so if baculoviruses are to be used more extensively, their limited host range needs to be expanded (Glick et al., 2010).

6. CONCLUSION

Due to "chemical pesticide" problems in India, there is an urgent need to promote environment-friendly 'biopesticides'. Moreover, recent government policies also favour biopesticides. Application of biopesticides has thus become an integral component of nutrient management system. Apart from these, they play a vital role in increasing the agricultural production.

Demand for biopesticides is increasing steadily in all parts of the world. Efficacy of biopesticides can be equal to or better than conventional products, especially for crops like vegetables, fruits, flowers and nuts, when they are used in integrated pest management systems. By combining safety and performance, biopesticides perform with high efficiency while providing the flexibility of minimum application restrictions, resistance management potential, superior residue and human and environmental safety benefits. As a natural agent, they pose little danger to the environment as compared to the dangerous, persistent and broad-spectrum chemical pesticides.

It is expected that in future they will play more important role in agriculture and forestry. Biopesticides clearly have a potential to play a significant role in development of future integrated pest management strategies. Hopefully, in the near future more rational approach will be gradually adopted towards biopesticides and fate of biopesticides will not be determined by short-term profits from chemical pesticides (Usta, 2013).

Baculoviruses have been confirmed to be safe in terms of their effect on beneficial insects and vertebrates. NPV is one of the important biopesticides, because it has less residual toxicity, is ecofriendly, has self-perpetuating nature and is compatible with many chemical pesticides. Hence, NPV can be implemented as one of major component in integrated pest management programme. But, scope is always there to develop quality control guidelines and methodologies, systematic registration policies, to identify effective stains and to develop UV (ultraviolet) resistant strains. In addition, the guidelines and training for implementation of biocontrol agents should be made available by competent agencies (Erayya et al., 2013).

References

Aronson AI, and Shai Y (2001). Why *Bacillus thuringiensis* insecticidal toxins are so effective: unique features of their mode of action. FEMS Microbiol Lett 195: 1–8.

Berliner E (1915). Über die Schlaffsucht der Mehlmottenraupe (*Ephestia kühniella* Zell.) und ihren Erreger *Bacillus thuringiensis* , n.sp., Z. Angewandte Entomologie 2: 29–56.

Blissard GW (1996). Baculovirus-insect cell interactions. Cytotechnology 20: 73–93.

Boyce FM, and Bucher NLR (1996). Baculovirus-mediated gene transfer into mammalian cells. Proc Natl Acad Sci USA 93: 2348–2352.

Braunagel SC, Russell WK, Rosas-Acosta G, Russell DH, and Summers MD (2003). Determination of protein composition of the occlusion-derived virus of *Autographa californica* nucleopolyhedrovirus. Proc Natl Acad Sci USA 100: 9797–9802.

Chakraborty S, Monsour C, Teakle R, and Reid S (1999). Yield, biological activity, and field performance of a wild-type Helicoverpa nucleopolyhedrovirus produced in *H. zea* cell cultures. J Invertebr Pathol 73: 199–205.

Erayya Jagdish J, Sajeesh PK, and Upadhyay V (2013). Nuclear Polyhedrosis Virus (NPV), A potential Biopesticide: A Review. Res J Agric For Sci, 1(8): 30–33.

Glick BR, Pasternal JJ, and Pattern CL (2010). Molecular Biotechnology: Principles and Applications of Recombinant DNA, 4th edition.

Gupta S, and Dikshit AK (2010). Biopesticides: An eco-friendly approach for pest control. J. Biopesticides 3: 186–188.

Huh NE, and Weaver RF (1990). Identifying the RNA polymerases that synthesize specific transcripts of the *Autographa californica* nuclear polyhedrosis virus. J Gen Virol 71: 195–202.

Ishiwata S (1901). On a kind of severe flacherie (sotto disease). Dainihon Sanshi Kaiho 114: 1–5.

Jehle JA, Blissard GW, Bonning BC, Cory JS, Herniou EA, Rohrmann GF, Theilmann DA, Thiem SM, and Vlak JM (2006). On the classification and nomenclature of baculoviruses: a proposal for revision. Arch Virol 151: 1257–1266.

Jem KJ, Gong T, Mullen J, and Georgis R (1997). Development of an industrial insect cell culture process for large scale production of baculovirus biopesticides. In: Maramorosch K and Mitsuhashi J (eds) Invertebrate Cell Culture: Novel Directions and Biotechnology Applications. New Hampshire: Science Publishers 173–180.

Kost TA, Condreay JP, and Jarvis DL (2005). Baculovirus as versatile vectors for protein expression in insect and mammalian cells. Nature Biotech 23: 567–575.

Krell PJ (1996). Passage effect of virus infection in insect cells. Cytotechnology 20: 125–137.

Kumar S, Chandra A, and Pandey KC (2008). *Bacillus thuringiensis* (Bt) transgenic crop: an environmentally friendly insect-pest management strategy. J Environ Biol 29: 641–653.

Lereclus D, Delécluse A, and Lecadet M (1993). Diversity of *Bacillus thuringiensis* toxins and genes. In: Entwistle PF, Cory JS, Bailey MJ, Higgs S (eds) *Bacillus thuringiensis* An Environmental Biopesticide: Theory and Practice, Wiley, Chilchester 37–69.

Marvier M, McCreedy C, Regetz J, and Kareiva P (2007). Meta-analysis of effects of Bt cotton and maize on nontarget invertebrates. Science 316: 1475–1477.

Mazid S, Kalita JS, and Rajkhowa RC (2011). A review on the use of biopesticides in insect pest management. Int J Sci Adv Technol 1: 169–178.

McIntosh AH, Rice WC, and Ignoffo CM (1987). Genotypic variants in wild type populations of Baculovirus. In: Maramorsoch K (ed), Biotechnology in Inverterbrate pathology & cell culture, Academic Press, New York, 305–325.

Monteiro F, Carinhas N, Carrondo MJT, Bernal V, and Alves PM (2012). Toward system-level understanding of baculovirus–host cell interactions: from molecular fundamental studies to large-scale proteomics approaches. Front Microbiol 3: 1–16.

Moscardi F (1999). Assessment of the application of baculoviruses for control of Lepidoptera. Ann Rev Entomol 44: 257–289.

Moscardi F (2007). A Nucleopolyhedrovirus for control of the velvetbean caterpillar in Brazilian Soybeans. In: Vincent C, Goethel MS, Lazarovits G (eds) Biological Control: A Global Perspective. Oxfordshire, UK, and Cambridge, USA: CAB International 344–352.

Pedrini MRS, Christian P, Nielsen LK, Reid S, and Chan LCL (2006). Importance of virus-medium interactions on the biological activity of wild-type *Heliothine* nucleopolyhedroviruses propagated via suspension insect cell cultures. J Virol Meth 136: 267–272.

Pedrini MRS, Nielsen LK, Reid S, and Chan LCL (2005). Properties of a unique mutant of *Helicoverpa armigera* single-nucleocapsid nucleopolyhedrovirus that exhibits a partial many polyhedra and few Polyhedra phenotype on extended serial passaging in suspension cell cultures. *In Vitro* Cell Devel Biol 41: 289–297.

Rhodes DJ (1996) Economics of baculovirus-insect cell production systems. Cytotechnology 20: 291–297

Roh JY, Choi JY, Li MS, Jin BR, and Je YH (2007). *Bacillus thuringiensis* as a specific, safe, and effective tool for insect pest control. J Microbiol Biotechnol 17: 547–559.

Schnepf E, Crickmore N, Van Rie J, Lereclus D, Baum J, Feitelson J, Zeigler DR, and Dean DH (1998). *Bacillus thuringiensis* and its pesticidal crystal proteins. Microbiol Mol Biol Rev 62: 775-806.

Shelly A, Wei W, and Pante N (2013). Baculovirus Nuclear Import: Open, Nuclear Pore Complex (NPC) Sesame. 5: 1885–1900.

Slavicek JM, Hayes-Plazolles N, and Kelly ME (2001). Identification of a *Lymantria* nucleopolyhedro virus isolate that does not accumulate few-polyhedra mutants during extended serial passage in cell culture. Biol Control 22: 159–168.

Steinhaus EA (1949). Principals of Insect Pathology. McGraw Hill, New York, 757.

Thakore Y (2006). The biopesticide market for global agricultural use. Ind Biotechnol 2: 194–208.

Usta C (2013). Microorganisms in biological pest control–A Review (Bacterial Toxin Application and Effect of Environmental Factors). Curr Prog Biol Res 287–317.

3

Rhizobacterial Efficacy for Sustainable Crop Productivity in Agroecosystems

Manoj Kaushal
Resilient Dryland Systems, International Crops Research Institute for the Semi-Arid Tropics (ICRISAT), Hyderabad, India
Email: *kaushal.mbg@gmail.com*

1. INTRODUCTION

Sustainable agricultural development is best defined as the management and conservation of agricultural resource base in an appropriate, economically viable and socially acceptable way. Thus, sustainable agriculture should ensure availability, access, utilization and stability of resources to present and future generations. However, ecosystem resilience has emerged as a key factor to reach the goal of sustainability. Resilience is defined as the capability of a system to anticipate, absorb, accommodate or recover from the effects of an undergoing change so as to ensure the preservation or restoration of its essential basic structures, identity and functions. There has been substantial increase in the productivity of the crops with the use of chemical inputs to provide nutrition to the crop and to control pests and diseases. However, these have several negative effects like development of pathogen resistance to the applied agents and their non-target environmental impacts, degradation of soil productivity, depletion of non-renewable sources of energy (Prasad et al., 2014) etc. Plant growth promoting rhizobacteria (PGPR) commonly called rhizobacteria, have received prominent attention because of their multifarious activities to improve plant growth and disease control ensuring crop yields in a sustainable way. The application of PGPR to crops offers an environment friendly approach of reducing chemical inputs (Prasad et al., 2005, 2015).

2. RHIZOBACTERIAL CLASSIFICATION AND THEIR PLANT GROWTH MECHANISMS

Rhizobacteria (PGPR) encompass all bacteria that inhabit rhizosphere and they promote plant growth either through direct mechanisms which include production of phytohormones, enhanced availability of nutrients, stress alleviation or by indirect mechanisms that include suppression of pathogens by antibiosis, synthesis of lytic enzymes and induced systemic resistance (ISR) (Glick, 2012). Bhattacharya and Jha (2012) functionally categorized rhizobacteria as (i) biofertilizers (a substance consisting of live microorganisms having biological nitrogen fixation and phosphorus solubilisation capabilities), (ii) phytostimulators (microorganism with phytohormonal production capability) and (iii) biopesticides (microorganisms controlling phytopathogens). However, based on their location, Gray and Smith (2005) characterized rhizobacteria as extracellular (ePGPRs) and intracellular PGPRs (iPGPRs). Intracellular PGPRs exist inside plant cells and are able to produce specialized nodular structures, whereas ePGPRs colonizing the rhizosphere, the rhizoplane or within the spaces of the root cortex, but not inside the cells. Inoculating crops with rhizobacterial strains has resulted in significant increase in growth and productivity.

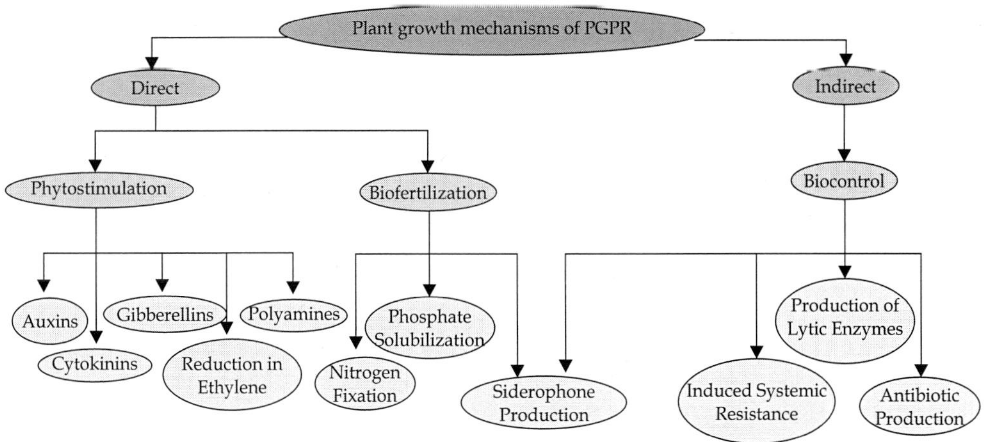

Fig. 3.1 Mechanisms of plant growth promotion exhibited by rhizobacteria

3. PHYTOHORMONAL PRODUCTION

Phytohormonal production by rhizobacteria is among most recognized mechanisms promoting plant growth. It has been reported that the PGPR can stimulate plant growth through production of auxins, gibberellins and cytokinins or by reducing high levels of endogenous ethylene in the plant (Glick, 2012). It has been reported that indole acetic acid (IAA) plays central regulatory roles in apical dominance, lateral root initiation and vascular differentiation. IAA production by PGPR causes modifications in root system architecture by increasing number of root tips and root surface area, thus increasing nutrient acquisition (Mantelin and Touraine, 2004). Different pathways are involved in the biosynthesis of IAA based on the chemical nature of intermediate molecules produced using tryptophan as precursor. However, in Gram-positive bacteria, the main pathway for IAA biosynthesis involves indole-3-pyruvic acid (IPA). PGPR can also stimulate the plant auxin route in an indirect manner. It has been observed that during root colonization by rhizobacterial strains (possessing nitrite reductase activity), NO is produced which results in lateral root formation as NO is involved in auxin signalling pathway (Molina-Favero et al., 2008). IAA production by rhizobacteria, results in increased root length, root surface (Egamberdieva and Kucharova, 2009). Increased root surface area enables the plant to access more nutrients from soil and thus contributes to plant growth promotion. Plant cell walls get loosened in response to rhizobacterial IAA thus causing an increase in amount of root exudates. Thus, the additional nutrient availability further supports the growth of rhizospheric bacteria (Glick, 2012). Many of the rhizobium species show ability to synthesize IAA and some studies have indicated that auxin plays a key role in the nodulation of legume crops and as a whole, in rhizobium-legume symbiosis. It has been proved that flavonoids (as inducers of nodulation genes) intensify IAA synthesis through rhizobiums. Moreover, it has been shown that compared with roots that don't have nodes, nodular roots contain higher levels of IAA hormone and it plays a role in the development of the root system and its maintenance.

Besides auxins, cytokinins and gibberellins producing rhizobacterial strains have been reported to cause plant growth. Production of gibberellins has been reported in *Achromobacter xylosoxidans, Acinetobacter calcoaceticus, Azospirillum* sp., *Azotobacter* sp., *Bacillus* sp., *Herbaspirillum seropedicae, Gluconobacter diazotrophicus* and rhizobia (Dodd et al., 2010). Fulchieri et al. (1993) showed that application of gibberellic acid on maize at a concentration similar to that produced by *Azospirillum*, promotes root growth due to increased root hair density in root zones.

Cytokinin stimulates cell divisions, cell enlargement, induce root hairs proliferation, but inhibit lateral root formation and primary root elongation. Cytokinin production (especially zeatin) has been reported in *Azospirillum brasilense, Bradyrhizobium japonicum, Bacillus licheniformis, Pseudomonas fluorescens,* and *Paenibacillus polymyxa*. Inoculating plants with rhizobacteria producing cytokinins stimulated shoot growth and reduced the root to shoot ratio (Arkhipova et al., 2007).

In the plant, the higher concentration of ethylene induces defoliation and other cellular processes that trammel crop yield. It has been investigated that certain PGPR strains possess the enzyme ACC deaminase which can cleave the plant ethylene precursor ACC to ammonia and α-ketobutyrate thereby lowers the level of ethylene thus promoting plant growth (Glick, 2012). In addition to a direct role of ethylene on plant growth, this hormone can also act as a virulence factor and a signaling molecule in induced systemic resistance (ISR) against pathogen attack (Van Loon, 2007).

Polyamines are aliphatic nitrogen compounds ubiquitous in bacteria, plants and animals, and are involved in various metabolic and hormonal pathways that regulate plant growth and development as well as plant responses under drought stress (Alcazar et al., 2010). Enhanced root growth caused by cadaverine (polyamine) production promoted growth in *Oryza* seedlings when inoculated with *A. brasilense* Az39 (Cassan et al., 2009). Wheat and maize plants inoculated with *Azospirillum brasilense* strain Az39, produced polyamines, such as spermidine and spermine (Perrig et al., 2007), and putrescine in culture.

4. BIOFERTILIZERS

Biofertilizers are considered an environment-friendly alternative to chemical fertilizers ensuring increased crop productivity. The well-studied PGPR's considered biofertilizers correspond to nitrogen fixation, utilization of insoluble forms of phosphorus and production of siderophores.

4.1 Nitrogen Fixation

Nitrogen is an essential nutrient for all forms of life on the earth. The prokaryotic microbes play an important role in conversion of atmospheric nitrogen (N_2) to available form for plant and microbial growth. Rhizobacteria improve the growth and nitrogen fixation by inducing the occupancy of introduced rhizobia in the nodules of the legume. Inoculating rice with *A. amazonense* increased grain dry matter and nitrogen accumulation. Promotion of growth by *A. amazonense* for rice plants was primarily a response to nitrogen fixation (Rodrigucs ct al., 2008). Study conducted by Kaushal and Kaushal (2013), reported that most efficient nitrogen fixing isolate *Bacillus pumilus* isolated from cauliflower rhizosphere showed highest nitrogenase activity (437.26 ηmole C_2H_4 h^{-1} mg^{-1} protein or about 109 mg of N_2 fixed/ha per day). Islam et al. (2010) also reported the Acetylene Reduction Array (ARA) of free living bacteria in the range of 1.8 ηmole C_2H_4 h^{-1} mg^{-1} protein to 2,844.7 26 ηmole C_2H_4 h^{-1} mg^{-1} protein.

4.2 Phosphate Solubilization

Phosphate solubilizing bacteria (PSB) can be a viable substitute to chemical phosphatic fertilizers as they help plants to acquire the available forms of phosphorus. Bacterial genera like *Azotobacter, Bacillus, Burkholderia, Pseudomonas* and *Rhizobium* are the most significant phosphate solubilizing bacteria (Bhattacharyya and Jha, 2012). PSB in general bring about mobilisation of insoluble phosphates, however their plant growth promoting activities, such as indole acetic acid (IAA), gibberellic acid, cytokinins, ethylene production, hydrogen cyanide (HCN) production, asymbiotic nitrogen fixation and resistance to soil-borne pathogens, etc., have been observed by Cattelan et al. (1999). PSB have been used since long as bio-inoculants for various crops due to their antagonistic and phosphate solubilizing properties. Citrate, lactate, and succinate are various organic acids produced by PSB that solubilise mineral phosphates. Gluconic acid and 2-ketogluconic acid are other active organic acids produced by rhizobacteria which consume the organic acids in root exudates and, therefore, indirectly may moderate solubilisation of P and other immobile elements, such as Fe and Mn.

4.3 Siderophore Production

Siderophores are low molecular weight, iron chelating ligands synthesized by microorganisms that suppress pathogenic microorganisms by limiting iron availability. Under iron-limiting conditions, microorganisms produce a range of iron chelating compounds or siderophores which have a high affinity for ferric ions. These bacterial iron chelators are thought to sequester the limited supply of iron available in the rhizosphere making it unavailable to pathogenic fungi, thereby restricting their growth. Recent studies clearly revealed that iron nutrition of plant influences the microbial community structure in rhizosphere. Iron competition in *Pseudomonads* has been intensively studied and role of pyoverdin siderophore produced by many *Pseudomonas* spp. has been clearly demonstrated in the control of fungal pathogen like *Pythium* and *Fusarium* spp. (Duijff et al., 1994). Another siderophore called pseuobactin produced by *Pseudomonads* was responsible for inducing systemic resistance against many fungal diseases (Leeman et al., 1996). *Bacillus subtilis* is the source of the first 2, 3-dihydroxybenzoyl containing substances which have been shown to stimulate uptake of iron. Sharma et al. (2003) suggested that siderophores produced by *Pseudomonas* spp. may be used by the bacteria (homologously) or in effecting plant nutrition (heterologously). The problem of iron non-availability particularly in calcareous soils may be overcome by incorporation of siderophore producing strains of fluorescent *Psuedomonads* (FLPs). Cantero et al. (2007) isolated bacterial strains showing dissimilatory Fe-reducing activity from rhizosphere of bean and maize plants. Isolates were identified as *Bacillus megaterium, Arthrobacter* sp. and *Stenotrophomonas maltophilia.* All the isolates showed Fe reduction in alkaline soil and were able to stimulate plant growth. Ahmad et al. (2008) also proved that production of siderophore by PGPR directly influence plant growth promotion. Chakraborty et al. (2006) isolated siderophore producing *Bacillus megaterium* from tea rhizosphere and reported that the bacterial inoculation has reduced disease incidence in tea.

5. BIOLOGICAL CONTROL

In modern input intensive agricultural production system, the indiscriminate use of agrochemicals is posing a chronic threat not only economically, but also ecologically to the entire ecosystem throughout the world. The mechanism of plant growth promoting bacteria as biocontrol agents has become increasingly better understood over the past two decades. Rhizosphere competence of biocontrol agents comprises effective root colonization combined with the ability to survive and proliferate along growing plant roots over a considerable time period, in the presence of indigenous microflora. Studies have shown that plant growth promoting rhizobacteria can be applied to a wide range of plants for the purpose of disease control and growth enhancement (Van Loon et al., 2007). The application of PGPR as biological agent has been recommended to control bacterial wilt of tomato caused by *Ralstonia solanecearum* (Guo et al., 2002), black rot of crucifers caused by *Xanthomonas campestris* pv. *campestris,* root rot of cauliflower caused by *Pythium ultimum* var. *ultimum* (Mariano et al., 2002) and dry rot disease caused by *Fusarium* spp. (Recep et al,. 2009). Anith et al. (2004) compared the effectiveness of different strains of *Pseudomonas* (*P. putida* 89B61) and *Bacillus* (*B. pumilus* SE34) to control bacterial wilt caused by *Ralstonia solanacearum* and recommended *P. putida* 89B61 application for effective control under greenhouse condition. Srivastav et al. (2004) recommended the use of phosphate solubilizing bacteria of *Pseudomonas* spp. against a number of plant pathogenic fungi. The application of *Pseudomonas fluorescens* to black pepper is known to enhance root proliferation and fibre root production in addition to act as biocontrol agent against foot rot caused by *Phytophthora capsici* (Paul and Sharma, 2006). The offensive plant growth promoting bacterial colonization and defensive retention of rhizosphere niches are enabled by production of allelo-chemicals; including antibiotics, iron chelating siderophores, biocidal volatiles, lytic enzymes and detoxification enzymes. Butyrolactones, zwittermycin A, kanosamine, oligomycin A, phenazine-1-carboxylic acid, pyoluteorin, pyrrolnitrin, viscosinamide, xanthobaccin, and 2,4-diacetyl phloroglucinol (2,4-DAPG) are various antibiotics produced by rhizobacterial strains. The 2,4-DAPG is one of the most efficient antibiotics (phloroglucinol compound) having antifungal, antibacterial and antihelmintic properties. It causes membrane damage to *Pythium* spp. and also displayed inhibitory effects to its zoospores. Pyrrolnitrin inhibits fungal respiratory chains. HCN is released as product of secondary metabolism by several microorganisms and affects sensitive organisms by inhibiting the synthesis of ATP-mediated cytochrome oxidase. Studies have shown the role of HCN produced by many rhizospheric bacteria in disease suppression of the pathogens. Various bacterial genera have been reported for HCN production, including species of *Alcaligenes, Aeromonas, Bacillus, Pseudomonas* and *Rhizobium*. *Bacillus* spp. produces various biotics, such as polymyxin, circulin and colistin, which are active against Gram-positive and Gram-negative bacteria, as well as an many pathogenic fungi. The growth and development of filamentous fungi can be repressed both *in vitro* and *in vivo* by lytic enzymes, such as chitinases and gluconases produced by *Bacillus* and *Pseudomonas* spp. (Maksimov et al., 2011). *Bacillus* species can exert biocontrol which is being characterized at the molecular level. The biosynthetic gene cluster responsible for the production of the antibiotic zwittermicinA in *Bacillus cereus* has been identified (Milner et al., 1999). *Pseudomonas* produces thiamine which acts as an important factor in quorum sensing dependent expression of biological control. Global regulation and quorum sensing are key factors involved in the biosynthesis of most anti-fungal metabolites anti-fungal metabolites (AFMs). The *gacS/gacA*gene directs global regulation which encodes a two-component regulatory system that senses an unknown signal(s). However, in quorum sensing involved by an Acyl-Homoserine Lactore (AHL) synthase, such as LuxI v leads to the production of *N*-acyl homoserine lactone (AHL) signal molecules. The AHL causes activation of a transcriptional regulator, such as LuxR which then stimulates gene expression (Bloemberg and Lugtenberg, 2001).

Induced systemic resistance (ISR) has been reported as one of the mechanisms by which PGPR reduces plant disease, through the manipulation of the host plant's physical and biochemical properties. Interaction of rhizobacteria with plant roots results in enhancement of plants defensive capacity against a broad spectrum of pathogens and pests, phenomenon called ISR (Lugtenberg and Kamilova, 2009). ISR involves jasmonic acid (JA) and ethylene (ET) signaling pathways within the plant and these hormones stimulate the host plant's defense responses against a variety of plant pathogens (Glick, 2012). Pieterse et al. (1998) reported that ISR triggered by *Pseudomonas fluorescens*WCS417r signals resistance responses through JA and ET dependent pathways. A regulator protein NPR1 is required for expression of ISR. Jetiyanov and Kloepper (2002) observed that PGPR mediate biological control indirectly by eliciting ISR against a number of plant diseases. Rhizobacterial elicited ISR has been demonstrated in many plant species, including *Arabidopsis,* carnation, cucumber, radish, tobacco and tomato (Van Loon et al., 2007). One of the consequences of ISR reported is the reinforcement of the cell wall through enhanced lignin synthesis and callose apposition. The bacterial components known to induce ISR are lipopolysaccharides (LPS), flagella, siderophores, cyclic lipopeptides, 2,4-diacetylphloroglucinol, homoserine lactones, and volatiles like, acetoin and 2,3-butanediol (Lugtenberg and Kamilova 2009). A large number of defense enzymes that have been associated with ISR include phenylalanine ammonia lyase (PAL), chitinase, β-1,3-glucanase, peroxidase (PO), polyphenol oxidase (PPO), superoxide dismutase (SOD), catalase (CAT), lipoxygenase (LOX), ascorbate peroxidase (APX) and proteinase inhibitors. These enzymes cause release of molecules that elicit the initial steps in induction of resistance (Van Loon et al., 2007).

6. STRESS ALLEVIATION

Rhizobacteria can provide protection to plants against the deleterious effects of environmental stresses, such as drought, flooding, high temperature and salinity. Different mechanisms are exhibited by rhizobacterial strains imparting resistance to plants against various stresses. The term induced systemic tolerance (IST) has been proposed for rhizobacterial induced physical and chemical changes that result in enhanced tolerance to abiotic stress (Yang et al., 2009). One of the mechanisms employed by rhizobacterial strains ensuring plant survival under stressed conditions involves the modification in content of bacterial phytohormones, such as auxins, gibberellins, cytokinins and ABA (Bresson et al., 2013). Production of IAA, gibberellins by PGPR strains displayed increased root length, root surface area and number of root tips, causing enhanced nutrient uptake thus improving plant growth under stress conditions (Egamberdieva and Kucharova, 2009). Drought and salinity disrupt photosynthesis and increases photorespiration, causing enhanced production of reactive oxygen species (ROS), such as the superoxide radical, hydrogen peroxide and hydroxyl radical. Excess ROS causes damages to proteins, DNA and lipids, so it is mandatory to regulate the ROS levels through the coordination of ROS production and ROS scavenging systems to combat oxidative damage occurring during stress conditions. A significant relation between drought stress and antioxidant enzyme activity has been reported. Inoculating maize plants with *Bacillus* species imparted protection against drought stress by reduced activity of antioxidant enzymes APX and Glutatheone peroxidase (GPX) (Vardharajula et al., 2011). Rhizobacterial strains possess the enzyme 1-aminocyclopropane-1-carboxylate (ACC) deaminase which cleaves the plant ethylene precursor ACC to ammonia and α-ketobutyrate, thereby causing reduced ethylene level during stressed conditions. Bacterial ACC deaminase confers tolerance against water deficit in tomato and pepper plants inoculated with *Achromobacter piechaudii* ARV8 (Mayak et al., 2004). Siddikee et al. (2010) reported that inoculation with 14 halotolerant bacterial strains ameliorated salt stress in canola plants through the reduction of ethylene production via ACC deaminase activity.

Rhizobacterial production of osmoprotectants such as sugars and amino acids is another acclimatization response observed under stressed conditions. Proline synthesis has been shown to be increased in abiotically stressed plants in the presence of beneficial bacteria, such as *Burkholderia*, as well as *Arthrobacter* and *Bacillus* (Sziderics et al., 2007). Plants treated with exopolysaccharide (EPS) producing bacteria showed resistance to water stress through improved soil structure and soil aggregation (Sandhya et al., 2009). EPS binds Na^+ ions which decrease its accumulation in plants and preventing their transfer to leaves thus alleviating salt stress in plants (Ashraf et al., 2004). *Bacillus subtilis* produces volatile organic compounds resulting in decreased expression of high-affinity K^+ transporter (HKT1) in roots, but its upregulation in the shoots, thus decreasing root Na^+ import, but facilitating Na^+ exclusion from the shoot by retrieving Na^+ from the xylem and facilitating root-to-root Na^+ recirculation (Zhang et al., 2008). Production of heat shock proteins (Berjak, 2006), dehydrins (Timmusk and Wagner, 1999) and volatile organic compounds (Ryu, 2004) have also been reported to impart tolerance to plants against abioic stresses.

Production of volatile organic compounds (VOCs) by rhizobacteria show antimicrobial and plant-growth modulating activities. VOCs promote plant growth by regulating auxin homeostasis in which the gene expression for auxin production was upregulated. In addition, microarray data revealed coordinated regulation of cell wall loosening enzymes that implicated cell expansion with *B. subtilis* GB03 exposure (Zhang et al., 2007). *Bacillus subtilis* and *B. amyloliquefaciens* produced 3-hydroxy-2-butanone and 2,3-butanediol which enhanced total leaf surface area and induced systemic resistance (ISR) of *A. thaliana* (Ryu et al., 2004).

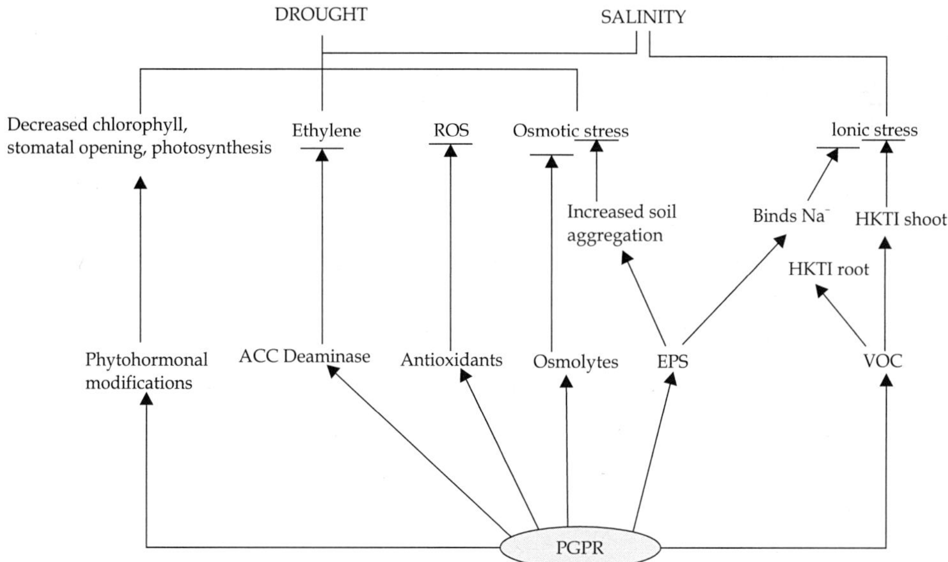

Fig. 3.2 Stress alleviating mechanisms of rhizobacteria

(Abbreviations: ROS, reactive oxygen species; *HKT1,* high affinity K^+ transporter; ACC, 1-aminocyclopropane-1-carboxylate; VOC, volatile organic compound; EPS, exopolysaccharides)

7. CONCLUSION

Rhizobacteria are efficient microbial competitors in the root zone to enhance plant growth directly through biofertilization, phytostimulation and indirectly by reducing soil-borne pathogens. The application of rhizobacteria to crops as inoculants is a trend for future as it substantially reduces the chemical inputs thus contributing towards increased crop yields. In addition, various rhizobacterial strains imparting stress tolerance have been identified. However, there is need to develop more effective and longer shelf-lived PGPR strains to achieve sustainable crop production in agroecosystems.

Acknowledgements

The author is thankful to ICRISAT and ICAR for supporting research work done.

References

Ahmad F, Ahmadand I, and Khan MS (2008). Screening of free living rhizospheric bacteria for their multiple plant growth promoting activities. Microbiol Res 163(2):173–181.

Alcazar R, Bitrian M, Bartels D, Koncz C, Altabella T, and Tiburcio AF (2010). Polyamines, molecules with regulatory functions in plant abiotic stress tolerance. Planta 231:1237–1249.

Anith KN, Momol MT, Kloepper JW, Marois JJ, Olson SM, and Jones JB (2004). Efficacy of plant growth promoting rhizobacteria, acibenzolar-S-methyl and soil amendment for integrated management of bacterial wilt of tomato. Plant Disease 88: 669–673.

Arkhipova TN, PrinsenE,Veselov SU, Martinenko EV, Melentiev AI, and Kudoyarova GR (2007). Cytokinin producing bacteria enhance plant growth in drying soil. Plant and Soil 292: 305–315.

Ashraf M, Berge SH, and Mahmood OT (2004). Inoculating wheat seedling with exopolysaccharide-producing bacteria restricts sodium uptake and stimulates plant growth under salt stress. Biol Fert Soils 40:157–162.

Berjak P (2006). Unifying perspectives of some mechanisms basic to desiccation tolerance across life forms. Seed Sci Res 16: 1–15.

Bhattacharyya PN, and Jha DK (2012). Plant growth-promoting rhizobacteria (PGPR): emergence in agriculture. World J Microbiol Biotechnol 28: 1327–1350.

Bloemberg GV, and Lugtenberg BJ (2001). Molecular basis of plant growth promotion and biocontrol by rhizobacteria. Curr Opin Plant Biol 4: 343–350.

Bresson J, Varoquaux F, Bontpart T, Touraine B, and Vile D (2013). The PGPR strain *Phyllobacterium brassicacearum* STM196 induces a reproductive delay and physiological changes that result in improved drought tolerance in *Arabidopsis*. New Phytol 200: 558–569.

Cantero EV, Calderon EH, Becerra CV, Meza JE, Lopez M, Cuevas RA, and Bucio JL (2007). Role of dissimilatory fermentative iron-reducing bacteria in Fe uptake by common bean (*Phaseolus vulgaris* L.) plants grown in alkaline soil. Plant Soil 291: 263–273.

Cassan F, Maiale S, Masciarelli O, Vidal A, Luna V, and Ruiz O (2009). Cadaverine production by *Azospirillum brasilense* and its possible role in plant growth promotion and osmotic stress mitigation. Eur J Soil Biol 45: 12–19.

Cattelan AJ, Hartel PG, and Fuhrmann JJ (1999). Screening of plant growth promoting rhizobacteria to promote early soybean growth. Soil Sci Soc Am J 63: 1670–1680.

Chakraborty U, Chakraborty B, and Basnet M (2006). Plant growth promotion and induction of resistance in *Camellia sinensis* by *Bacillus megaterium*. J Basic Microbiol 34: 925–929.

Dodd IC, Zinovkina NY, Safronova VI, and Belimov AA (2010). Rhizobacterial mediation of plant hormone status. Annal Appl Biol 157: 361–379.

Duijff BJ, Bakker PAHM, and Schippers B (1994). Suppression of *Fusarium* wilt of carnation by *Pseudomonas putida* WCS358 at different levels of disease incidence and iron availability. Biocontrol Sci Techn 4: 279–288.

Egamberdieva D, and Kucharova Z (2009). Selection for root colonizing bacteria stimulating wheat growth in saline soils. Biol Fert Soil 45: 561–573.

Fulchieri M, Lucangeli C, and Bottini R (1993). Inoculation with *Azospirillum lipoferum* affects growth and gibberellin status on corn seedling roots. Plant Cell Physiol 34: 1305–1309.

Glick BR (2012). Plant growth-promoting bacteria: mechanisms and applications. Scientifica.Article ID 963401, 15 pages,http://dx.doi.org/10.6064/2012/963401.

Gray EJ, and Smith DL (2005). Intracellular and extracellular PGPR: commonalities and distinctions in the plant–bacterium signalling processes. Soil Biol Biochem 37:395–412.

Guo JH, Qi HY, and Li SM (2002). Biocontrol efficiency of three PGPR strains admixture of pepper bacterial wilt. Bacterial Wilt Newsletter 17:3.

Islam MR, Ahamed R, Rahman MO, Akbar MA, Al-Amin M, Alam KD, and Lyzu F (2010). In vitro antimicrobial activities of four medicinally important plants in Bangladesh. Eur J Sci Res 39: 199–206.

Jetiyanov J, and Kloepper JW (2002). Mixture of plant growth promoting rhizobacterial for induction of systemic resistance against multiple plant diseases. Biol Control 24: 285–291.

Kaushal M, and Kaushal R (2013). Screening and characterization of plant growth promoting rhizobacteria isolated from rhizosphere of cauliflower (*Brassica oleracea* var. *botrytis* L.). Afr J Microbiol Res 7(17): 1657–1663.

Leeman M, Van Peet JA, Den Quden FM, Henisbroek M, Bakker PAHM, and Schippers B (1996). Iron availability affects induction of systemic resistance to *Fusarium* wilt of radish by *Pseudomonas fluorescens*. Phytopathol 86: 149–155.

Lugtenberg B, and Kamilova F (2009). Plant growth promoting rhizobacteria. Annu Rev Microbiol 63: 541–556.

Maksimov IV, Abizgil'dina RR, and Pusenkova LI (2011). Plant growth promoting rhizobacteria as alternative to chemical crop protectors from pathogens (Review). Appl Biochem Microbiol 47: 333–345.

Mantelin S, and Touraine B (2004). Plant growth-promoting rhizobacteria and nitrate availability: impacts on root development and nitrate uptake. J Expt Bot 55: 27–34.

Mariano RLR, Luna CL, and Souto Maior AM (2002). Production of biocontrol agent for crucifers black rot disease. Braz J Chem Eng 19(2): 133–140.

Mayak S, Tirosh T, and Glick BR (2004). Plant growth promoting bacteria that confer resistance to water stress in tomato and pepper. Plant Sci 166: 525–530.

Milner JI, Stohl EA, and Handelsman J (1999). Zwittermicin A biosynthetic cluster. Gene 237: 403–411.

Molina-Favero C, Creus CM, Simontacchi M, Puntarulo S, and Lamattina L (2008). Aerobic nitric oxide production by *Azospirillum brasilense* Sp245 and its influence on root architecture in tomato. Mol Plant Microbe Interact 21: 1001–1009.

Paul D, and Sharma YR (2006). Plant growth promoting rhizhobacteria (PGPR)-mediated root proliferation in black pepper (*Piper nigrum* L.) as evidenced through GS Root software. Arch Phytopathol Plant Protect 39: 1–4.

Perrig D, Boiero ML, Masciarelli OA, Penna C, Ruiz OA, Cassan FD, and Luna MV (2007). Plant-growth-promoting compounds produced by two agronomically important strains of *Azospirillum brasilense*, and implications for inoculant formulation. Appl Microbiol Biotechno l75: 1143–1150.

Pieterse CMJ, Van Wees SCM, Van Pelt JA, Knoester M, Laan R, Gerrits H, Weisbeek PJ, and Van Loon LC (1998). A novel signaling pathway controlling induced systemic resistance in Arabidopsis. Plant Cell 10: 1571–1580.

Prasad R, Garg AP, and Varma A (2005). Interaction of medicinal plants with plant growth promoting rhizobacteria and symbiotic fungi. In: Podila GK, Varma A (eds), Basic Research and Applications of Mycorrhizae. IK International Pvt Ltd. New Delhi, India, 1: 363–40719.

Prasad R, Kumar V, and Prasad KS (2014). Nanotechnology in sustainable agriculture: present concerns and future aspects. African Journal of Biotechnology 13(6): 705–713.

Prasad R, Kumar M, and Varma A (2015). Role of PGPR in soil fertility and plant health. In: Egamberdieva D, Shrivastava S, Varma A (eds)Plant Growth-Promoting Rhizobacteria and Medicinal Plants, Springer International Publishing Switzerland 247–260.

Recep K, Fikrettin S, Erkol D, and Cafer E (2009). Biological control of the potato dry rot caused by Fusarium species using PGPR strains. Biol Control 50(2):194–198.

Rodrigues EP, Rodrigues LS, de Oliveira ALM, Baldani VLD, Teixeira KRD, Urquiaga S, and Reis VM (2008). *Azospirillum amazonense* inoculation: Effects on growth, yield and N_2-fixation of rice (*Oryza sativa* L.). Plant Soil 302: 249–261.

Ryu CM (2004). Bacterial volatiles induce systemic resistance in *Arabidopsis*. Plant Physiol 134: 1017–1026.

Ryu CM, Farag MA, Hu CH, Reddy MS, Kloepper JW, and Pare PW (2004). Bacterial volatiles induce systemic resistance in *Arabidopsis*. Plant Physiol 134: 1017–1026.

Sandhya V, Ali SkZ, Grover M, Reddy G, and Venkateswarlu B (2009). Alleviation of drought stress effects in sunflower seedlings by exopolysaccharides producing *Pseudomonas putida* strain P45. Biol Fert Soil 46: 17–26.

Sharma A, Johri BN, Sharma AK, and Glick BR (2003). Plant growth promoting bacterium Pseudomonas spp. strain GRP3 influences iron acquisition in mung bean (*Vigna radiate* L. Wilzeck). Soil Biol Biochem 35: 887–894.

Siddikee MA, Chauhan PS, Anandham R, Han GH, and Sa T (2010). Isolation, characterization, and use for plant growth promotion under salt stress, of ACC deaminase - producing halotolerant bacteria derived from coastal soil. J Microbiol Biotechnol 20: 1577–1584.

Srivastav S, Yadav KS, and Kundu BS (2004). Prospects of using phosphate solubilizing *Pseudomonas* as biofungicide. Ind J Microbio l44: 91–94.

Sziderics AH, Rasche F, Trognitz F, Sessitsch A, and Wilhelm E (2007). Bacterial endophytes contribute to abiotic stress adaptation in pepper plants (*Capsicum annuum* L.). Can J Microbiol 53: 1195–1202.

Timmusk S, Wagner EGH (1999). The plant growth-promoting rhizobacterium *Paenibacillus polymyxa* induces changes in *Arabidopsis thaliana* gene expression: a possible connection between biotic and abiotic stress responses. Mol Plant Microbe Inter 12: 951–959.

Van Loon LC (2007). Plant responses to plant growth-promoting rhizobacteria. Eur J Plant Pathol 119: 243–254.

Vardharajula S, Zulfikar Ali S, Grover M, Reddy G, and Bandi V (2011). Drought-tolerant plant growth promoting *Bacillus* sp. effect on growth, osmolytes and antioxidant status of maize under drought stress. J Plant Inter 6: 1–14.

Yang J, and Kloepper JW, Ryu CM (2009). Rhizosphere bacteria help plants tolerate abiotic stress. Trends Plant Sci 14: 1–4.

Zhang H, Kim MS, Krishnamachari V, Payton P, Sun Y, Grimson M, Farag MA, Ryu CM, Allen R, Melo IS, and Pare PW (2007). Rhizobacterial volatile emissions regulate auxin homeostasis and cell expansion in *Arabidopsis*. Planta 226: 839–851.

Zhang H, Kim MS, Sun Y, Dowd SE, Shi H, and Pare PW (2008). Soil bacteria confer plant salt tolerance by tissue-specific regulation of sodium transporter HKT1. Mol Plant Microbe Interact 21: 737–744.

4

Heavy Metal Resistance Bacterial Strains from Agricultural Polluted Sites of Northern India

TANVEER BILAL[1], NAVREET KAUR[2], BISMA MALIK[1], REIAZ UL REHMAN[1] AND MANOJ KUMAR[3*]

[1]Department of Bioresources, University of Kashmir, Srinagar–190006
[2]School of Biotechnology, Lovely Professional University, Phagwara, Punjab, India
[3]Amity Institute of Microbial Technology, Amity University Uttar Pradesh, Noida–201303
*Corresponding author e-*mail: manojjnu@gmail.com, pztanveer@gmail.com, bisma767@gmail.com and reiazrehman@yahoo.co.in*

ABSTRACT

Heavy metals play a pivotal role in the metabolic processes of the biota. However, due to their excessive use and dumping at high concentrations, they become toxic to both humans and microorganisms. Bacteria have evolved uptake and efflux mechanisms to adapt in heavy metals contaminated environments and thus represent a potential source for bioremediation processes. Bacteria capable of accumulating heavy metals were isolated from soil samples from northern India (Punjab) and identified by standard biochemical tests. Out of the different isolates, four were capable of growing in the presence of arsenic at varying concentrations (1.15-4.62 mg/kg), nickel (2.96-38.3 mg/kg), cadmium (0-18.72 mg/kg) and chromium (7.715-33.28 mg/kg) which were sampled as "heavy metal resistant bacteria" mentioned as HMR1, HMR2, HMR3, HMR4 respectively. The growth inhibitory assay results for each of the bacterial isolates from the tested heavy metals were performed. The maximum tolerance of HMR1 against arsenic was 500 µg/ml, HMR2 against cadmium was 700 µg/ml, HMR3 against chromium was 250 µg/ml, of HMR4 against nickel was 175 µg/ml. Pure cultures were raised from different samples (HMR1 to HMR4) and subjected for fatty acid methyl esters analysis to confirm them as *Bacillus, Stenotrophomonas*.

1. INTRODUCTION

Industrialization and mal-agricultural practices have raised the level of toxic heavy metals in the environment thus degrading the quality of soil (Pirzadah et al., 2014). At low concentrations, metal ions play a vital role in biological systems, often serving as an important component in the enzymatic reactions (Gort et al., 1994; Buffoni and Ignesti, 2000; Cooper et al., 2000). However, above certain threshold concentrations, these cause extreme toxicity not only to the humans and environment but also to the microorganisms (Pirzadah et al., 2014). Heavy metals generally exert an inhibitory action on microorganisms by blocking essential functional groups, displacing essential metal ions or modifying the active conformations of biological molecules (Doelman et al., 1994; Bruins et al., 2000; Gikas et al., 2009). Thus, microorganisms have acquired a variety of mechanisms for their survival under metal stressed conditions. These include transport across the cell membrane, biosorption to the cell walls and entrapment in extracellular capsules, extracellular precipitation, complexation and oxidation-reduction reactions (Dopson et al., 2003). However, the extent and efficiency of bioremediation varies noticeably with the metal as well as with the microorganisms. Most of the times, these resistance mechanisms are located on the plasmids but in some organisms these have been revealed on chromosomal genes (Zolgharnein et al., 2007). The mechanism involved in gaining resistance against these toxic heavy metals is the long-term exposure to heavy metals or widespread use of antibiotics. Generally, when microorganisms are subjected to heavy metal ions, resistance comes mostly in plasmid-encoded bacteria. Resistance genes encode genetic information of microorganisms that is changed by external or internal stimuli. It has been reported that *S. aureus* has one main mechanism for adapting to high levels of environmental copper via increased oxidative stress resistance (Baker et al., 2010). Many microorganisms possess the ability to secrete antioxidant enzymes such as superoxide dismutase that helps them to provide tolerance to these toxic metal ions by employing redox reactions (Singh et al., 2013). In addition to this, there are certain cysteine-rich proteins known as metallothionein present in microorganisms to detoxify these toxic heavy metals (Chudobova et al., 2015). However, regulation of expression is probably caused by metal binding to the transcription factor MTF-1, although the information about expression is currently still insufficient (Babula et al., 2012). Many bacterial strains contain genetic determinants of resistance to heavy metals such as, arsenic (*arsC*), cadmium (*cadA*) and chromium (*chrA*) (Gladysheva et al., 1994; Nucifora et al., 1989; Cooksey, 1994). Microorganisms possess the capability to uptake these toxic heavy metals either actively (bioaccumulation) and/or passively (adsorption) or biotransformation of toxic heavy metals to less toxic forms (Ahemad, 2014; Malik et al., 2014). Among numerous bioremediation methods, the biosorption phenomenon seems to be more versatile for commercial scale compared to the bioaccumulation process, as microbes will require addition of nutrients for their active uptake of heavy metals, which enhances the biological oxygen demand or chemical oxygen demand in the waste. Moreover, it is quite difficult to maintain a healthy population of microorganisms due to heavy metal toxicity and other environmental factors (Dilek et al., 1998). Therefore, microorganisms isolated from contaminated sites exhibit more tolerance to these toxic metal ions than non-polluted areas. As the cell wall of microbes is composed of polysaccharides, lipids as well as proteins offer various functional groups such as hydroxyl, amino, carboxylate and phosphate groups that, in turn, bind to metal ions (Scott et al., 1992). A novel technology has recently been reported for obtaining highly efficient biosorbents from endophytes, a hyperaccumulator, which is more convenient than the traditional method of obtaining biosorbents (Xiao et al., 2010). In another investigation genetic diversity of endophytic bacteria from the copper-tolerant species of *Elshotzia spliendens* and *Commelina communis*, reported enhanced dry weights of roots and aboveground tissues compared to uninoculated plants (Sun et al., 2010). The metal tolerance ability of these microorganisms can be manipulated in plethora of ways to facilitate several biotechnological applications (Cavicchioli and Thomas, 2002). These can be used for any kind of bio-mining of expensive metals, directly on ores or by recovering metals from effluents of industrial processes (Acevedo, 2000). Bioremediation of heavy metals using these microorganisms has received a great deal of attention in recent years as it offers a cheaper alternative to chemical technology and is an eco-friendly option (Volesky, 2001; Ray and Ray, 2009, Malik, 2004). Strategies for bioremediation of toxic metals include mainly the microbe-mediated enzymatic reduction of toxic compounds (Torres et al., 2003 and De et al., 2006). Based on the potential applications of metal resistant bacteria, researchers have been exploring the microbial world with the utmost efforts. Heavy metal resistance in microorganisms from different habitats has been reported by several groups (Sharma et al., 2000; Shi et al., 2002; Chien et al., 2008; Singh et al., 2010). A plethora of microbial genera that includes *Bacillus, Enterobacter, Escherichia, Pseudomonas* and also some yeasts and fungi were found to be adapted to such toxic environment (Ezzouhri et al., 2009; Abdelatey et al., 2011). Although, numerous studies have been reported from the past 30 years on such aspects which provided the base to capture the genetic potential to upgrade the environmental disorders. The present study is specifically meant for contaminated industrial sites which are treated as one of the major threats to the masses. At the same time, this study will give some additional insights pertaining to the heavy metal sensitive flora to be cultivated in targeted area (northern India: Punjab).

2. SOURCES OF HEAVY METALS

Heavy metals are either present naturally in the environment from pedogenetic processes of weathering of parent materials or generated through anthropogenic sources. Currently, with the use of numerous anthropogenic activities like mining, smelting, electroplating, use of pesticides and synthetic fertilizers as well as bio-solids (viz., composts and municipal solid waste), atmospheric deposition, etc., (Wuana et al., 2011) (Fig. 4.1).

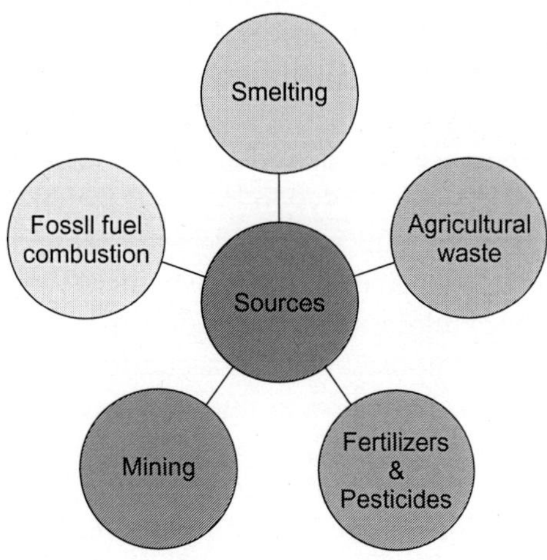

Fig. 4.1 Different sources of heavy metals

3. MOLECULAR MECHANISMS INVOLVED IN BIOREMEDIATION PROCESS

Numerous mechanisms involved in removal of heavy metals by employing use of microorganisms are known. Brim et al. (2003) reported that in a genetically engineered bacterium *Deinococcus geothemalis*, Hg reduction occurs at elevated temperatures due to the expression of mer operon from *E.* coli coded for Hg^{2+} reduction. By means of genetic engineering mercury-resistant bacteria *Cupriavidus metallidurans* strain MSR33 was engineered by introducing a pTP6 plasmid that provided genes (merB and merG) regulating Hg biodegradation along with the synthesis of organomercurial lyase protein (MerB) and mercuric reductase (MerA) (Rojas et al., 2011). Similarly, engineered *Pseudomonas* strain with the pMR68 plasmid with novel genes (*mer*) made that strain tolerant to mercury heavy metal (Sone et al., 2013). There are two distinct mechanisms for Hg degradation by bacteria (*Klebsiella pneumonia* M426) which include mercury volatilization by reduction of Hg (II) to Hg (0) and mercury precipitation as insoluble Hg due to volatile thiol (H2S) (Essa et al., 2002). Another bacterium such as *Deinococcus radiodurans* considered a radiation-tolerant bacterium that has a natural tendency to reduce Cr (IV) to Cr (III) was engineered by incorporating cloned genes of *tod* and *xyl* operons of *Pseudomonas putida* to degrade toluene completely (Brim et al., 2006). Microbial metabolites like metal bound coenzymes and siderophores mainly involved the degradation pathway (Penny et al., 2010).

4. DISCUSSION

4.1. Physico-chemical Analysis of Samples

As metal-resistant bacteria play an essential role in detoxifying these toxic metal ions thus plays a pivotal role in bioremediation process. Bioremediation technology has gained a worldwide importance to maintain the quality of soil besides it is a sustainable and a cost-effective technology. With this consideration, the present study was carried to isolate and characterize distinct metal-resistant bacteria from northern India (Punjab). Physio-chemical characterization was done by measuring pH that ranges from 7.0-8.5 for each sample was taken into consideration. The maximum alkalinity was observed in the sludge sample collected from site 2 (pH 8.5) followed by the grit sample from site 3 (pH 8.0) and minimum was noted for the sludge sample from site 1 (pH 7.0). The obtained moisture content values ranges from 10 to 25%. The observed moisture content values fall within the standard limits of Standards Association of Australia (*AS 1289 B1.1-1977*) which is common for all the studied samples. Variation was noted in the moisture content values of the sludge samples from the two different sites. The sample from site 1 showed the maximum moisture level (25%) whereas, 10% moisture content was noted for the sludge from site 2. The reason behind such variability in the moisture content is only because of the periodic existence of the sludge. The sludge was collected from site 2 after 7 days of exposure. On the other hand, the same sample was collected fresh from site 1 directly from the bioreactor under operation. The variation observed in the moisture content from different sites was found to have no correlation with the growth of microorganisms. In contrast, there is wide evidence reporting the correlation of microbial growth and moisture content (Yeager and Ward, 1981). Table 4.1 depicts the results of elemental analysis of various samples based on ICP-AES. Besides the presence of crude elements, some of the heavy metals that include arsenic, cadmium, cobalt, chromium, manganese, nickel, lead, copper, iron and żinc were also detected in the samples used in this study. Sample 1 showed higher proportion of all the metals detected in the samples as compared to the other two samples. The level of arsenic observed in the samples ranged between 2.70-4.62 mg/kg of the sample. The concentration of arsenic was found to be maximum for sample 1 (4.62 mg/kg) and it is minimum in sample 2 (2.7 mg/kg). The level of cadmium and chromium ions present in the studied isolates also showed the same pattern as observed for arsenic. The amount of cadmium in the studied samples ranged from 0-18.72 mg/kg. The sludge sample from site 2 showed the absence of cadmium. Shukla et al. (2007) also recorded similar value (<0.001 mg/l) for cadmium concentration in the effluent sample from aeration tank of common effluent treatment plant. The level of chromium detected in the soil during the studies was much higher (7.72-33.28 mg/kg) for all the samples, as compared to the safe limit prescribed by World Health Organization (WHO) in 2005 which is 0.007 mg/kg of chromium. It indicates that contamination of heavy metals in the samples used in our study was alarming. The above evidence regarding the presence of heavy metals in the studied samples indicates the perspective of bacterial species to be highly reciprocated in terms of gaining resistance.

Table 4.1: Elemental composition of effluents from contaminated sites determined by ICP-AES analysis

Element	Sample 1*	Sample 2*	Sample 3*
Arsenic	4.62	2.7	3.67
Calcium	30210	3904	26085
Cadmium	18.72	0	16.37
Cobalt	4.16	4.525	4.165
Chromium	32.93	20.15	33.28
Copper	168.9	53.55	136.5
Iron	10360	9350	8395
Potassium	2070	2012	2490
Magnesium	7952	2551	7150
Manganese	160.4	173.9	165.5
Sodium	2160	348.9	1143
Nickel	38.3	11.55	36.34
Phosphorus	2231	428.3	2872
Lead	258	23.85	182.3
Sulphur	4106	707	3077
Zinc	637	149	452

4.2 Isolation and Purification of Metal-Resistant Bacterial Species

Primary screening of colonies was performed using nutrient agar medium supplemented with heavy metals. Three isolates were selected in the secondary screening found resistant against arsenic, cadmium and chromium. The isolates were designated as HMR-2 from site 1 which was found resistant against arsenic, HMR-3 from site 2 exhibits resistance against cadmium and HMR-5 from site 3 was found to be resistant against chromium (Fig. 4.2). All the isolates showed high resistance to heavy metals with the minimum inhibitory concentration (MIC) for heavy metals ranging from 200 µg/ml to 850 µg/ml. Similar inhibitory response was observed for both the cases on semi-solid media as well as in nutrient broth. MIC test indicated the highest resistance against arsenic for the isolate HMR-2 and the lowest resistance against chromium for isolates HMR-5. Separate growth studies of three isolates (HMR2, HMR3 and HMR5) were performed using nutrient broth supplemented with different concentrations of arsenic, cadmium and chromium. In the present study, the increased concentration of supplemented metals is proportionally declining the growth of cells based on MIC study. In control regime, the immediate growth was observed irrespective of the late growth phase in lower concentration of metals which was found very prominent up to 12 hours of lag phase. MICs in various metal ions have been tested for several bacterial strains in similar fashion (Keramati et al., 2011).

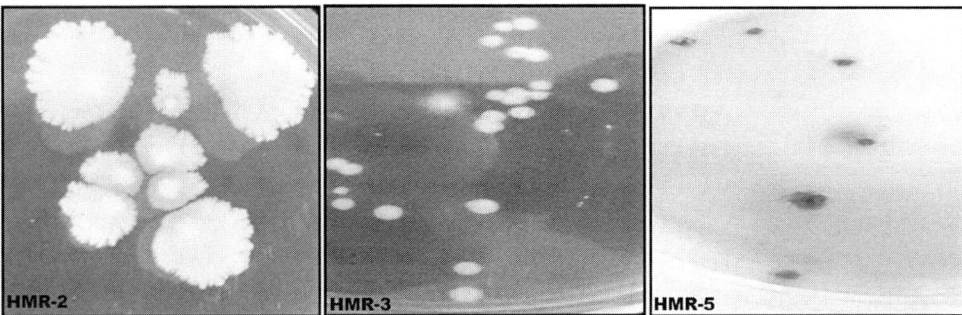

Fig. 4.2 Bacterial isolates resistant against respective heavy metals

4.3 Biochemical Characterization

HMR2 and HMR3 were identified as rod-shaped Gram-positive bacteria while HMR5 was identified as Gram-negative bacteria. The HMR-2 and HMR-3 showed resistance up to 850 µg/ml of arsenic and 750 µg/ml of cadmium respectively. HMR-5 was found resistant upto 250 µg/ml of chromium. A series of optical density was observed for all the samples ranging 0.01 to 0.6 at 600 nm. All the three isolates were characterized by different biochemical tests as per the standard profiling criteria (Table 4.2). Based on the biochemical analysis, the bacterial isolates were found to be *Bacillus* and *Serratia* species, where *Bergey's Manual of Determinative Bacteriology* was taken as reference for identification.

Table 4.2: Biochemical characterization of the bacterial isolates

Strains	Biochemical tests														
	A	B	C	D	E	F	G	H	I	J	K	L	M	N	O
HMR-1	-/+/-/+	-	+	-	-	+	-	+	+	+	-	+	+	-	+
HMR-2	-/+/-/+	-	+	-	-	+	-	+	+	+	-	+	+	-	+
HMR-3	-/-/+/+	-	+	-	+	+	-	-	+	+	-	+	+	-	-

*A = IMViC test, B = Coagulase test, C = Catalase test, D = Urease test, E = Motility test, F = Nitrate test, G = Gelatin liquefaction, H = H_2S production, I = Starch hydrolysis, J = Caesin hydrolysis, K = Mannitol salt agar, L = Glucose utilization, M = Sucrose utilization, N = Lactose utilization, O = TSI test

4.4 16S rDNA Gene Amplification

The primers amplified the 16S rDNA gene successfully for all the three strains, although no conspicuous variations in the size of rDNA gene products among the isolates were discernible. However, the size of the PCR amplified 16S rDNA gene product of all the isolates investigated in this study was approximately 1500 bp corresponding to DNA marker band (1 kb).

4.5 Analysis of Sequencing Results

Comparative analysis of the sequences obtained (Fig. 4.3) with already available database showed that the bacterial isolates were close to the members of genus *Bacillus* and *Serratia* species. The sequence obtained for the sample S1 showed a high similarity to the genus *Bacillus*. It was inferred on the basis of phylogenetic tree constructed for the sample S1 (Fig. 4.4a) that this bacterial isolate contains sequences identical to the *Bacillus cereus*, *Bacillus thuringenesis* and *Bacillus anthracis*. The sequence homology is more to *Bacillus cereus*. So, it can be concluded without any ambiguity that this bacterial isolate belongs to the genus *Bacillus*. The interpretation for the sequence obtained for the sample S2 was also done in a similar manner. Detailed analysis of the phylogenetic tree (Fig. 4.4b) revealed homology of this bacterial isolate to the genus *Bacillus*. Sequences identical to the species *Bacillus anthracis*, *Bacillus thuringenesis* and *Bacillus cereus* were found. Although, different phylogenetic trees were obtained for the two isolates (Sample S1 and S2), however, both of these were found to be closely related to the *Bacillus* species. Similarly, an evolutionary relationship was built by constructing a phylogenetic tree (Fig. 4.4c) for the sequence obtained for the sample S3. It showed homology to the genus *Serratia*. Similarity was found more to the *Serratia marcescans*.

>Sample-S1
**TCGAGCGAATGGATTAAGAGCTTGCTCTTATGAAGTTAGCGGCGGACGGGTGAGTA
ACACGTGGGTAACCTGCCCATAAGACTGGGATAACTCCGGGAAACCGGGGCTAATA
CCGGATAACATTTTGAACCGCATGGTTCGAAATTGAAAGGCGGCTTCGGCTGTCAC
TTATGGATGGACCCGCGTCGCATTAGCTAGTTGGTGAGGTAACGGCTCACCAAGGC
AACGATGCGTAGCCGACCTGAGAGGGTGATCGGCCACACTGGGACTGAGACACGG
CCCAGACTCCTACGGGAGGCAGCAGTAGGGAATCTTCCGCAATGGACGAAAGTCT
GACGGAGCAACGCCGCGTGAGTGATGAAGGCTTTCGGGTCGTAAAACTCTGTTGT
TAGGGAAGAACAAGTGCTAGTTGAATAAGCTGGCACCTTGACGGTACCTAACCAGA
AAGCCACGGCTAACTACGTGCCAGCAGCCGCGGTAATACGTAGGTGGCAAGCGTTA
TCCGGAATTATTGGGCGTAAAGCGCGCGCAGGTGGTTTCTTAAGTCTGATGTGAAA
GCCCACGGCTCAACCGTGGAGGGTCATTGGAAACTGGGAGACTTGAGTGCAGAAG
AGGAAAGTGGAATTCCATGTGTAGCGGTGAAATGCGTAGAGATATGGAGGAACACC
AGTGGCGAAGGCGACTTTCTGGTCTGTAACTGACACTGAGGCGCGAAAGCGTGGG
GAGCAAACAGGATTAGATACCCTGGTAGTCCACGC**

>Sample-S2
**AGTCGAGCGAATGGATTAAGAGCTTGCTCTTATGAAGTTAGCGGCGGACGGGTGA
GTAACACGTGGGTAACCTGCCCATAAGACTGGGATAACTCCGGGAAACCGGGGCT
AATACCGGATAACATTTTGAACCGCATGGTTCGAAATTGAAAGGCGGCTTCGGCTG
TCACTTATGGATGGACCCGCGTCGCATTAGCTAGTTGGTGAGGTAACGGCTCACCA
AGGCAACGATGCGTAGCCGACCTGAGAGGGTGATCGGCCACACTGGGACTGAGAC
ACGGCCCAGACTCCTACGGGAGGCAGCAGTAGGGAATCTTCCGCAATGGACGAAA
GTCTGACGGAGCAACGCCGCGTGAGTGATGAAGGCTTTCGGGTCGTAAAACTCTG
TTGTTAGGGAAGAACAAGTGCTAGTTGAATAAGCTGGCACCTTGACGGTACCTAAC
CAGAAAGCCACGGCTAACTACGTGCCAGCAGCCGCGGTAATACGTAGGTGGCAAG
CGTTATCCGGAATTATTGGGCGTAAAGCGCGCGCAGGTGGTTTCTTAAGTCTGATG
TGAAAGCCCACGGCTCAACCGTGGAGGGTCATTGGAAACTGGGAGACTTGAGTGC
AGAAGAGGAAAGTGGAATTCCATGTGTAGCGGTGAAATGCGTAGAGATATGGAGG
AACACCAGTGGCGAAGGCGACTTTCTGGTCTGTAACTGACACTGAGGCGCGAAAG
CGTGGGGAGCAAACAGGATTAGATACCCT**

>Sample-S3
**CACAGGGGAGCTTGCTCCCCGGGTGACGAGCGGCGGACGGGTGAGTAATGTCTG
GGAAACTGCCTGATGGAGGGGGATAACTACTGGAAACGGTAGCTAATACCGCATAA
CGTCGCAAGACCAAAGAGGGGGACCTTCGGGCCTCTTGCCATCAGATGTGCCCAG
ATGGGATTAGCTAGTAGGTGGGGTAATGGCTCACCTAGGCGACGATCCCTAGCTGG
TCTGAGAGGATGACCAGCCACACTGGAACTGAGACACGGTCCAGACTCCTACGGG
AGGCAGCAGTGGGGAATATTGCACAATGGGCGCAAGCCTGATGCAGCCATGCCGC
GTGTGTGAAGAAGGCCTTCGGGTTGTAAAGCACTTTCAGCGAGGAGGAAGGTGGT
GAGCTTAATACGCTCATCAATTGACGTTACTCGCAGAAGAAGCACCGGCTAACTCC
GTGCCAGCAGCCGCGGTAATACGGAGGGTGCAAGCGTTAATCGGAATTACTGGGC
GTAAAGCGCACGCAGGCGGTTTGTTAAGTCAGATGTGAAATCCCCGGGCTCAACC
TGGGAACTGCATTTGAAACTGGCAAGCTAGAGTCTCGTAGAGGGGGGTAGAATTC
CAGGTGTAGCGGTGAAATGCGTAGAGATCTGGAGGAATACCGGTGGCGAAGGCGG
CCCCCTGGACGAAGACTGACGCTCAGGTGCGAAAGCGTGGGGAGCAAACAGGATT
AGATACCCTGGTAGTCCACGCTG**

Fig. 4.3 PCR amplification profile of the bacterial isolates

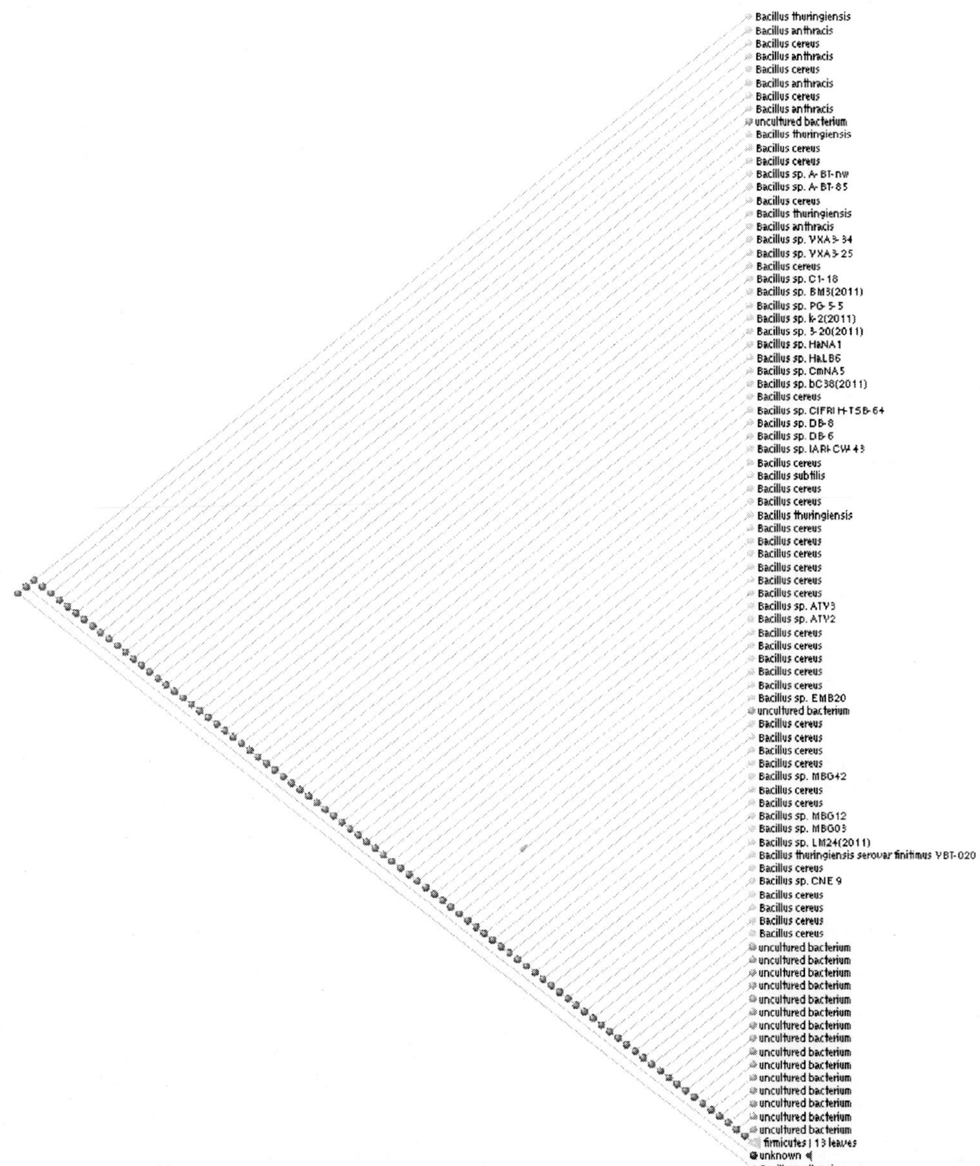

Fig. 4.4a Phylogram depicting the evolutionary relationship of sample S1 with *Bacillus* species

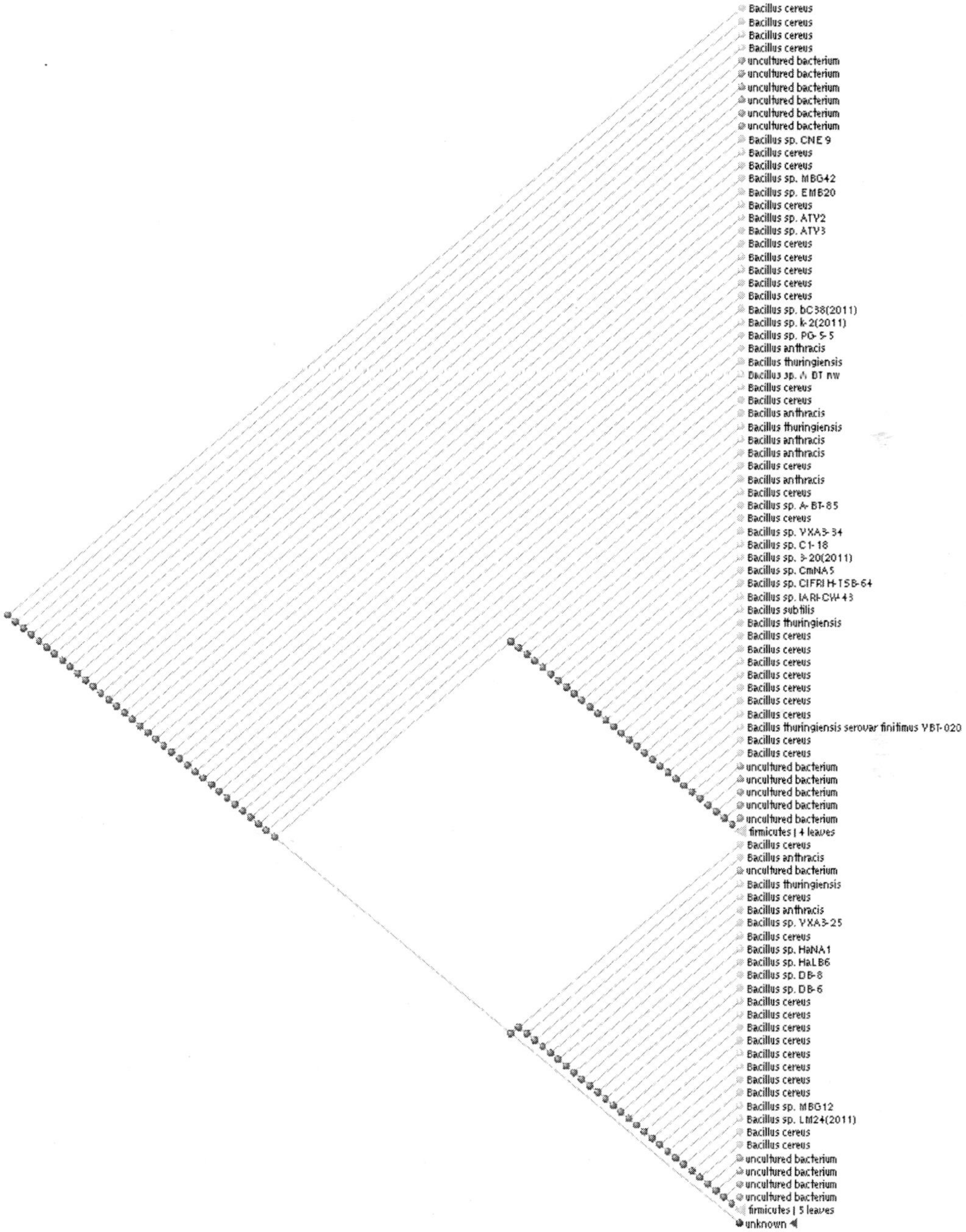

Fig. 4.4b Phylogram depicting the evolutionary relationship of sample S2 with *Bacillus* species

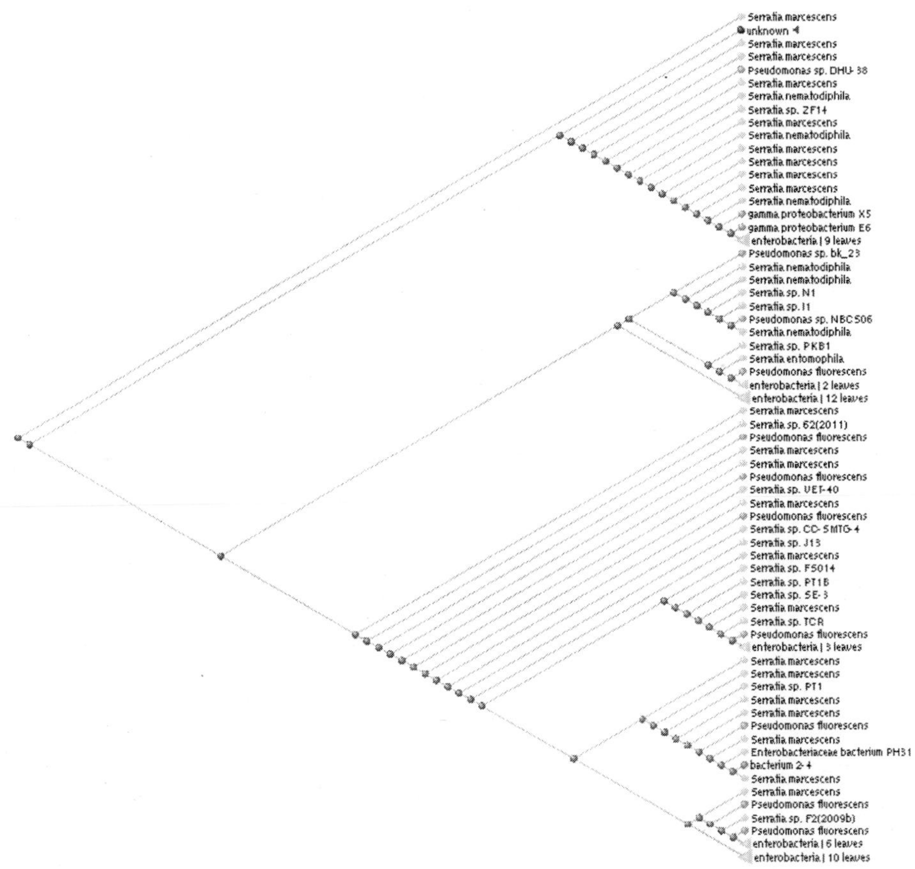

Fig. 4.4c Phylogram depicting the evolutionary relationship of sample S3 with *Serratia* species

5. CONCLUSION

Microcosm experiments were performed for native sludge, sewage dumps and effluents collected from different contaminated sites of industry of north India. Using advance technology and serial inhibitory test, the microbial composition and activity were analyzed for contaminated soil containing arsenic (ranges from 1.15–4.62 mg/kg), nickel (ranges from 2.96–38.3 mg/kg), cadmium (ranges from 0–18.72 mg/kg) and chromium (ranges from 7.715–33.28 mg/kg) which were sampled as "heavy metal-resistant bacteria". Isolates from different soil composition reflects the potential survival against the toxicity of heavy metals and got validated as the remediation house. Bacterial isolates will further be kept for molecular analysis; a subject of discussion can describe the fate of heavy metals for this particular phenomenon.

References

Abdelatey LM., Kkhalil WKB, Aali TH, Mahrou KF (2011). Heavy metal resistance and gene expression analysis of metal resistance genes in Gram-positive and Gram-negative bacteria present in Egyptian soils. Journal of Applied Sciences in Environmental Sanitation 6(2):201-211.

Acevedo F (2000). The use of bioreactors in biomining processes. EJ Electronic Biotechnol 3:184-94.

Ahemad (2014). Remediation of metalliferous soils through the heavy metal resistant plant growth promoting bacteria: Paradigms and prospects. Arab J Chem http://dx.doi.org/10.1016/j.arabjc.2014.11.020.

Babula P, Masarik M, Adam V, Eckschlager T, Stiborova M, Trnkova L, et al. (2012). Mammalians'metallothioneins and their properties and functions. Metallomics 4:739–50.

Baker J, Sitthisak S, Sengupta M, Johnson M, Jayaswal RK, Morrissey JA (2010). Copper stress induces a global stress response in *Staphylococcus aureus* and represses sae and agr expression and biofilm formation. Appl Environ Microbiol 76:150–60.

Brim H, Osborne JP, Kostandarithes HM, Fredrickson JK, Wackett LP, Daly MJ (2006). *Deinococcus radiodurans* engineered for complete toluene degradation facilities Cr(IV) reduction. Microbiol 152: 2469–2477.

Brim H, Venkateshwaran A, Kostandarithes HM, Fredrickson JK, Daly MJ (2003). Engineering *Deinococcus geothermalis* for bioremediation of high temperature radioactive waste environments. App Environ Microbiol 69: 4575–4582.

Bruins MR, Kapil S, Oehme FW (2000). Microbial resistance to metals in the environment. Exoc Environ Safe 45: 198-207.

Buffoni F, Ignesti G (2000). The copper containing amine-oxidase: biochemical aspects and functional role. Mol Genet Metab 71:559-64.

Cavicchioli R, Thomas T (2002). Extremophiles. 2nd Edition.Encyclopedia of Microbiology, Volume 2, Academic Press, San Diego.p.317-37.

Chien MF, Huang CC, Kusano T, Endo G (2008). Facilities for transcription and mobilization of an exon-less bacterial group II intron nested in transposon TnMERI1. Gene 408: 164-71.

Chudobova D, Dostalova S, Ruttkay-Nedecky B, Guran R, Rodrigo MAM, Tmejova K, Krizkova S, Zitka O, Adam V, Kizek R (2015). The effect of metal ions on Staphylococcus aureus revealed bybiochemical and mass spectrometric analyses. Microbiological Research 170: 147–156.

Cooksey DA (1994). Molecular mechanisms for copper resistance and accumulation in bacteria. FEMS Microbiol Rev 14:381-86.

Cooper RA, Knowles PF, Brown DE, McGuirl MA, Dooley DM (2000). Evidence for copper and 3,4,6-trihydroxyphenylalanine quinine cofactors in an amine oxidase from the Gram negative bacterium *Escherichia coli* K12. Biochem J 288: 337-40.

De J, Sarkar A, Ramaiah N (2006). Bioremediation of toxic substances by mercury resistant marine bacteria. Ecotoxicol 15:385-389.

Dilek FB, Gokcay CF, Yetis U (1998). Combined effects of Ni (II) and Cr (VI) on activated sludge. Water Research 32(2): 303-312.

Doelman P, Jansen E, Michels M, Van Til M (1994). Effects of heavy metals in soil on microbial diversity and activity as shown by the sensitivity-resistance index, an ecologically relevant parameters. Biol Fert Soil 17:177-184.

Dopson M, Baker-Austin C, Koppineedi PR, Bond PL (2003). Growth in sulfidic mineral environments: metal resistance mechanisms in acidophilic micro-organisms. Microbiol 149(8):1959-70.

Essa AMM, Macaskie LE, Brown NL (2002). Mechanisms of mercury bioremediation. Biochem. Soc. Trans. 30:672–674.

Ezzouhri L, Castro E, Moya M, Espinola F, Lairini K (2009). Heavy metal tolerance of filamentous fungi isolated from polluted sites in Tangier, Morocco. Afr J Microbiol Res 3(2):35-48.

Gikas P, Sengor SS, Ginn T, Moberly J, Peyton B (2009). The effects of heavy metals and temperature on microbial growth and lag. Glob Nest J 11(3):325-332.

Gladysheva TB, Oden KL, Rosen BP (1994). Properties of the arsenate reductase of plasmid R773. Biochem 33:7288-93.

Gort AS, Howell LD, Haewood J (1994). Responses of diverse heterotrophic bacteria to elevated copper concentrations. Can J Microbiol 40(5):408-11.

Keramati P, Hoodaji M, Tahmourespour A (2011). Multi-metal resistance study of bacteria highly resistant to mercury isolated from dental clinic effluent. Afr J Microbiol Res 5(7): 831-837.

Malik A (2004). Metal bioremediation through growing cells.Environ Inter 30:261-278.

Malik B, Pirzadah TB, Tahir I, Dar TH, Rehman RU (2014). Recent Trends and Approaches in Phytoremediation. In: K.R. (eds.) Hakeem et al. Soil Remediation and Plants, Prospects and ChallengesISBN: 978-0-12-799937-1 pp, 131-142.

Nucifora G, Chu L, Misra TK, et al. (1989). Cadmium resistance from *Staphylococcus aureus* plasmid pI258 cad A gene results from a cadmium-efflux ATPase. Proc Natl Acad Sci USA 86:3544–48.

Penny C, Vuilleumier S, Bringel F (2010). Microbial degradation of tetrachloromethane: Mechanisms and perspectives for bioremediation. FEMS Microbiol. Ecol. 74: 257–275.

Pirzadah TB, Malik B, Tahir I, Kumar M, Varma A, Rehman RU (2014). Phytoremediation: An Eco-Friendly Green Technology for Pollution Prevention, Control and Remediation. In: K. R. Hakeem et al. (eds.) Soil Remediation and Plants, Prospects and Challenges ISBN: 978-0-12-799937-1 pp, 107-122.

Ray SA, Ray MK (2009). Bioremediation of heavy metal toxicity-with special reference to chromium. Al Ameen J Med Sci 2(2):57-63.

Rojas LA, Yanez C, Gonzalez M, Lobos S, Smalla K, Seeger M (2011). Characterization of the metabolically modified heavy metal-resistant *Cupriavidusn metallidurans* strain MSR33 generated for mercury bioremediation. PLoS One-e17555.

Scott JA, Karanjkar AM (1992). Repeated cadmium biosorption by regenerated Enterobacter aerogenes biofilm attached to activated carbon. Biotechnol. Lett. 14:737–740.

Sharma KP, Frenkel A, Balkwill LD (2000). A new *Klebsiella planticola* strain (cd-1) grows anaerobically at high cadmium concentrations and precipitates cadmium sulphide. Appl Environ Microbiol 66: 3083-87.

Shi W, Bischoff M, Turco R, Konopka A (2002). Long-term effects of chromium and lead upon the activity of soil microbial communities. Appl Soil Ecol 21:169-77.

Shukla OP, Rai UN, Singh NK, Dubey S, Baghel VS (2007). Isolation and characterization of chromate resistant bacteria from tannery effluent. J Environ Biol 28(2):399-403.

Singh P, Raghukumar C, Parvatkar RR, Mascarenhas-Pereira MBL (2013). Heavy metal tolerance in the psychrotolerant Cryptococcus sp isolated from deep-sea sediments of the Central Indian Basin. Yeast 30:93–101.

Singh V, Chauhan PK, Kanta R, Dhewa T, Kumar V (2010). Isolation and characterization of Pseudomonas resistant to heavy metals contaminants. Inter J Pharm Sci Rev Res 3(2):164-66.

Sone Y, Mochizuki Y, Koizawa K, Nakamura R, Pan-Hou H, Itoh T, Kiyono M (2013). Mercurial resistance determinants in *Pseudomonas* strain K-62 plasmid pMR68. AMB Express 3, Article 41.

Sun LN, Zhang YF, He LY, Chen ZJ, Wang QY, Qian M, Sheng XF (2010). Genetic diversity and characterization of heavy metal-resistant-endophytic bacteria from two copper-tolerant plantspecies on copper mine wasteland. Bioresour. Technol 101:501–509.

Torres E, Bustos-Jaimes I, Le Borgne S (2003). Potential use of oxidative enzymes for the detoxification of organic pollutants. Appl Catal B-Environ 46:1-15.

Volesky B (2001). Detoxification of metal-bearing effluents biosorption for the next century. Hydrometallurgy 44:203-216.

Wuana RA, Okieimen FE (2011). Heavy metals in contaminated soils: A review of sources, chemistry, risks and best available strategies for remediation. ISRN Ecol, Article 20.

Xiao X, Luo S, Zeng G, Wei W, Wan Y, Chen L, Guo H, Cao Z, Yang L, Chen J et al. (2010). Biosorpiton of cadmium by endophytic fungus (EF) *Microsphaeropsis* sp. LSE10 isolated from cadmium hyperaccumulator *Solanum nigrum* L. Bioresour. Technol. 101:1668–1674.

Yeager JG, Ward RL (1981). Effects of moisture content on long-term survival and regrowth of bacteria in wastewater sludge. Appl. Environ. Microbiol. 41(5): 1117-1122.

Zolgharnein H, Azmi MLM, Saad MZ, Mutalib AR, Mohamed CAR (2007). Detection of plasmids in heavy metal resistance bacteria isolated from the Persian Gulf and enclosed industrial areas. Iranian J Biotech 5(4):232-39.

5

Microbial Metagenomics: Herbicides for Weed Management in Sustainable Agriculture

Priyanku Teotia[1], Manoj Kumar[2] and Vivek Kumar[2]*

[1]Department of Botany, CCS University, Meerut, India

[2]Amity Institute of Microbial Technology, Amity University Sector-125, Noida, India

*Corresponding author Email: *vivekbps@gmail.com*

1. INTRODUCTION

Metagenomics literally means "beyond the genome". Metagenomics is the study of metagenomes, genetic material recovered directly from environmental samples, without culturing the microbes on Petri plate. Like genomics itself, metagenomics is both a set of research techniques, comprising many related approaches and methods, and a research field. In Greek, *meta* means "after or beyond." In its approaches and methods, metagenomics circumvents the unculturability and genomic diversity of most microbes, the biggest roadblocks to advances in clinical and environmental microbiology. Meta in the first context recognizes the need to develop computational methods that maximize understanding of the genetic composition and activities of communities so complex that they can only be sampled, never completely characterized. In the second sense, that of a research field, meta means that this new science seeks to understand biology at the aggregate level, transcending the individual organism to focus on the genes in the community and how genes might influence each other's activities in serving collective functions. Individual organisms remain the units of community activities, of course, and we anticipate that metagenomics will complement and stimulate research on individuals and their genomes. In the next decades, we expect that the top-down approach of metagenomics, the bottom-up approach of classical microbiology, and organism level genomics will merge. We will understand communities, and the collection of communities that forms the biosphere, as a nested system of systems of which humans are a part and on which human survival depends. In some situations, it will be possible to apply the new understanding to problems of urgency and importance (Shabana and Mohamed, 2005).

Metagenomics in either sense will probably never be circumscribed tightly by a definition, and it would be undesirable to attempt to so limit it now, but the term includes cultivation-independent genome-level characterization of communities or their members, high throughput gene level studies of communities with methods borrowed from genomics, and other "omics" studies.

2. METAGENOMICS FOR WEED MANAGEMENT: STATUS

All plants and animals have closely associated microbial communities that make necessary macro- and micro-nutrients, vitamins, growth hormones available to their hosts. These functions are conducted within complex communities' intricate, balanced and integrated entities that adapt swiftly and flexibly to environmental change. But historically, the study of microbes has focused on single species in pure culture, so understanding of these complex communities lags behind for understanding of their individual members. We know enough, however, to confirm that microbes, as communities, are the key players in maintaining environmental stability (Sauerborn et al., 2007).

Weeds cause serious ecological problems and may even alter the process of ecosystem by displacing the native flora and fauna. In the past 20 year, biocontrol research has uncovered an abundance of microorganisms that suppress weeds across the full spectrum of plant growth stages. Although there has been increasing interest in Europe over the last few decades in trying to harness the potential of biological control as a management tool for weeds. The incentives to use alternative technology are based on a number of interrelated factors, not least the dramatic rise in popularity of organic products (linked to increasing environmental awareness) and a powerful antiGMO lobby. Combine this with new legislation to remove some chemical herbicides from the market or to restrict their usage, as well as soaring development costs for new products, and the opportunities for biological control have never been better. These developments included commercialization of, and increasing potential for, microbial herbicide use, particularly in agricultural systems in North America (Chandramohan and Charudattan, 2003).

Here, we analyse the reasons for the limited uptake, and the challenges or constraints facing weed biocontrol from two different approaches: classical and inundative. For classical biological control against invasive alien weeds, Europe has lagged behind than other continents (e.g., Australasia, North America) to the point that there have been no introductions for those that are in the pipeline still need to clear considerable legislative hurdles. These issues are highlighted for past and ongoing projects. For inundative biological control against indigenous or naturalized weeds, the use of products based on plant pathogens (bioherbicides) the constraints are largely technological and commercial rather than bureaucratic, e.g., stability and efficacy; costs of production and registration; and limited markets (Sands and Pilgeram, 2009).

The concept of utilizing microbial herbicides has been explored for more than a quarter century, but there remain many challenges to overcome before they can be widely used in agricultural, range and forest lands, or waterways. Those challenges include improving the efficacy of the microbial activity, survival of microorganisms, persistence of the suppressive compounds, delivery systems, determining host range, and avoiding injury to non-target organisms. Other considerations are interactions with chemical herbicides, regulations, commercialization and mass production, and economic feasibility. Biological herbicides represent a means to reduce dependence on synthetic herbicides; focus on ecologically grounded methods of management; reduce weed seed bank populations through environment-friendly practices; and potentially reduce costs of weed control in crop production, rangeland restoration, forestry and aquatic systems (Bailey et al., 2010; Kennedy and Stubbs, 2007).

3. METAGENOMICS AND DEVELOPMENT OF BIOHERBICIDES

There are some very tough weeds. They compete with valuable crop and forage plants and threaten many native plant species. Many of these weeds have been very expensive and difficult, if not impossible, to control with more traditional methods. In some instances, the chemicals used for control are nonselective compounds, which will also damage nontarget plants and may leach out of sandy or gravelly soils, or compounds that give top growth control only. In addition, because of leaching, chemicals cannot be used on weeds that grow close to bodies of water. Many problem weeds are abundant in rangeland and along riverbanks and gullies, so the use of non-chemical control methods, such as mowing or cultivation can be limited by the topography and size of these areas. These difficulties have prompted scientists to investigate a third alternative: the biological control of weeds (Bailey and Mupondwa, 2006). Biological control covers two key concepts: the deliberate use of a weed's "natural enemies" to suppress its population and the use of these live organisms to maintain this lower population density. A weed's natural enemies may be arthropods, bacteria or fungi. These "control agents" feed upon or cause disease in the weed, thereby limiting its growth, reproduction and spread (Casella et al., 2010).

The invisible members of a microbial community can differ vastly in their biochemical activities and interactions, not only between species global occurrences of herbicide-resistant weed populations have increased the demand for development of new herbicides targeting novel mechanisms of action. Metagenomic approaches to natural drug discovery offer potential for isolating weed suppressive compounds from microorganisms. In past research, traditional techniques entailed isolating compounds from living organisms, whereas metagenomic approaches involve extracting fragments of DNA from soil and exploring for compounds of interest produced by the transformed hosts. Several herbicidal compounds have been isolated from soil bacteria through culturing methods and have led to the development of popular herbicides, such as glufosinate (Mortensen et al., 2012). In this chapter, we talk about the emergence of metagenomic approaches for weed management in the context of natural product discovery using traditional culture dependent isolation and the more recent culture independent methods. The same techniques can be used to isolate herbicide resistance genes. Adoption of metagenomic approaches in pest management research can lead to novel control strategies in cropping and landscape systems. Advances in molecular biology in the past decade have allowed researchers to discover novel plant microbial relationships relevant to weed control (Rector, 2008). The expanding field of metagenomics allows both microbial identity and function to be determined by extracting DNA and RNA directly from the environment, which circumvents the need to isolate and culture microorganisms (Handelsman, 2004). Direct isolation of DNA from the environment increases the diversity of microbial strains recovered. Metagenomic studies have shown that a single gram of soil holds from 2000 to over 52,000 microbial species or genomes (Gans et al., 2005; Schloss and Handelsman, 2006). This estimate far exceeds the number of microbial species found using traditional culturing techniques. The massive diversity of bacterial and fungal genomes in soil provides a potentially vast pool of genes that code for the production of herbicidal compounds and herbicide resistance. As a consequence, two approaches to weed management can be developed using metagenomic techniques: isolation of novel herbicides produced by vector hosts expression; and identification of herbicide resistance genes in vector hosts exposed to high levels of an herbicide.

Both approaches consist of different functional screening assays, but the initial construction of the metagenomic libraries (clones containing fragments of the metagenomes) are identical. Incorporating metagenomic approaches to a weed scientist's toolbox is feasible today as commercial kits for DNA extraction and cloning are widely available. For example, DNA extraction kits are available for different environmental samples, including soil, water, plant, and stool. Likewise, several vector host kits are provided through a multitude of companies optimizing for different expression systems and insert lengths (Ma et al., 2002).

In the end, it is possible to view ecosystems themselves as biological units with their own genetic repertoires and to sidestep consideration of individual species. Such understanding can be achieved only with methods that go beyond the pure culture and single whole-genome approaches that have dominated microbial genomics. We must move directly to the genes, to defining environments by the potential and realized biochemical and geochemical activities of the genes that are there, and the complex patterns of interactions within and between cells that regulate their responses to changes in their physical and biological surroundings. We must do this while recognizing that except in restricted environments and specialized consortia with limited numbers of genetically homogeneous constituents. We will be dealing with enormous amounts of data that will represent an incomplete sampling of the genetic diversity present. In short, we must adopt the methods of metagenomics (Kniffin et al., 2013).

4. MICROBIAL HERBICIDES DEVELOPMENT

Microbial derived herbicides, especially microorganism secondary metabolites, are a new kind of microbial herbicide to control weeds, which are always phytotoxins. Microorganisms can produce a lot of metabolites whose characteristics are diversity in structure and biological activity, and easily degraded. These bioactive components invade into the host plant, causes pathogenicity, destroy their structure and lead them to produce necrotic lesions or chlorotic halo. Now biologists and agriculturalists are paying more attention to natural products that have herbicidal activity, for these natural phytotoxins have a specific target and new different chemical structures which are difficult to synthesize by common pesticide synthesis method. They are less poisonous to most of mammalian systems, easily degraded and so far result in no biological disaster compared to chemical herbicides (Charudattan, 2005). These phytotoxins are very different in chemical structure and size. Some of them are polypeptides, terpenes, macrocylic and bakelite (Ghosheh, 2005). They are also different in their host plant speciality; they either apply to a single species or to one kind of plant. Others are not only specific to one kind of host plant, but also to a lot of plants.

5. BACTERIAL PHYTOTOXINS

Most of the bacteria with an ability to produce toxins are *Pseudomonas, Erwinia, Xanthomonas, Streptomyces, Corynebacterium* and some are non-fluorescent pseudomonads (Kremer 1993). *Pseudomonas syringae* pv. phaseolicola is a bacterial plant pathogen which causes halo blight disease and localized death on common bean (*Phaseolus vulgaris* L) and kudzu (*Pueraria lobata*), whose toxin was called phaseolotoxin. Once it infects plant root, it will spread to shoot terminus, then causes stunting, chlorosis and even causes foliage necrotic lesions (Sauerborn et al., 2007). Brady and Clardy (2000) have purified the phytotoxin, which was inhibitory to downy brome (*Bromus tectorum*) and was a complex consisting of at least two polypeptides, a chromophore, fatty acid esters, and a lipopolysaccharide matrix. Further purification of this compound resulted in near complete loss of phytotoxin because purification procedures could damage the phytotoxin (van Elsas et al., 2008). Wenzel and Müller (2005) evaluated effects of a crude preparation of microbial genome on various physical processes in the roots of downy brome seedlings. It was found that cell division, respiration, and synthesis of protein, RNA and DNA were not inhibited or only slightly inhibited, whilst disruption of lipid synthesis and membrane integrity were significant, which might account for inhibition of root elongation. Other phytotoxins from *Pseudomonas* have different physiological mechanisms. Phaseolotoxin, a tripeptidefrom *Pseudomonas syringae* pv. phaseolicola, inhibits arginine synthesis by competing with carbamoyl phosphate for the binding site on ornithine carbamoyltransferase (Caesar, 2003).

Further studies showed that this toxin could specifically increase K^+ efflux and H^+ATPase activity and was revealed to have a possible role in Ca^{2+} transport (Uchiyama and Miyazaki, 2010). The majority of biological herbicides developed so far are mycoherbicides; however, several bacterial herbicides are under development as well. A screening procedure for a bacterial biological control agent has been developed. The screening procedure takes advantage of the soil microbial community and the diversity within. This screening includes sampling the soil and plant material at the peak of population growth and when suppression naturally occurs. An initial assay can separate the weed suppressive microorganisms from the bulk of the native population. Further multiple screenings on various plants are suggested to get rid of those isolates that may also inhibit beneficial plants. While inhibition of the target weed is critical, non-host range must be determined early in the process to ensure the product progresses through the registration process and eventually becomes marketable. Both steps need to be thorough to ensure that no lesser known, but economically important, host is detected later in the process. Souissi and Kremer (1998) utilized a multiple well plate procedure with leafy spurge (*Euphorbia esula* L.) callus to rapidly determine phytotoxicity of rhizobacterial isolates, and Vidal et al. (2004) have developed a successful method to produce yellow starthistle (*Centaurea solstitialis* L.) calli for bioassay screening of biocontrol pathogens, such as *Phoma exigua*. A quick substitute bioassay will reduce the length of time needed for the initial screenings. In many cases, a quick screen is not possible (Kremer, 2005), but automating data recording can often make data collection easier.

6. BIOLOGICAL CONTROL IDENTIFICATION AND INTRODUCTION IN ENVIRONMENT

6.1 Identifying Target Weeds

To be considered a good candidate for biological control, a weed should be non-native should present in numbers and densities greater than in its native range and numerous enough to cause environmental or economic damage. The weed should also be present over a broad geographic range and have a few or not at all beneficial qualities. Should also have taxonomic characteristics sufficiently distinct from those of economically important and native plant species. For classical biocontrol, the weed should occur in relatively undisturbed areas to allow for the establishment of biological control agents. Cultivation, mowing and other disturbances can have a destructive effect on many arthropod biocontrol agents. Inundative biocontrol agents, such as bacteria and fungi are less sensitive to these types of disturbances so may be used in cropland.

6.2 Biocontrol Identification and Specialization Assessing

Scientists observe weeds in their areas of origin and collect the naturally occurring microorganisms producing biological herbicides hindering and affecting the growth of plants and affecting their survival. The screening procedure takes advantage of the soil microbial community and the diversity within. This screening includes sampling the soil and plant material at the peak of population growth and when the suppression naturally occurs. An initial assay can separate the weed suppressive microorganisms from the bulk of the native population.

In metagenomics, the aim is to understand the composition and operation of complex microbial consortia in environmental samples through sequencing and analysis of their DNA. Similarly, metatranscriptomics and metaproteomics target the RNA and proteins obtained from such samples. Technological advances in next generation sequencing methods are fuelling a rapid increase in the number and scope of environmental sequencing projects. These organisms are subjected to a multi-level screening process to assess their host range and their effect on the weed. These screening efforts do two things: they ensure the safety of any valuable crop, forage or native plant species that the agents may encounter when released, and they assess the efficacy of the agent.

Assessing the host range (how specific the potential biocontrol agent is to a particular plant) is probably the most important step in this process. Over very long periods, some plants and herbivores have evolved to form very close associations. Plants have developed a number of defences, such as toxic chemicals, that plant eating organisms (herbivores) must overcome. Some herbivores have evolved the ability to bypass only certain host plants defences, such that they cannot feed or develop on anything else. To find out how specialized a particular agent is, scientists collect and expose them to a wide assortment of plants. These plants include crop and forage species as well as species native to the intended release area, especially if the species are close relatives of the weed. Screening potential biological control agents ensures that only those with a very narrow host range (i.e., those that represent no threat to crop, forage or native species) are released.

6.3 Controlled Release of Agent

All biological control agents must be approved under the Plant Protection Act and are reviewed by an advisory panel of Agriculture and Environment ministries before they can be released. Following government approval, the agents are released on their target weeds at selected experimental sites, which are closely monitored. Data from these sites help to assess both the agents potential for survival under field conditions and their potential to cause damage to the target weed.

6.4 Final Release of Agent

If the agents survive our harsh climate and damage or suppress the weeds in the controlled tests, they can be released on other weed infestations. The agents are released on as many varied sites as agent numbers, time and resources will permit. The agents are then carefully monitored. Release on a large variety of sites allows scientists to find the habitats best suited to each species.

6.5 Monitoring Release Sites

Release sites should be monitored periodically to assess the size of the biological control agent population and the effect of the agent on the weed. Monitoring release sites is very important to determine if agents have established on a site. Some insect biocontrol agents may need two to five years before their populations increase to a sufficient size to have a visible impact on weed numbers.When the classical control agent is establishing on a weed infestation and increasing its numbers, the site should not be moved or disturbed. Herbicide may be used along the boundaries of the weed infestation to help contain the weed while the biological agent is increasing in population and spread (Kennedy and Stubbs, 2007).

6.6 Maintaining Control Agent

If a biological control agent is released on a site with favourable conditions, its population can grow quickly. An important point to remember is that it can take years before a biocontrol agent's population reaches this size. The agents should not be collected for redistribution before this time. Since the weeds are never completely eradicated, a small population of microorganisms stays on the few remaining plants, microbes that cannot be supported by the weed population at this lower level either move to new stands or die.

7. MICROBIAL HERBICIDES DEVELOPMENT CHALLENGES

It is becoming widely accepted, that a sole dependence of weed control strategies on chemical herbicides is unfortunate and that alternative and complementary control options should be considered. Hence, the overall intention of new control strategies is to suppress problematic weed populations in an ecologically and economically feasible manner by integrating a number of different methods (Einhorn and Brandau, 2000). Metagenomics microbiological weed control represents an eligible strategy that can be integrated into such a system. Thereby biocontrol methods should be considered components of an overall integrated weed management in order to achieve efficiency and to reduce herbicide application to the minimum possible extent. The biological control of weeds by plant pathogens has gained acceptance as a practical, safe, and environmentally beneficial method applicable to agroecosystem (Boyetchko, 2005). The application of plant pathogens comes especially into consideration for parasitic weeds, for weeds difficult to control via chemical means, or for smallscale and specialized crops where the development of specific chemical solutions is too expensive (Auld and Morin, 1995; Zdor et al., 2005). Other instances where plant pathogens may be preferable over chemical herbicides are aquatic weeds to avoid chemical contamination of water and situations where a close relation between weeds and crops account for a high degree of specificity, such as grass weeds in cereal production (Auld and Morin, 1995; Duke et al., 2000).

7.1 Biological Challenges

It is desirable for a bioherbicide to act relatively quickly and have sufficient efficacy to control weeds. Unfortunately, many of the weed pathogens discovered may provide only partial control of only one weed species, even under ideal conditions (Charudattan, 2005). Host specificity is related to the basic biology of the pathogen and to host variability (Chutia et al., 2007). Biological constraints including host variability and resistance, as well (Auld and Morin, 1995), phyllosphere of plants, a requirement for more than 12 hour of dew period for severe infection by a pathogen, has been reported for several potential bioherbicides.

7.2 Environmental Limitations

Environmental factors influence formulation performance of bioherbicides as inoculum production depends on sporulation of the formulation. In the application of bioherbicides, environmental conditions prevailing in the phyllosphere of plants are frequently hostile for biological control agents (Brady, 2007; Donato et al., 2010). For foliar biocontrol agents, environmental factors that influence plant infection and disease development are temperature, free moisture or dew period, and protection against UV irradiation. For soil applied biocontrol agents, physical and chemical characteristics of soil, moisture, and temperature regimens, as well as microbial competition are to be monitored (Duke, 2012).

7.3 Mercantile Limitations

Some reasons why biocontrol agents have met with limited commercial success are difficulty of production, sensitivity to UV light and desiccation, requirements of high humidity for infection, insufficient performance over a wide range of environmental conditions, and lack of appropriate formulation (Craig et al., 2009). Formulations should be used to improve product stability, bioactivity, and delivery (i.e., ability to mix and spray the product) as well as to integrate the biopesticide into a pest management system (Culliney, 2005) including good market potential, ease of production and application, adequate product stability and shelf life during transportation as well as in storage and guaranteed propagule viability and efficacy over the long term (Delmont et al., 2011).

7.4 Technological Limitations

Several technological limitations have been identified that could prevent the widespread use of bioherbicides, pathogenical strains, formulation method and the interaction of these two parameters significantly affect the shelf life of the formulations at room temperature (Hynes and Boyetchko, 2006; Hillis et al., 2011). Other important characteristics of a successful formulation are convenience of use, compatibility with end user equipment and practices, and effectiveness at rates consistent with agricultural practices (Hugenholtz and Tyson, 2008). The inclusion of novel synergists in bioherbicide formulations could take them past the point of research, and into the development of efficacious, reliable, and economical products for the marketplace (McMahon et al., 2012). All these parameters need to be taken into consideration when developing an appropriate microbial herbicide formulation.

8. OUTLOOK AND CONCLUSION

The wealth of genetic potential of microorganisms on this earth is boundless. There have been many investigations of potential products for weed management. Some have been successful at suppressing weeds in the field and a select few are marketed products that now reduce weed infestations. Further studies are needed to continue to search for additional tools to combat weeds. Numerous microbial candidates exist, and preliminary research into biological characterization has been conducted on these candidates for several decades. Despite all of this research and expense poured into development of microbial biological control agents, only a few have been successful and fewer still have persisted in the marketplace. Many candidates have failed, and often for one of multiple common reasons; production problems, lack of stabilization of high titers following fermentation, lack of adequate shelf life of formulations under warehouse temperatures, lack of an economic viable delivery system, or loss of virulence of the product before reaching the target. Therefore, there is a critical need to better understanding of the mode of action of metagenomic microbial herbicides involved in host-pathogen interactions which consequently leads to enhance the virulence of pathogen and/or suppress the host plant's defence. In fact, there is a substantial difference in studying the mode of action between chemical herbicides and microbial herbicides. In case of chemical herbicides, the active ingredient is a chemical substance, whereas in microbial herbicides, the active ingredient is secondary metabolites produced by the microbes in the phytosphere. Though bioherbicides will not solve all the environmental and weed management problems associated with synthetic herbicides, nor will replace the current or future arsenal of synthetic herbicides but their role will probably be complimentary components in successful weed management systems, and in the discovery of novel phytotoxins with new chemistries and new molecular sites of action.

Continued research on these areas is important in order to fully understand interactions of microorganisms and plants (crops and weeds), and to discover new phytopathogenic microorganisms or microbial phytotoxins useful as bioherbicides. The potential that biological control agents as a weed management tool can offer is often overlooked because of: (i) perceptions about its novelty and unproven track record, (ii) the perceptions about the performance of biological control agents relative to herbicide weed control, (iii) the familiarity of using traditional weed control methods, or (iv) the concerns over its environmental safety (i.e., non-target impacts of classical biological control agents). However, it should be noted that only a small fraction of newly tested herbicide chemistries ever make it to market as new herbicides for numerous reasons. As a result the metagenomic approach is gaining momentum. New bioherbicides will find place in irrigated lands, wastelands as well as in parasite weeds or resistant weed control. Research on synergy test of pathogens and herbicides developmental technology, microbial and fungal toxins, and application of biotechnology, especially genetic engineering is required. Thus, microbes have a place in integrated ecological based weed management and their potential is just being realized.

References

Auld BA, and Morin L (1995). Constraints in the development of bioherbicides. Weed Technol 9: 638–652.

Bailey KL, and Mupondwa EK (2006). Developing microbial weed control products: commercial, biological, and technological considerations. In: Singh HP, Batish DR, and Kohli RK (eds.) Handbook of Sustainable Weed Management. The Haworth Press Inc., Binghamton, NY. pp. 431–473.

Bailey K, Boyetchko S, and Langle T (2010). Social and economic drivers shaping the future of biological control: A canadian perspective on the factors affecting the development and use of microbial biopesticides. Biological control: theory and application in pest management. 52(3): 221–229.

Boyetchko SM (2005). Biological herbicides in the future. In: Ivany JA (ed) Weed Management in Transition. Topics in Canadian Weed Science (Vol. 2). Sainte-Anne-de-Bellevue, Quebec. So Canadian Weed Science Society-Societecanadiėnne de malherbologie. pp. 29–47.

Brady SF (2007). Construction of soil environmental DNA cosmid libraries and screening for clones that produce biologically active small molecules. Nat Protocol 2:1297–1305.

Brady SF, and Clardy J (2000). Long-chain N-acyl amino acid antibiotics isolated from heterologously expressed environmental DNA. J Am Chem Soc 122:12903–12904.

Caesar AJ (2003). Synergistic interaction of soilborne plant pathogens and root-attacking insects in classical biological control of an exotic rangeland weed. Biol Contr 28: 144–153.

Casella F, Charudattan R, and Vurro M (2010). Effectiveness and technological feasibility of bioherbicide candidates for biocontrol of green foxtail (*Setaria viridis*). Biocontrol Sci Technol 20(9-10): 1027–1045.

Chandramohan S, and Charudattan R (2003). A multiple-pathogen system for bioherbicidal control of several weeds. Biocontr Sci Technol 13:199–205.

Charudattan R (2005). Ecological, practical, and political inputs into selection of weed targets: what makes a good biological control target? Biological control: theory and application in pest management. 35(3): 183–196.

Chutia M, Mahanta JJ, Bhattacharya N, Bhuyan M, Boruah P, and Sarma TC (2007). Microbial herbicides for weed management. Plant Pathol J 6: 210–218.

Craig JW, Chang FY, and Brady SF (2009). Natural products from environmental DNA hosted in *Ralstonia metallidurans*. ACS Chem Biol 4: 23–28.

Culliney TW (2005). Benefits of classical biological control for managing invasive plants. Crit Rev Plant Sci 24: 131–150.

Delmont TO, Robe P, Clark I, Simonet P, and Vogel TM (2011). Metagenomic comparison of direct and indirect soil DNA extraction approaches. J Microbiol Methods 86: 397–400.

Donato JJ, Moe LA, Converse BJ, Smart KD, Berklein FC, McManus PS, and Handelsman J (2010). Metagenomic analysis of apple orchard soil reveals antibiotic resistance genes encoding predicted bifunctional proteins. Appl Environ Microbiol 76: 4396–4401.

Duke SO (2012). Why have no new herbicide modes of action appeared in recent years? Pest Manag Sci 68: 505–512.

Duke SO, Dayan FE, Romagni JG, and Rimando AM (2000). Natural products as sources of herbicides: current status and future trends. Weed Res 40: 99–111.

Einhorn G, and Brandau J (2006). Influence of microorganisms on weed and crop seeds? J Plant Dis Prot 20: 317–324.

Gans J, Wolinsky M, and Dunbar J (2005). Computational improvements reveal great bacterial diversity and high metal toxicity in soil. Science 309: 1387–1390.

Ghosheh H (2005). Constraints in implementing biological weed control: A review. Weed Biol Manag 5(3): 83–92.

Handelsman J (2004). Metagenomics: application of genomics to uncultured microorganisms. Microbiol Mol Biol Rev 68: 669–685.

Hillis DG, Fletcher J, Solomon KR, and Sibley PK (2011). Effects of ten antibiotics on seed germination and root elongation in three plant species. Arch Environ Contam Toxicol 60: 220–232.

Hugenholtz P, and Tyson GW (2008). Microbiology: metagenomics. Nature 455: 481–483

Hynes RK, and Boyetchko SM (2006) Research initiatives in the art and science of biopesticide formulations. Soil Biol Biochem 38: 845–849.

Kennedy A, and Stubbs T (2007). Management effects on the incidence of jointed goatgrass inhibitory rhizobacteria. Biological control: theory and application in pest management. 40 (2): 213–221.

Kniffin JK, Carver SM, and DiTommaso A (2013). Advancing weed management strategies using metagenomic techniques. Weed Sci 61:171–184.

Kremer JR (2005). The Role of bioherbicides in weed management. Bio pest Int 3: 127–141.

Kremer RJ (1993). Management of weed seed banks with microorganisms. Ecol Appl 3: 42–52.

Ma J, Xu L, and Wang S (2002). A quick, simple, and accurate method of screening herbicide activity using green algae cell suspension cultures. Weed Sci 50: 555–559.

McMahon MD, Guan C, Handelsman J, and Thomas MG (2012). Metagenomic analysis of *Streptomyces lividans* reveals host-dependent functional expression. Appl Environ Microbiol 78: 3622–3629.

Mortensen DA, Egan JF, Maxwell BD, Ryan MR, and Smith RG (2012). Navigating a critical juncture for sustainable weed management. Biosci 62: 75–84.

Rector BG (2008). Molecular biology approaches to control of intractable weeds: new strategies and complements to existing biological practices. Plant Sci 175: 437–448.

Sands D, and Pilgeram A (2009). Methods for selecting hypervirulent biocontrol agents of weeds: why and how. Pest Manage Sci 65(5): 581–587.

Sauerborn J, Mueller-Stoever D, and Hershenhorn J (2007). The role of biological control in managing parasitic weeds. Crop Prot 26: 246–254.

Schloss PD, and Handelsman J (2006). Toward a census of bacteria in soil. PLOS Comput Biol 2: 786–793.

Shabana Y, and Mohamed Z (2005). Integrated control of water hyacinth with a mycoherbicide and phenylpropanoid pathway inhibitor. Biocontrol Sci Technol 15 (7): 659–669.

Souissi T, and Kremer R (1998). A rapid microplate callus bioassay for assessment of rhizobacteria for biocontrol of leafy spurge (*Euphorbia esula* L.). Biocontrol Sci Technol 8: 83–92.

Uchiyama T, and Miyazaki K (2010). Product-induced gene expression, a product-responsive reporter assay used to screen metagenomic libraries for enzyme-encoding genes. Appl Environ Microbiol 76: 7029–7035.

vanElsas JD, Speksnijder A, and van Overbeek LS (2008). A procedure for the metagenomics exploration of disease-suppressive soils. J Microbiol Methods 75: 515–522.

Vidal K, Guermache F, and Widmer T (2004). *In vitro* culturing of yellow starthistle (*Centaurea solstitialis*) for screening biological control agents. Biological Control: Theory and Application in Pest Management. 30(2): 330–335.

Wenzel SC, and Müller R (2005). Recent developments towards the heterologous expression of complex bacterial natural product biosynthetic pathways. Curr Opin Biotechnol 16: 594–606.

Zdor RE, Alexander CM, and Kremer RJ (2005). Weed suppression by deleterious rhizobacteria is affected by formulation and soil properties. Commun Soil Sci Plant Anal 36: 1289–1299.

6

Biodegradation of Petroleum Hydrocarbon Pollutants in Marine Environment

Sunita J. Varjani
Science and Technology Department
Gujarat National Law University
Gandhinagar, Gujarat, India
Email: *drsvs18@gmail.com*

1. INTRODUCTION

Petroleum products play an important role in modern society. Petroleum pollution with its attendant ecological consequences is a major environmental issue (Jha et al., 2008). Petroleum industry activities, such as manufacturing, transport, processing, handling and storage of crude oil leads to environmentsal pollution. Oil spill is a hazardous situation in which spillage of oil may be either marine or terrestrial (Varjani, 2014). Oil spills create devasting effects on environment and life forms. There is an urgent need to find the protective and preventive measures to avoid spillage (Ferradji et al., 2014). Physical, chemical and biological technologies have been developed to control oil spills. Although physical and chemical methods are the first response option, they rarely achieve complete cleanup of oil spills. The commonly used technologies for remediation are expensive and can lead to incomplete decomposition of contaminants (Varjani and Upasani, 2012; Ferradji et al., 2014). Bioremediation by biodegradation of hydrocarbon pollutants exploits the ability of microorganisms to degrade and/or detoxify organic contaminants, has been established as an efficient, economic, versatile and environmentally sound treatment (Varjani et al., 2015).

In the past decade, a great deal of research concerning biodegradation of hydrocarbons has been done and much is known about microbial mechanisms and pathways involved. Many reviews have been published outlining general mechanisms of hydrocarbon degradation (Watkinson and Morgan, 1990; Salleh et al., 2003; Speight and Arjoon, 2012; Jaekel et al., 2013). Diverse group of bacteria, fungi, cyanobacteria and algae are involved in hydrocarbon degradation (Chaillan et al., 2004). They are widely distributed in marine, freshwater and soil habitats, i.e., they are omnipresent. Biodegradation of petroleum components is 'energetically favourable' (Speight and Arjoon, 2012).

This chapter aims to provide updated overview of bioremediation as well as phytoremediation of petroleum hydrocarbons and will throw a light to find out way for development of clean and green environment.

2. CRUDE OIL COMPOSITION

Crude oil (naturally occurring raw oil) or petroleum (crude oil and its refined products) is produced by thermal decay of buried organic material over millions of years. Crude oil once extracted from subsurface is transported to refineries where it undergoes distillation to produce various petroleum products that are widely used in day-to-day life (Speight 2007). Petroleum hydrocarbons can be defined as a complex mixture of constituents in varying proportions, mainly composed of carbon and hydrogen. Crude oil can be categorized into four broad fractions (a) saturates (aliphatics), (b) aromatics (ringed hydrocarbons), (c) resins and (d) asphaltenes (Balba et al., 1998; Zheng et al., 2001; Stroud 2007; Speight 2007; Jain et al., 2011).

Aliphatic hydrocarbons are saturated or unsaturated and linear or branched open-chain structures, such as n-alkanes, iso-alkanes, cyclo-alkanes (naphthenes), terpenes and steranes (Venosa and Zhu, 2003; Stroud, 2007). Aromatics are ringed hydrocarbon molecules (a) monocyclic aromatic hydrocarbons (MAHs), viz., BTEX (benzene, toluene, ethylbenzene and xylenes) and (b) polycyclic aromatic hydrocarbons (PAHs). PAHs contain more than one benzene ring, such as naphthalene (two-ring), phenanthrene and anthracene (three-ring), pyrene and chrysenes (four-ring), fluorenthene and benzo[a] pyrene (five-ring) (Macaulay and Rees, 2014).

Resins and asphaltenes contain numerous polar functional groups formed with N, S, O and trace metals such as Ni, V, Fe (Balba et al., 1998; Zheng et al., 2001; Speight 2007). Resins are smaller molecules than asphaltenes and contain aromatic compounds with long alkyl chain (Bardon et al., 1996). Resins are structurally similar to surface-active molecules in crude oil and act as peptizing agents keeping asphaltenes in suspension thereby promoting the stability of crude oil (Parra-Barraza et al., 2003).

3. OIL SPILL INCIDENTS

An oil spill is a condition in which a liquid petroleum hydrocarbon is released unnaturally into the environment and causes pollution. Oil spill term is usually applied to marine ecosystem. However, it is also reported that if oil is released into ocean or coastal waters it is referred to as marine oil spill and if oil is spilled on land it is known as terrestrial oil spills (Varjani and Srivastava, 2015). Oil spills mainly occur due to discharge from oil tankers, offshore and drilling activities, spill of refined petroleum products and waste oil, fuels used in large ships such as bunker (Varjani et al., 2015).

Oil spill incidents occur worldwide. Some of them are Exxon Valdez spill (1989), Nakhodka spill (1997), Erik spill (1999), Prestige spill (2002), Gopalpur beach spill (2010) and Mumbai coast spill (2011). Major and minor oil spills in Indian waters since 2005 along with its location and vessel involved are depicted in Table 6.1. The devasting effects of these oil spills have increased concerns of environmentalists, chemists, biotechnologists and engineers (Brooijmans et al., 2009; Rai et al., 2011; Chaerun, 2012). The Exxon Valdez spill was the largest and most hazardous in USA (USEPA, 2009).Various marine oil spills are extensively reviewed by Rai et al. (2011). Oil spill at mumbai cost (2011) is represented in Fig. 6.1.

Table 6.1: Minor and major oil spills in Indian water (Since 2005)

Date	Position/Location	Vessel Involved
June '05	Vishakhapatnam Port	MV Jinan VRWD-5
September '05	Off Vishakhapatnam	MV Royal Ocean 2
May '06	Oyster Rocks, Karwar	MV Ocean Saraya
August '06	Great Nicobar Island	Bright Artemis
October '07	Off Jakhau, Gujarat	MV Star Leikanger
August '09	South Gujarat and North Maharashtra	Not Known
April '10	Gopalpur in Orissa	MV Malavika
January '11	Arabian Sea, Mumbai	Mumbai-Uran pipeline spill
October '13	ONGC Pipeline, Uran, Mumbai	Mumbai High-Uran pipeline rupture

Source: Blue Waters (2013), Indian Coast Guard; Blue Waters (2014) Indian Coast Guard.

Fig. 6.1 Oil spill of Mumbai coast

(*Source:* http://www.deccanherald.com)

3.1 Impacts of Oil Incidents with Concern to Agriculture

Marine oil spills can cause serious, long-term environmental and human health problems (Brooijmans et al., 2009). The carcinogenic, mutagenic, neurotoxic and immunotoxic petroleum pollutants released through spill(s) pose serious threat to ecosystem and living beings (Fingas, 2001). Various toxic organic compounds when introduced in natural environment during oil extraction (during seismic work, oil spill and several other activities) changes the geo-chemical composition of soil, river and other components of environment. This, in turn, affects agriculture and leads to a drastic decline in output for both farming and fishing activities affecting livelihood of people (Stanley, 1990). Oiled birds suffer from developmental abnormalities which may results in loss of eggs or even death (Varjani, 2014). Oil spill threatening turtle nesting site at Gopalpur, Orissa, India is shown in Fig. 6.2.

Fig. 6.2 Oil spill threatens turtle nesting site (Gopalpur, Orissa, India) (*Source*: projectsmileindia.blogspot.com)

Farming is one of the basic activity of the people in India. Large-scale spills that kill herbaceous vegetation quickly however its devasting effect on trees and shrubs can be seen after a month of spill. Small scale spills (1 to 5% oil) usually have much less severe effects on plant community and oil components of spill are biodegraded (Redmon et al., 2010). Oil spills reduce oxygen absorption of water thereby cause lesser oxygen dissolution under oil spills than the deep sea levels. As oil spill occurs, oil spreads on farmlands and in water bodies. Crude oil seeps into ground and is taken up by the roots of plants (Fingas, 2001; Inoni et al., 2006). Oil spilled on agricultural land can impact soil fertility (a) cause poor growth of plants, (b) affects germination of seeds and (c) pollute groundwater resources (Fingas, 2001; Oyem and `Oyem, 2013). Udo and Fayemi (1975) stated that oil spills can change soil properties, can increase in water holding capacity, loss of soil structure, introduction of reduced conditions by exclusion of air from soil, and production of hydrogen sulphide, which adversely affect plant growth. However, extent of damage depends on concentration and composition of spilled oil (Jha et al., 2008). Numerous field and laboratory studies have revealed that oil on leaves of plants, penetrates the leaf and interferes with its functioning chiefly by reducing transpiration and photosynthesis which leads to chlorophyll deficiency and quick death. If oil pollution is light leaves become yellow and drop soon after, but under heavy contamination, complete shedding of leaves takes place (Inoni et al., 2006). Oil spills can affect various trees, such as pineapple, banana, palm and corn. Oil spillage affects food crop production and productivity in terms of quantity and quality (Udo and Fayemi, 1975; Dublin et al., 1998;). Polluted groundwater with toxic oil constituents when consumed by animals and humans can have detrimental effects (Varjani, 2014).

4. BIOREMEDIATION BY MICROORGANISMS AND PHYTOREMEDIATION

Bioremediation is a treatment process that uses microorganisms (yeast, fungi, or bacteria) to breakdown or degrade hazardous substances into less toxic or nontoxic substances. The ability to utilize hydrocarbons is widely distributed among diverse microbial populations (Atlas, 1981). It is an emerging technology for environmental protection. Apart from ecofriendly nature and economical feasibility bioremediation by microbes also has some other advantages like minimization of site destruction and permanent elimination of pollutant(s). Some disadvantages of bioremediation by microorganisms are toxicity of pollutants, production of unknown byproducts and biodegradation of some recalcitrant components, etc. (Erdogan and Karaca, 2011). Interest in potential of bioremediation technology increased after its substantial success in cleanup of oil spill due to Exxon Valdez oil tanker disaster at Prince William Sound, Alaska (Bragg et al., 1994). The microbial degradation of hydrocarbons appears to be a promising tool to control petroleum pollution. Biodegradation is a primary mechanism of bioremediation in which oleophilic microbes eliminate hydrocarbon pollutants from the environment (Leahy and Colwell, 1990; Macaulay and Rees, 2014).

Phytoremediation is a technology that uses plants to manage wide variety of environmental pollutants from ecosystems polluted with hydrocarbons and other organic as well as inorganic pollutants (Das and Chandran, 2011). However, Ighovie and Ikechukwu (2014) have defined phytoremediation as the use of plants and/or associated microorganisms to remove or render harmful material harmless or to detoxify oil polluted sites. There are different mechanisms by which phytoremediation is carried out such as, hydraulic control, phytovolatilization, rhizoremediation and phytotransformation (Diab, 2008; Das and Chandran, 2011). Advantages of phytoremediation over physico-chemical methods include cost effectiveness, aesthetic advantages and long-term applicability. Though phytoremediation can be applied for remediation of numerous contaminated sites not much information is available for pollutant fate and transformation pathways. Very less data is available on contaminant removal rates and effciencies directly attributable to plants under field conditions (Das and Chandran, 2011). Grasses are used as primary remediation agents which are generally planted between rows of trees to provide soil stabilization and protection against wind-blown dust that could move contaminants offsite (White et al., 2006; Das and Chandran, 2011). Major species involved in phytoremediation are grasses, legumes, trees and several other monocots and dicots (White et al., 2006; Ighovie and Ikechukwu 2014). Diab (2008) reported that growth of both plant and microorganism (of rhizosphere) is must for effective and succeesful phytoremediation. Microbial degradation in rhizosphere area might be significant mechanism for removal of some petroleum hydrocarbon pollutants such as PAHs. The reason for this may be hydrophobic nature of PAHs (Miya and Firestone, 2001).

5. MICROBIAL DEGRADATION OF PETROLEUM HYDROCARBONS

Microorganisms play an important role in maintaining of ecosystem. They are used to bio-monitor global ecosystems and cleanup of environment (Atlas, 1981). There are some important hydrocarbon utilizing genera, such as *Achromobacter, Acinetobacter, Arthrobacter, Aspergillus, Brevibacterium, Cellulomonas, Corynebacterium, Flavobacterium, Micrococcus, Nocardia, Ochrobactrum, Pseudomonas, Stenotrophomaonas, Vibrio, Amorphoteca, Fusarium, Graphium, Neosartoria, Paecilomyces, Penicillium, Sporobolomyces, Talaromyces* and *yeast* (*Candida, Pichia, Rhodotorula* and *Yarrowia*) (Boguslawska-Was and Dabrowski, 2001;Varjani and Upasani, 2013).

Microscopic photograph (40×) of oleophilic hydrocarbon degrading colony of bacterial isolates on Bushnell-Hass agar plate with crude oil is shown in Fig. 6.3. Oil-droplets present in the vicinity of bacterial colonies indicate oleophilic nature. These hydrocarbon degrading bacterial isolates were obtained from crude oil contaminated soil samples of sites located in Ahmedabad asset of ONGC. GenBank accession no. of 16S rDNA sequences of these isolates are KC713609 and KC713610, respectively (Varjani and Upasani, 2012; Varjani, 2014).

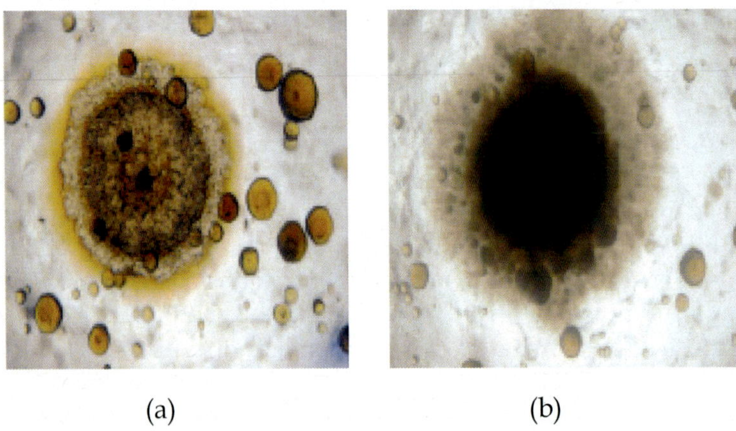

Fig. 6.3 Microscopic photograph (40×) demonstrating oleophilic nature of the HUBC isolate(s) on Bushnell-Hass agar plate with crude oil (*Source:* Varjani and Upasani, 2012; Varjani, 2014)

6. FACTORS AFFECTING BIODEGRADATION OF PETROLEUM HYDROCARBONS

Microbial communities are highly sensitive to their environment and can rapidly respond to changes in their surrounding conditions. Biodegradation rates are influenced by (a) characteristics of pollutant, (b) accessibility of nutrients, (c) nature, quantity and type of microorganisms and (d) physical, chemical and environmental factors, such as pH, water content, nutrients amendment, temperature, oxygen and bioavailability of pollutants (Varjani et al., 2014). Therefore, it is important to consider all these factors before selecting any alternative for remediation process (Atlas, 1981; Kauppi et al., 2011). Nutrient availability as well as bioavailability and biosurfactant are reviewed here.

6.1 Nutrient Availability

Availability of nutrients play an important role in performance of microbes and their degradation activity. Though hydrocarbons are excellent source of carbon and energy, they do not contain significant concentrations of other nutrients (such as N and P) required for microbial growth. If organic material has less amount of nitrogen in relation to carbon then microorganism will utilize soil nitrogen for further decomposition and soil nitrogen will be immobilized which will not be available for plant use (Oyem and Oyem, 2013). Adjustment of carbon/nitrogen/phosphorus/potassium (C-N-P-K) ratios by addition of urea, phosphate, N-P-K fertilizers, ammonium and phosphate salts accelerates biodegradation of crude oil or gasoline in soil and groundwater (Leahy and Colwell, 1990). The most adequate C:N:P to promote microbial growth is 100:10:1 (Beškoskiet al., 2011; Dias et al., 2012).

6.2 Bioavailability and Biosurfactant

Availability of hydrocarbon as substrate is a major problem faced by microorganisms. Bioavailability is the amount of a substance that is physico-chemically accessible to microorganisms. It plays a crucial role in efficient biodegradation to determine overall success of a bioremediation process (Chaudhry et al., 2005). Three modes of hydrocarbon transport to microbial cells are (a) interaction of cells with hydrocarbon dissolved in aqueous phase, (b) direct contact of cells with large hydrocarbon drops and (c) interaction of cells with "solubilized / pseudosolubilized" or "accommodated" hydrocarbon droplets much smaller than cells (Goswami and Singh, 1991). Pollutant bioavailability depends on its chemical properties (hydrophobicity, volatility), soil properties, environmental conditions and biological activity (Pilon-Smits, 2005). However, solubility and dissolution rates play a key role for bioavailability. Hydrophobic, less water-soluble molecules sorb more strongly to soil components than hydrophilic molecules (Jain et al., 2011).

Hydrocarbon degraders may have different mechanisms to increase absorbance of hydrocarbons; secretion of biosurfactant or bioemulsifier is one of them (Whang et al., 2008). Biosurfactant plays a crucial role to enhance bioavailability of hydrocarbon pollutant and it is very well studied. These surface-active agents may stimulate dissolution or desorption rates, solubilization or even emulsification of hydrocarbons (Mulligan, 2005; Dias et al., 2012).

7. CONCLUSION

Oil pollution of marine ecosystem can cause long term and/or short term effects on environment and life forms. Oil floats on top of water which decreases light penetration into water which limits photosynthesis of marine plants and phytoplankton, affecting human life directly and indirectly. Biological methods for oil spill cleanup are preferred over other conventional methods due to its advantages on environment and economy. Biodegradation rate of hydrocarbons from oil spills depends on local environmental conditions and microbial communities. Oil spills either marine or terrestrial, it affects agriculture adversely. Very less data is available on phytoremediation of pollutant removal rates and its fate as well as transformation pathways. Successful phytoremediation can be achieved by proper growth of both plant and rhizospheric-microorganisms. In the present scenario, bioremediation of petroleum hydrocarbons is preferred over phytoremediation because of its positive effects as discussed in this chapter. Application of appropriate and sufficient inorganic NPK fertilizer to restore carbon to nitrogen ratio (C/N) to the optimum required to stimulate and sustain microbial activity. Stimulating growth of indigenous microbes by cultivating soil to distribute nutrient and lime as well as to aerate treatment zone which will help to develop sustainable agriculture. The government of India should be strict towards the oil companies to follow environmental laws and regulations as well as a penalty plan established, i.e., oil companies whose activities cause excessive pollution.

Acknowledgements

I am grateful to Professor Dr. Bimal Patel, Vice Chancellor, Gujarat National Law University (GNLU), Gandhinagar, Gujarat for granting permission to perform research work. I express my gratitude to Academic and Administrative staff of GNLU for their kind support.

References

Atlas RM (1981). Microbial degradation of petroleum hydrocarbons: an environmental perspective. Microbiol Rev 45: 180–209.

Balba MT, Al-Awadhi N, and Al-Daher R (1998). Bioremediation of oil-contaminated soil: Microbiological methods for feasibility assessment and field evaluation. J Microbiol Methods 32: 155–164.

Bardon C, Barre L, Espinat D, Guille V, Li MH, Lambard J,Ravey JC, Rosenberg E, and Zemb T (1996). The colloidal structure of crude oils and suspensions of asphaltenes and resins. Fuel Sci Tech Int 14: 203– 242.

Beškoski VP, Gojgic-Cvijovic G, Milic J, Ilic M, Miletic S, Šolevic T, and Vrvic MM (2011). Ex situ bioremediation of a soil contaminated by Mazut (heavy residual fuel oil) A field experiment. Chemosphere 83: 34–40.

Blue waters (2013). Indian Coast Guard, Volume XIV, Issue 1, Indian Coast Guard. http://www.indiancoastguard.nic.in/IndianCoastGuard/NOSDCP/NOSDCP%20 Publications_files/blue%20water_Jan_13.pdf (Last accessed: 23.05.2015).

Blue waters (2014). Indian Coast Guard. Volume XV, Issue 1, http://www. indiancoastguard. nic.in/indiancoastguard/Blue_water/BLUE%20WATERS%20 JAN%202014.pdf(Last accessed: 23.05.2015).

Boguslawska-Was E, and Dabrowski W (2001). The seasonal variability of yeasts and yeast-like organisms in water and bottom sediment of the Szczecin Lagoon. Int J hygiene and Env Health 203: 451–458.

Bragg JR, Prince RC, Harner EJ, and Atlas RM (1994). Effectiveness of bioremediation for the Exxon Valdez oil spill. Nature 368: 413–418.

Brooijmans RJW, Pastink MI, and Siezen RJ (2009). Hydrocarbon-degrading bacteria: the oil-spill clean-up crew. Microbial Biotechnology 2: 587–594.

Chaerun SK (2012). The Nakhodka oil spill and its bioremediation. J Life Sci Tech 1: 29–33.

Chaillan F, LeFlèche A, Bury E, Phantavong YH, Grimont P, Saliot A, and Oudot J (2004). Identification and biodegradation potential of tropical aerobic hydrocarbon degrading organisms. Research in Microbiol 155(7): 587–595.

Chaudhry Q, Blom-Zandstra M, Gupta S, and Joner EJ (2005). Utilizing the synergy between plants and rhizosphere microorganisms to enhance breakdown of organic pollutants in the environment. Env Sci Pollu Res 12: 34–48.

Das N, and Chandran P (2011). Microbial degradation of petroleum hydrocarbon contaminants: an overview. Biotechnology Research International, Article ID 941810, 2011:1–13 doi:10.4061/ 2011/941810.

Diab EA (2008). Phytoremediation of crude oil contaminated desert soil using the rhizosphere effects. Global J. Env. Res. 2: 66–73.

Dias RL, Ruberto L, Hernandez E, Vazquez SC, Lo-Baldo A, Panno MTD, and Cormack WPM (2012). Bioremediation of an aged diesel oil-contaminated Antarctic soil: Evaluation of the "on site"biostimulation strategy using different nutrient source. Int Biodeter Biodegr 75: 96–103.

Dublin Green WF, Nwankwo JN, and Irrechukwu DO (1998). Effective regulation and management of HSE issues in the petroleum industry in Nigeria. 1998 society of petroleum engineers' International conference on health safety and environment in oil and gas exploration and production. Caracas, Venezuela. p. 9.

Erdogan E, Karaca A (2011). Bioremediation of crude oil polluted soils. Asian J Biotech 3: 206–213.

Ferradji FZ, Mnif S, Badis A, Rebbani S, Fodil D, Eddouaouda K, and Ferradji SS (2014). Naphthalene and crude oil degradation by biosurfactant producing *Streptomyces* spp. isolated from Mitidja plain soil (North of Algeria). Int Biodeter Biodegr 86: 300–308.

Fingas M (2001). The Basics of Oil Spill Cleanup. CRC Press LLC, USA.

Goswami P, and Singh HD (1991) Different modes of hydrocarbon uptake by two *Pseudomonas* species. Biotechnol Bioeng 37: 1–11

Ighovie ES, and Ikechukwu EE (2014). Phytoremediation of crude oil contaminated soil with *Axonopus compressus* in the Niger Delta region of Nigeria. Natu Resour 5: 59–67.

Inoni OE, Omotor DG, and Adun FN (2006). The effect of oil spillage on crop yield and farm income in Delta state, Nigeria. J Eur Agri 7: 41–48.

Jaekel U, Musat N, Adam B, Kuypers M, Grundmann O, and Musat F (2013). Anaerobic degradation of propane and butane by sulfate-reducing bacteria enriched from marine hydrocarbon cold seeps. ISMEJ 7: 885–895.

Jain PK, Gupta VK, Gaur RK, Lowry M, Jaroli DP, and Chauhan UK (2011). Bioremediation of petroleum oil contaminated soil and water. Res J Env Toxico 5: 1–26.

Jha MN, Levy J, and Gao Y (2008). Advances in remote sensing for oil spill disaster management:state-of-the-art sensors technology for oil spill surveillance. Sensors 8: 236–255.

Kauppi S, Sinkkonen A, and Romantschuk M (2011). Enhancing bioremediation of diesel-fuel-contaminated soil in a boreal climate: comparison of biostimulation and bioaugmentation. Int Biodeter Biodegr 65: 359–368.

Leahy JH, and Colwell R (1990). Microbial degradation of hydrocarbons in the environment. Microbiol Rev 54: 305–315.

Macaulay BM, and Rees D (2014). Bioremediation of oil spills: a review of challenges for research advancement. Annals of Env Sci 8: 9–37.

Miya RK, and Firestone MK (2001). Enhanced phenanthrene biodegradation in soil by slender oat root exudates and root debris. J Env Quality 30: 1911–1918.

Mulligan CN (2005). Environmental applications for biosurfactants. Environ Pollut 133:183–198.

Oyem ILR, and Oyem IL (2013). Effects of crude oil spillage on soil physico-chemical properties in Ugborodo community. IJMER 3: 3336–3342.

Parra-Barraza H, Hernández-Montiel D, Lizardi J, Hernández J, Urbina RH, and Valdez MA (2003). The zeta potential and surface properties of asphaltenes obtained with different crude oil/n-heptane proportions. Fuel 82: 869–874.

Pilon-Smits E (2005). Phytoremediation.Annu Rev Plant Biol 56: 15–39.

Rai N, Pandey IP, and Joshi K (2011). Impacts and management of oil spill pollution along the Indian coastal areas. J Ind Res Tech 1: 119–126.

Redmon LA, McFarland ML, Feagley SE, and Miller TD (2010). Oil spills: Potential impacts and management strategies for agricultural land. Agr Life Ext. 4: 1–4.

Salleh AB, Ghazali FM, Rahman RNZA, and Basri M (2003). Bioremediation of petroleum hydrocarbon pollution. Indian J Biotechnol 2: 411–425.

Speight JG (2007). The chemistry and technology of petroleum, 5th edn., volume 114, CRC, Taylor & Francis, Boca Raton, Florida, USA.

Speight JG, and Arjoon KK (2012). Bioremediation of petroleum and petroleum products. Wiley, Hoboken, New Jersey, USA (DOI: 10.1002/9781118528471).

Stanley WR (1990). Socio-economic impact of oil in Nigeria. Geography Journal 22: 67–79.

Stroud JL, Paton GI, and Semple KT (2007). Microbe-aliphatic hydrocarbon interactions in soil: implications for biodegradation and bioremediation. J ApplMicrobiol 102: 1239–1253.

Udo EJ, and Fayemi AA (1975). The effect of oil pollution of soil on germination growth and nutrient uptake of corn. J Environ Qual. 4(4): 537-540.

U.S. EPA (2009). Hazardous Waste Report.http://www.epa.gov/osw/inforesources/data /br09/br2009rpt.pdf (Last accessed: 23.05.2015).

Varjani SJ (2014). Hydrocarbon degrading and biosurfactants (bio-emulsifiers) producing bacteria from petroleum oil wells. PhD Thesis, Kadi Sarva Vishwavidyalaya, Gandhinagar, India.

Varjani SJ, Rana DP, Bateja S, Upasani VN (2013). Isolation and screening for Hydrocarbon Utilizing Bacteria (HUB) from petroleum samples. Int J Curr Microbiol App Sci 2: 48–60.

Varjani SJ, Rana DP, Jain AK, Bateja S, and Upasani VN (2015). Synergistic *ex-situ* biodegradation of crude oil by halotolerant bacterial consortium of indigenous strains isolated from on shore sites of Gujarat, India. Int Biodeter Biodegr 103: 116–124.

Varjani SJ, and Srivastava VK (2015). Green technology and sustainable development of environment. Renew Res J 3: 244–249.

Varjani SJ, Thaker MB, and Upasani VN (2014). Optimization of growth conditions of native hydrocarbon utilizing bacterial consortium "HUBC" obtained from petroleum pollutant contaminated sites. I JAR 4: 69–71.

Varjani SJ, and Upasani VN (2012). Characterization of hydrocarbon utilizing *Pseudomonas* strains from crude oil contaminated samples. Int J Scient Comput 6: 120–127.

Varjani SJ, and Upasani VN (2013). Comparative studies on bacterial consortia for hydrocarbon degradation. IJIRSET 2: 5377–5383.

Venosa DA, and Zhu X (2003). Biodegradation of crude oil contaminating marine shorelines and freshwater wetlands. Spill Sci Technol B 8: 63–178.

Watkinson RJ, and Morgan P (1990). Physiology of aliphatic hydrocarbon-degrading microorganisms. Biodegradation 1: 79–92.

Whang LM, Liu PW, Ma CC, and Cheng SS (2008). Application of biosurfactants, rhamnolipid, and surfactin, for enhanced biodegradation of diesel-contaminated water and soil. J Haz Mat 151: 155–163.

White PM, Wolf DC, Thoma GJ, and Reynolds CM (2006). Phytoremediation of alkylated polycyclic aromatic hydrocarbons in crude oil contaminated soil. Water air soil pollut 169: 207–220.

Zheng J, Behrens SH, BorkovecM, and Powers SE (2001). Predicting the wettability of quartz surfaces exposed to dense nonaqueous phase liquids. Env Sci Tech 35: 2207–2213.

7

Biochar and Its Role in Soil Fertility Management

Charu Gupta[1*], Dhan Prakash[1] and Sneh Gupta[2]
[1]Amity Institute for Herbal Research & Studies
Amity University, UP, Sector-125, Noida–201313, India
[2]Department of Zoology, RG (PG) College, Meerut (UP)–250001, India
*Corresponding author Email: *charumicro@gmail.com*

1. INTRODUCTION

Biochar first came into broad public awareness through the example of the Amazon, where the hypothesis is that Amazonian inhabitants added biochar along with other organic and household wastes over centuries to modify the surface soil horizon into a highly productive and fertile soil called Terra Preta, which is in direct contrast to the typical weathered Oxisol soils in close proximity. Biochar is exciting to many people because of its role in such soil-building processes (Wilson, 2014). Biochar is a solid material obtained from the carbonisation of biomass. Biochar may be added to soils with the intention to improve soil functions and to reduce emissions from biomass that would otherwise naturally degrade to greenhouse gases. Biochar also has appreciable carbon sequestration value.

Biochar is a stable carbon (C) compound created when biomass (feedstock) is heated to temperatures between 300 and 1000°C, under low (preferably zero) oxygen concentrations. Biochar, due to its aromatic structure and long mean residence time in soil (more than 100 years) has the potential for long-term carbon sequestration in the soil. Biochar can be produced from a variety of biomass feed stocks, but is generally designated as biochar only if it produces a useable co-product for soil improvement. The efficiency and effectiveness of the process of its creation and use can vary and the specific biomass sources used can affect the characterization and stability of the biochar. In a research, application of biochar with inorganic fertilizers and biofertilizer increased the soil fertility of maize in Alfisols (Gokila and Baskar, 2015).

Biochar is generally produced through thermal conversion (pyrolysis). As agricultural waste and wood is made up of cellulose, hemicelluloses and lignin; these building units are produced into volatile products through the thermal conversion process. Thus, biochar (black charred carbon) is a product of thermal decomposition of biomass (pyrolysis).

Recently, the application of biochar in mitigation strategy for sequestering recalcitrant carbon into agricultural soils that improve soil fertility has also been well discussed (Glaser et al., 2002). Modern pyrolysis techniques allow energy production from syngas (mainly carbon monoxide, hydrogen and methane and other hydrocarbons) and/or liquid-fuel production, while simultaneously producing different types of biochar (Laird et al., 2009). In addition, biochar composition depends and differs in their material properties (aldehyde concentration, aromaticity, cation exchange capacity, pH, nutrient contents, porosity, energy density) depending on feedstock and pyrolysis conditions (Amonette and Joseph, 2009; Downie et al., 2009).

Studies have shown that biochar is stable and hence suitable for carbon sequestration, promotes plant growth in poor soils particularly when combined with nitrogen-fertilizers (Blackwell et al., 2009) and also reduces nutrient leaching (Major et al., 2010). Moreover, studies have also shown that the addition of biochar also improves the cation exchange capacity of soil as the functional groups are oxidized (Liang et al., 2006; Cheng et al., 2008).

The mechanisms involving chemical and physical properties of biochar particles affecting water movement through soil, remove pollutants, alter microbial communities and reduce emissions of greenhouse gases is yet to be unraveled. It is believed that biochar can help farmers with poor soils. Different types of biochar have unique potential to mitigate some of the greatest soil-health constraints to crop productivity, for example, in highly weathered and sandy soils. Part of the difficulty is that biochar can be produced from all kinds of biomass and at different temperatures and speeds, which leads to huge variation in the substance and in results. Researchers are also studying the biochar's influence on microbial activity in soil. Microbes typically act as a community, for example, many pathogenic bacteria attack plant's roots only when they have sufficient numbers to overwhelm the host's immune response. Biochar can inhibit this by binding to the signaling molecules that bacterial cells secrete to coordinate their activity.

Biochar application to soil has resulted in agronomic benefits, including improved water-holding capacity. However, limited studies exist quantifying different biochar roles in soil water retention, especially when the soil is drier than field capacity. Studies have indicated that biochar may not generally improve water retention at all water contents with one-time application rates. The combined application of biochar with 100 percent recommended dose of fertilizer and biofertilizer increased the nutrient use efficiency and soil fertility status of the soil (Brantley et al., 2015).

It is believed that biochar can help reverse carbon dioxide levels in the atmosphere, improve overall soil quality and raise soil's water retention ability. It may also help filter toxic chemicals in the soil. Moreover, adding biochar to 10 percent of the world's croplands would store 29 billion tons of carbon dioxide equivalent. This roughly equals the world's annual greenhouse emissions. US soils alone can absorb up to 330 million tons of carbon annually with better carbon management practices. The addition of biochar radically improves the soil fertility and allows production of healthier crops.

2. ROLE OF SOIL CARBON IN BIOCHAR FORMATION

There are two main sources of soil carbon: organic and inorganic. Organic forms can be further be divided into "recalcitrant carbon" (resistant to decay) like humus, and "labile carbon," (carbon is both bioavailable and physically accessible to microbes). The organic carbon pool includes both the living bodies and the dead, decomposing bodies of bacteria, fungi, insects and worms, along with plant debris and manure. Inorganic carbon consists of carbonate ions, such as salts like calcium carbonate, and dolomite minerals, mostly in the form of rocks and sand.

It is believed that nature generates biochar in the process of naturally occurring wildfires in forests (Goldberg, 1985). Prairie fires can also generate a lot of biochar (Old, 1965). This kind of natural charcoal is present in some of the most valuable agricultural soils in the world: the carbon-rich Chernozems of the Russian steppe and the Mollisols of the US Midwestern prairie states (Skjemstad et al., 2002; Glaser and Amelung, 2003). Recently, scientists (Mao et al., 2012) have looked more closely at the Mollisols, and found that they contain charcoal that is structurally comparable to char in the Terra Preta soils and much more abundant than previously thought (40–50% of organic C).

3. FORMATION OF BIOCHAR

Biochar is a mixture of both organic and mineral carbon, depending on the conditions of formation. Biochar is made by heating biomass under the exclusion of air. This process is called pyrolysis, which includes the drying of the biomass and the subsequent release of flammable vapors. Technically, this can be done by many different methods. Some methods use a retort, which is a closed vessel that is externally heated. Heat is transferred through the metal vessel and vapors pass out of a vent where they can be burnt and help heat the retort. Gasification is another method that supplies enough air to burn the vapors, but prevents the complete combustion of the biomass material by excluding air from the charcoal zone, thus preserving the biochar (Wilson, 2014).

The resulting charcoal resembles a blackened, shrunken version of the original biomass. It has very little hydrogen and oxygen. Microscopically, it inherits much of the structure of the original biomass. The only difference is that the material now has been converted from lignin, cellulose and hemicellulose to some allotropes of carbon. There is a collection of disjointed graphite crystals based on hexagonally shaped carbon rings, with some leftover hydrogen and oxygen attached, along with minerals (ash) that were in the original feedstock. These hexagonal carbon compounds are fused carbon rings. Fused carbon rings are also called "aromatic" carbon. They are very stable and difficult to degrade. The more the biomass is heated, the more of these fused carbon rings are created (Keiluweit et al., 2010). The rings hook up with each other to form layers of discontinuous, rumpled sheets (graphite crystals).

Biochar becomes more mineral-like with heating. This mineral transformation creates the skeletal structure that looks like a carbon sponge. While the mineral, fused–carbon ring structure is hardly biodegradable. The recondensed vapors that can be found in the biochar pores and on its surfaces are less aromatic and are more biodegradable and can thus be considered organic phases of the biochar.

The fused carbon rings are also responsible for the electrical activation of the biochar carbon sponge. Fused carbon rings form a special bond with each other that allows electrons to move around the molecule producing electrical properties like those in graphene sheets and carbon nanotubes. Depending on the pyrolysis temperature and resulting arrangement of atoms, biochar can be an insulator, a semiconductor or a conductor of electricity. Electrically active fused carbon rings also support "redox" or oxidation and reduction reactions that are important to soil biochemistry, by acting as both a source and sink of electrons. In soils, microorganisms use aromatic carbon both as an electron donor and as an electron acceptor during metabolic chemical reactions. Biochar seems to not only serve as an electron buffer for redox reactions (Klupfel et al., 2014), but it also helps bacteria swap electrons among themselves, improving their metabolic efficiency as a microbial community (Kappler et al., 2014).

With its pores and its electrical charges, biochar is capable of both absorption and adsorption. The larger pores absorb water, air and soluble nutrients like a normal sponge. The surfaces of biochar adsorb materials by electrochemical bonds, working like an electric sponge. The amount of porosity depends mostly on the feedstock material, particle size, and the highest treatment temperature (HTT) (Brewer et al., 2014). At temperatures approaching 1000°C, pores begin to collapse and melt. For this reason, HTT is a key variable to know when specifying a biochar for a particular purpose. Porosity will also depend on the feedstock, with high ash feedstocks like grass reacting quite differently to heating than low ash feedstocks like wood or bamboo. For wood feedstocks, porosity typically peaks at an HTT of about 750°C (Downie et al., 2009).

Biochar is also effective as a polycyclic aromatic hydrocarbon (PAH) adsorbent and could be used to clean up PAH and other substances from soil (Chen and Chen, 2009).

4. SUBSTRATES FOR BIOCHAR PRODUCTION

In a study Rajkovich et al. (2012) investigated the effects of over 30 different biochars produced from corn (*Zea mays*) stover, nut shells, woodchips, manures, and sawdust bedding at four different pyrolysis temperatures on corn growth and reported differences between feedstock source, pyrolysis temperature, application rate on corn biomass, plant growth, nitrogen concentration, and total nitrogen uptake.

Novak et al. (2009) reported that greater biochar production temperatures could produce more alkaline pH, greater ash contents, and greater surface areas in the biochar and could result in increased soil pH. Biochar has also been observed to increase soil-water retention (Karhu et al., 2011). Laird et al. (2010) reported greater water retention at gravity-drained equilibrium and −100 and −500 kPa water potentials in a Hapludoll from Iowa with mixed hardwood biochar.

Novak et al. (2012) reported varied results depending on the biochar feedstock used; switch grass (*Panicum virgatum*) biochar maximized soil-water retention when applied at a 40 $Mgha^{-1}$ application rate incorporated to a depth of 15 cm to loamy sand compared to poultry litter, pecan (*Carya illinoinensis*) shell, and peanut (*Arachis hypogaea*) hull biochars by utilizing an incubation and leaching experiment to determine the percentage of water retained in pots.

A similar experimental design was used to test soil-water retention at −5 and −60 kPa with switch grass biochar produced at 500°C resulting in the greatest soil-water retention (Novak et al., 2012).

Birch (*Betula* spp.) biochar added to a silt loam in Finland increased soil water-holding capacity by 11% compared to the unamended control (Karhu et al., 2011). Based on moisture release curve data from a laboratory experiment with loamy sand, only the addition of peanut hull biochar applied at 88 $Mgha^{-1}$ increased soil water-holding capacity compared to lower biochar rates and various woodchip biochar types (Gaskin et al., 2007).

In a Brazilian sandy clay loam, sugarcane (*Saccharum* spp.) filter cake biochar addition led to greater water retention within the plant available water range, even though drier soils were also evaluated (Eykelbosh et al., 2014).

In yet another study, the effect of biochar on the properties of loamy sand Spodosol with different fertility level in the ploughed horizon has been studied in a 60-day laboratory experiment. It has been found that the application of biochar resulted in a significant increase of the soil-water content in the range of soil-water potential from -5 to -50 kPa, a significant decrease of the total shrinkage of the soil with high level of fertility after three wetting-drying cycles, and the accumulation of a significantly higher content of nitrates in the soil. At water content of 21 percent, less favorable conditions for the denitrification development were formed in the soil with high level of fertility than in the medium-fertile soil, which resulted in lower cumulative N_2O emission from the former. The addition of biochar resulted in no significant changes in the cumulative emission of CO_2 from the studied soil, but significantly reduced the emission of N_2O from the soil at the simultaneous addition of biochar and clover residues (Rizhiya et al., 2015).

In another study, it was observed that direct use of poultry manure on agricultural lands may cause environmental concerns, so there is a need to establish the suitability of the application of biochar derived from poultry manure for calcareous soil chemical properties and plant growth. The purpose of this study was to evaluate the effects of processed poultry manure (0, 5, 10 and 20 g/kg) and its biochar (0, 2.5, 5, 10 and 20 g/kg) on soil chemical properties of a calcareous soil and growth of bean (*Phaseolus vulgaris*) and maize (*Zea mays*) plants. In the incubation experiment, both processed poultry manure (PPM) and biochar decreased pH and the concentration of plant-available Fe of soil, but increased plant-available P, Zn, Cu and Mn concentrations. PPM and biochar increased the concentrations of exchangeable cations (K, Ca and Mg) in soil. PPM and biochar applications increased the growth of maize and bean plants. PPM and biochar resulted in increased concentrations of N, P, K, Ca, Fe, Zn, Cu and Mn in bean plants. In maize plants, PPM and biochar applications increased the N, P, K, Zn, Cu and Mn, but decreased the Ca and Mg concentrations. Results of this study reveal that poultry manure biochar can be used effectively for agricultural purposes (Inal et al., 2015).

5. EFFECTS OF BIOCHAR ON SOIL MICROBIOTA

Biochar effects on soil microbial abundance and community structure are keys for understanding the biogeochemical cycling of nutrients and organic matter turnover, but are poorly understood, particular in tropical areas. A study was conducted in a greenhouse experiment in which biochars produced from four different feedstocks [sewage sludge (B1), deinking sewage sludge (B2), Miscanthus (B3) and pine wood (B4)] was added at a rate of 3 percent (w/w) to two tropical soils (an Acrisol and a Ferralsol) planted with proso millet (*Panicum miliaceum* L.). The interactive effect of the addition of earthworms was also addressed. For this purpose soil samples were utilized from pots with or without the earthworm *Pontoscolex corethrurus*, which is a ubiquitous earthworm in tropical soils. Phospholipid fatty acid (PLFA) measurements showed that biochar type, soil type and the presence of earthworms significantly affected soil microbial community size and structure. In general, biochar addition affected fungal, but not bacterial populations. Overall, biochars rich in ash (B1 and B2) resulted in a marked increase in the fungi to bacteria ratio, while this ratio was unaltered after addition of biochars with a high fixed carbon content (B3 and B4). This study remarked the contrasting effect that both, biochar prepared from different materials and macrofauna, can have on soil microbial community. Such changes might end up with ecosystem-level effects (Paz-Ferreiro et al., 2015).

In another study, it was found that biochar-soil interaction enhances soil C storage via the processes of OM sorption to biochar and physical protection (Zimmerman et al., 2011). The interactive effects of biochar addition to soil on CO_2 evolution (priming) were evaluated by comparing the additive CO_2 release expected from separate incubations of soil and biochar with that actually measured from corresponding biochar and soil mixtures. In general, C mineralization was greater than expected (positive priming) for soils combined with biochars produced at low temperatures (250 and 400°C) and from grasses, particularly during the early incubation stage (first 90 d) and in soils of lower organic C content. It contrast, C mineralization was generally less than expected (negative priming) for soils combined with biochars produced at high temperatures (525 and 650°C) and from hard woods, particularly during the later incubation stage (250–500 d). Measurements of the stable isotopic signature of respired CO_2 indicated that, for grass biochars at least, it was predominantly pyrogenic C mineralization that was stimulated during early incubation and soil C mineralization that was suppressed during later incubation stages. It is hypothesized that the presence of soil organic matter (OM) stimulated the co-mineralization of the more labile components of biochar over the short term (Zimmerman et al., 2011).

The effect of biochar for soil amendment as a means to improve soil fertility and to mitigate climate change is less studied as compared to its effects on soil chemical properties. A review of the literature reveals a significant number of early studies on biochar-type materials as soil amendments either for managing pathogens, as inoculant carriers or for manipulative experiments to sorb signaling compounds or toxins. However, no studies exist in the soil biology literature that recognize the observed large variations of biochar physico-chemical properties. This shortcoming has hampered insight into mechanisms by which biochar influences soil microorganisms, fauna and plant roots. Additional factors limiting meaningful interpretation of many datasets are the clearly demonstrated sorption properties that interfere with standard extraction procedures for soil microbial biomass or enzyme assays, and the confounding effects of varying amounts of minerals. In most studies, microbial biomass has been found to increase as a result of biochar additions, with significant changes in microbial community composition and enzyme activities that may explain biogeochemical effects of biochar on element cycles, plant pathogens, and crop growth. Yet, very little is known about the mechanisms through which biochar affects microbial abundance and community composition. The effects of biochar on soil fauna are even less understood than its effects on microorganisms, apart from several notable studies on earthworms. It is clear, however, that sorption phenomena, pH and physical properties of biochars, such as pore structure, surface area and mineral matter play important roles in determining how different biochars affect soil biota. Observations on microbial dynamics lead to the conclusion of a possible improved resource use due to co-location of various resources in and around biochars. Sorption and thereby inactivation of growth-inhibiting substances likely plays a role for increased abundance of soil biota. No evidence exists so far for direct negative effects of biochars on plant roots. Occasionally, observed decreases in abundance of mycorrhizal fungi are likely caused by concomitant increases in nutrient availability, reducing the need for symbionts. In the short term, the release of a variety of organic molecules from fresh biochar may in some cases be responsible for increases or decreases in abundance and activity of soil biota (Lehmann et al., 2011).

6. BIOCHAR AND SEAWEEDS

Recently, some commercially cultivated seaweed is also being developed and used for biochar production for their use in soil amelioration. Seaweeds biomass from *Sargassum, Saccharina* and *Undaria* (brown seaweeds), *Eucheuma, Gracilaria* and *Kappaphycus* (red seaweeds) offers a good feedstock for the production of nutrient-rich biochar for soil amelioration. The properties of biochar depend on the origin of seaweeds.

The pH of seaweed biochar varies from 7–11 (neutral to alkaline) thereby allowing the broad-spectrum application in diverse soil types. The other characteristics of seaweed biochar are relatively low carbon content and surface area, essential trace elements like nitrogen (N), phosphorus (P) and potassium (K) and exchangeable cation like potassium.

Seaweed biochar is a unique material for soil amelioration that is consistently different to biochar derived from lignocellulosic feedstock. Blending of seaweed and lignocellulosic biochar could provide a soil ameliorant that combines a high fixed C content with a mineral-rich substrate to enhance crop productivity (Roberts et al., 2015).

The other advantage of using seaweeds in comparison to lignocellulosic feedstock is that it yields high proportions of biochar with lower in C content than lignocellulosic biochar. Besides seaweed biochar has a high exchangeable nutrient content and cation exchange capacity (CEC) (Bird et al., 2011). The high CEC of seaweed biochar makes it highly effective in the retention of nutrients in agricultural soils. It also makes it an effective biosorbent of metals from wastewater effluents. Seaweed biochar can also be modified to increase its affinity for problematic oxyanionic contaminants, such as selenium (Se) that otherwise show limited affinity for passive biosorbents (Kidgell et al., 2014a,b; Roberts et al., 2015). Despite the diverse applications of seaweed biochar, there have been a few studies that consider the production of biochar from a range of seaweed feedstocks.

In a recent study, seaweed biochar was produced from six species of seaweed and each of these species was collected from two geographically distinct locations that are representative of the dominant sources of each biomass feedstock. A clear characteristic of all seaweed biochars was the high concentrations of N, P and K. N ranged from 0.3–2.8 percent, P from 0.5–6.6 g kg^{-1}, and K from 5.1–119 g kg^{-1} (Roberts et al., 2015).

7. BIOCHAR AND COMPOST

Both biochar and compost provide similar benefits.

- Both improve tilth and reduces soil bulk density
- Increase soil water holding capacity
- Become more stable by combining with clay minerals
- Increase cation exchange capacity
- Improve fertilizer utilization, by reducing leaching from the root zone
- Retain minerals in plant available form
- Support soil microbial life and biodiversity
- Help plants to resist diseases and pathogens
- Help plants to grow better in high salt situations
- Add humus carbon to the soil carbon pool, reducing the atmospheric carbon pool

The difference is that unlike biochar, compost is quickly broken down by microbial action in soil depending primarily on climate. Biochar lasts at least ten times longer in most soils. Secondly, biochar has important synergistic effects when added to compost. Researchers find that biochar makes faster, more nutrients rich, more biologically diverse and more humified, stable compost (Wilson, 2014).

8. PROPERTIES OF BIOCHAR: ORGANIC AND INORGANIC COMPOSITION

Biochar composition can be broadly divided into relatively recalcitrant C, labile or leachable C and ash. The greatest chemical difference between biochar and other organic matter is the larger proportion of aromatic carbon and, specifically, the occurrence of fused aromatic C structures, in contrast to other aromatic structures of soil organic matter such as lignin (Schmidt and Noack, 2000). This fused aromatic structure of biochars in itself can have varying forms, including amorphous C, which is dominant at lower pyrolysis temperatures, and turbostratic C, which forms at higher temperatures (Keiluweit et al., 2010; Nguyen et al., 2010). It is clear that the nature of these C structures is the chief reason for the high stability of biochars (Nguyen et al., 2010). The chemical stability of a large fraction of a given biochar material means that microorganisms will not be able to readily utilize the C as an energy source or the N and possibly other nutrients contained in the C structure. However, depending on the type of biochar, a fraction may be readily leached and therefore is mineralizable (Lehmann et al., 2009) and in some cases has been shown to stimulate microbial activity and increase abundance (Steiner et al., 2008a).

The third major component comprises minerals that are present as ash inclusions in biochar. These minerals include several essential macro- and micronutrients for biological uptake and, therefore, represent valuable resources in the soil food web. Additionally, the presence of these elements during pyrolysis plays a role in the biochar chemical structure to the extent that they are incorporated into the aromatic structure or that organo-metal reactions are thermodynamically favorable at high temperatures (Lehmann et al., 2011).

9. ROLE OF BIOCHAR IN SOIL FERTILITY MANAGEMENT

9.1 Biochar Keeps Compost Moist and Aerated, and Promotes Biological Activity

The composting process is governed by various physical parameters that are subject to alteration by the addition of biochar materials as bulking agents. Some of the parameters that affect compost are: aeration, moisture content, temperature, bulk density, pH, electron buffering and the sorptive capacity of bulking agents. Water and air are both held in biochar pore spaces and voids, and the spaces between particles. Moisture is also the vehicle for bringing dissolved organic carbon, nitrogen and other plant nutritive compounds into contact with biochar surfaces where they can be captured. Biochar's stable carbon matrix accepts electrons from decomposing organic compounds, buffering electric charges that might otherwise impair microbial activity and be responsible for the production of greenhouse gases like methane and hydrogen sulfides.

All these properties of biochar promote microbial activity in compost. For instance, Steiner et al. (2011) tested 5 percent and 20 percent additions of pine chip biochar to poultry litter compost, and found that the addition of 20 percent biochar caused microbial respiration to peak earlier and at a higher level than either the 5 percent or 0 percent biochar treatments.

9.2 Biochar Increases Nitrogen Retention

When nitrogen-containing biomass is decayed, large amount of ammonia is released. Numerous studies have shown that biochar is effective to retain nitrogen in soils (Steiner et al., 2008b; Clough et al., 2013). Several studies have also shown that biochar enhances nitrogen retention in compost, reducing emissions of ammonia and increasing total nitrogen retention by as much as 65 percent (Steiner et al., 2010; Chen et al., 2010; Huang and Xue, 2014). The ammonia retention ability of biochar can actually improve during the composting process. During composting, biochar is subjected to an accelerated aging process thereby making biochar surfaces get oxidized and enriched by carboxylic (acid) functional groups. The latter more than doubled at the end of the composting period, thereby improving the capacity to exchange cations like ammonia.

9.3 Biochar Improves Compost Maturity and Humic Content

Several researchers have studied the effects of biochar on the timing and results of compost maturation and found that adding biochar to compost reduced the amount of dissolved organic carbon (labile carbon) in mature compost while increasing the fraction of stable humic materials (stable carbon). Following the addition of 2 percent biochar to compost, Jindo et al., (2012) recorded a 10 percent increase in carbon captured by humic substance extraction and a 30 percent decrease of water-soluble, easily degradable carbon. They also found an increase of fungal species diversity in the mature biochar compost as compared to the control and proposed that these fungi were responsible for the increased humification. Another study by Zhang et al. (2014) found that sewage sludge composted with wood biochar had up to 30 percent more humic substances than the control.

9.4 Biochar Compost Improves Plant Growth

Several researchers have experimented with various combinations of compost and biochar added as separate amendments (Fisher and Glaser 2012; Liu et al., 2012). These studies found improved plant growth response when biochar was added to soil along with compost.

Recently, study in Germany looked instead at biochar composted together with other materials. Schulz et al. (2013) tested six different amounts of biochar in compost, from 0 to 50 percent by weight, and also three different application rates of each compost type. Using oats in greenhouse pots on two different substrates (sandy soil and loamy soil), they found that plant growth increased with increasing application rates of each type of biochar compost, since the amount of deliverable nutrients was increased, at least by the compost fraction. However, they also discovered that plant growth was increased as the amount of biochar in the compost increased. The biochar may either have improved nutrient retention during the composting process with subsequent enhancement of nutrient delivery to plants, or it promoted plant growth through some other mechanism. However, the researchers confirmed that synergistic effects could be achieved by adding biochar to composts.

9.5 Biochar as Inoculant Carrier

Due to the positive effects of biochar on microorganism abundance and reproduction rates, biochar as an additive to commercial mycorrhizal inoculum, or even as a carrier material seems promising. Biochar-type materials have been suggested as inoculant carriers for substituting peat (non-renewable) that is expensive and emits greenhouse gas (Beck, 1991). Three aspects play a role for evaluating the suitability of the carrier material:

the survival of the inoculants during storage; the survival in the soil; and the inoculation efficiency.

Biochar remains in the soil and positively influence abundance of the inoculant organisms as shown for microbial biomass in general and possibly inoculation efficiency. The first scientific study was reported in 1957 that suggested adding biochar-type residues from vegetation fires to soil significantly enhanced nodulation of subterranean clover in a yellow podzolic soil (Hely et al., 1957). When biochar was mixed with peat, press mud, compost or farm yard manure, these inoculant carriers were observed to increase rhizobial cell counts, viability from 30 days to three months and nodulation in soybean and pigeon pea (Tilak and Subba Rao, 1978). Khavazi et al. (2007) used biochar to adjust the pH of the inoculant carrier material, which was as effective as other carrier materials and ensured survival of *Bradyrhizobium japonicum* for more than six months at an acceptable level. It is reasonable to assume that biochars may be designed specifically for certain inoculants and possibly soil conditions.

9.6 Biochar as Pathogen Antagonists

Biochar also suppresses plant diseases. The mechanism of suppression is similar to other organic amendments such as composts.

Several principal mechanisms have been proposed like (1) a direct release of inhibitors of plant pathogens; (2) the promotion of microorganisms that act antagonistic to pathogens, such as parasites, through production of antibiotics, or by successful competition for nutrients; (3) improved plant nutrition and vigor, leading to enhanced disease resistance; and (4) activation of plant defense mechanisms (induced systemic resistance) by enhancing certain microorganisms. Any and all of these four mechanisms may also be applicable to biochar. In addition, (5) the known strong sorption of organic compounds onto biochar may modify signaling between plant and pathogens, or (6) affect the mobility and activity of the pathogen itself (Noble and Coventry, 2005).

There are numerous studies reported to support the antagonistic nature of biochar against plant pathogens. A decrease in *Fusarium* infection of asparagus was found after addition of biochar made by fast pyrolysis of wood powder (Elmer and Pignatello, 2011). Also, infection of tomato with another soil-borne disease, bacterial wilt (*Ralstonia solanacearum*), was significantly reduced by adding wood biochar in some experiments and consistently by adding biochar made from municipal biowaste (Nerome et al., 2005). The disease suppression improved with greater application rates of up to 40 percent (v/v), with benefits persisting beyond 90 days after planting. Disease severity of powdery mildew (*Leveillula taurica*) significantly decreased after biochar was added to both a sandy soil and an organic potting mix (Elad et al., 2010), suggesting that biochar acts differently from other organic matter with respect to the studied disease. The studied biochar produced from citrus wood using traditional charcoal-making techniques was found to induce systemic resistance also to another foliar fungal pathogen, *Botrytis cinerea* (gray mold) on pepper and tomato and to the broad mite pest (*Polyphagotarsonemus latus*) on pepper (Elad et al., 2010).

9.7 Effect of Biochar on Nutrient Transformation

Numerous studies have shown that biochars have significant effects on microbially mediated transformation of nutrients in soil. In forest soils, nitrification was increased by biochar additions to soil (Ball et al., 2010). In addition, activity of alkaline phosphatase, aminopeptidase and N-acetylglucosaminidase was found to increase with biochar applications (Bailey et al., 2010; Jin 2010). The activity of alkaline phosphatase increased by 615 percent and aminopeptidase by 15 percent on increasing the rate of corn biochar application to Alfisol (Jin, 2010). Plant uptake of N and P and growth of fine roots and root hairs into biochar pores stimulate the production of organic N and P mineralizing enzymes (Lehmann et al., 2011).

9.8 Biochar as Pollution Wrangler

Although role of biochar in agriculture is well established; researchers are now looking at other applications. Research has shown that biochar can bind to heavy metals in soil, which helps to keep them from reaching plants or entering water supplies. That has attracted the notice of the US Environmental Protection Agency, other agencies, and companies seeking to reclaim land formerly used in mining. At the Hope Mine near Aspen, Colorado, biochar added in 2010 helped to neutralize the impacts of decades old mine refuse by immobilizing the metals and increasing the amount of water held on the slope thereby reducing the opportunity for contaminated water to become run-off. It also helped to spur plant growth on the formerly barren hillside, according to the Aspen Center for Environmental Studies.

Biochar is also showing promise in cleaning up polluted water, at a much cheaper replacement for activated charcoal, which is used at sites ranging from treatment plants to areas that are heavily contaminated with toxic chemicals. Biochar particles have a relatively large surface area, which expands even further in water, providing a vast number of sites for contaminants to bind with. This type of pollution remediation may be particularly beneficial in countries that lack full water-treatment systems. It could also help to remove antibiotics or chemical wastes, which are difficult to strip out with conventional water treatments.

Scientists have even explored biochar's potential for treating fluids used in oil and gas drilling, and as a component of print toners and paint products (Cernasky, 2015).

10. ANALYSIS OF BIOCHAR IN SOIL

Biochar is used with increasing frequency as a soil amendment because of its potentially beneficial effects on soil carbon sequestration, crop yield, nutrient leaching and greenhouse gas emissions. Simple methods for the analysis of biochar in soil are currently unavailable. In a study, the "loss on ignition" method was adopted for this purpose. The technique requires knowledge of the proportions of both biochar and biochar-free soil that are lost on ignition. One can use values determined prior to the amendment of the soil with biochar, assuming that the values do not change after biochar is incorporated in the soil. The technique accurately determined a wide range of biochar concentrations in field soil (Koide et al., 2011).

11. EFFECTS OF TEMPERATURE AND PROCESSING CONDITIONS ON BIOCHAR CHEMICAL PROPERTIES

There have been limited studies of how the pyrolysis process and activation conditions affect the chemical properties of biochar and how these properties alter soil carbon (C) and nitrogen (N) transformations when used as an amendment. One study compared the chemical properties of biochars produced through slow pyrolysis at 200, 400 and 600°C, in the presence or absence of steam and CO_2 activation at 800°C. Quantitative solid-state ^{13}C nuclear magnetic resonance spectroscopy and elemental analysis were used to evaluate processing condition effects on biochar chemical properties. NMR data showed that biochars produced at 400 and 600°C converted (>82%) of labile C constituents to aromatic C structures, which increased their recalcitrance. Also the later pyrolysis temperatures increased biochar cation exchange capacity (CEC) and pH, however, exposure to steam and CO_2 activation decreased their CEC. Amendment with biochars produced at 400 and 600°C and activated biochars significantly decreased soil inorganic N. This study highlights the importance of production conditions for designing biochars for use as amendments to sequester soil C and N, by promoting the formation of stable soil organic matter and by increasing retention of soil inorganic N. From a soil amendment perspective, this study suggests that activation of biochar did not benefit soil C and N transformations (Zhang et al., 2015).

12. BIOCHAR: PROSPECTS OF COMMERCIALIZATION

Today biochar is available in limited amounts, used mostly as a component of potting soil in greenhouses and as a soil amendment for high-value horticultural crops and trees. The most economical sources for biochar are agricultural and forest residues (i.e., chaff, stover, bark, chips and slash); wood residues being the most common feedstock.

The logistics of wood harvesting provides economic advantages. Compared to field-by-field processing of agricultural crops, wood is hauled and processed at centralized saw or chip mill plants. Wood residues can thereby be obtained fairly easily and cheaply. In addition, wood is drier than most agricultural crops, making wood industry residues less expensive to transport. The western pine bark beetle infestation is creating huge new volumes of wood, and the U.S. Forest Service estimate that as much as 1,00,000 trees fall each day in Northern Colorado and southern Wyoming (USFS, 2011).

A 2013 report by the International Biochar Initiative (IBI) provided a snapshot of the current state of the biochar economy. The report found that a majority of the global market share comes from the US (65%), followed by Europe (25%), Asia (7%), and Africa (3%). Ninety percent (90%) of commercial activity is focused on small specialty retail markets mostly nursery and garden centers. A small percentage (10%) is focused on larger scale markets, such as agriculture or land remediation projects. It has been reported that

Japan has a growing commercial biochar industry, with approximately 15,000 tons traded annually (Okimori et al., 2003).

In 2011, 94,159 tons of compost, a common soil amendment, was sold at a price of $30–50 per ton (USDA ERS, 2011). The price per ton for biochar is far higher and the volumes far lower. Companies responding to the IBI survey reported sales of only 911 tons in 2013, sixty-five percent (597 tons) of which was produced in North America. The average price for biochar in the US was $1.29 per pound or $2,580 per ton.

In the near term, due to price point, biochar will likely only be used in high-end specialty markets, but there is growing interest in developing biochar into new products that take advantage of its unique chemical properties. For example, today wastewater treatment facilities purchase thousands of tons of activated carbon annually to help absorb potential contaminants and reduce odors. Biochar has been shown to be a cost-effective alternative.

13. CONCLUSION

Biochars are carbon-rich soil additives derived from agricultural and other plant waste that enhances soil fertility and crop productivity. Biochars can also be produced from human sewage (an under utilized resource that is rich in soil nutrients and carbon). The combined application of biochar with 100 percent recommended dose of fertilizer and biofertilizer increased the nutrient use efficiency and soil fertility status of the soil. Biochar keeps compost moist and aerated and promotes biological activity, increases nitrogen retention, improves compost maturity and humic content, improves plant growth, can also be used as an inoculant carrier and as a pathogen antagonist. Recently, biochar is also finding its application in pollution treatment (bioremediation).

Acknowledgements

Authors are grateful to Dr. Ashok K. Chauhan, Founder President, Amity Group of Institutions and Mr. Atul Chauhan, Chancellor, Amity University UP, Noida for the encouragement, research facilities and financial support.

References

Amonette JE, and Joseph S (2009). Characteristics of biochar: microchemical properties. In: Lehmann J, Joseph S (eds) Biochar for Environmental Management-Science and Technology. Earthscan, London, pp. 33–52.

Bailey VL, Fansler SJ, Smith JL, and Bolton JrH (2010). Reconciling apparent variability in effects of biochar amendment on soil enzyme activities by assay optimization. Soil Biol Biochem 43: 296–301.

Ball PN, MacKenzie MD, DeLuca TH, and Holben WE (2010). Wildfire and charcoal enhance nitrification and ammonium-oxidizing bacteria abundance in dry montane forest soils. J Environ Qual 39: 1243–1253.

Beck DP (1991). Suitability of charcoal-amended mineral soil as carrier for *Rhizobium* inoculants. Soil Biol Biochem 23: 41–44.

Bird MI, Wurster CM, de Paula Silva PH, Bass AM, and de Nys R (2011). Algal biochar-production and properties. Biores Technol 102: 1886–1891.

Blackwell P, Riethmuller G, and Collins M (2009). Biochar application to soil. In: Lehmann J, Joseph S (eds) Biochar for Environmental Management: Science and Technology. Earthscan, London, pp. 207–226.

Brantley KE, Brye KR, Savin MC, and Longer DE (2015). Biochar source and application rate effects on soil water retention determined using wetting curves. Open Journal of Soil Science 5: 1–10.

Brewer CE, Chuang VJ, Masiello CA, Gonnermann H, Gao X, Dugan B, Driver LE, Panzacchi P, Zygourakis K, and Davies CA (2014). New approaches to measuring biochar density and porosity. Biomass Bioenergy 66: 176–185.

Cernansky R (2015). Agriculture: State-of-the-art soil. Nature 517(7534): 258–260.

Chen B, and Chen Z (2009). Sorption of naphthalene and 1-naphthol by biochars of orange peels with different pyrolytic temperatures. Chemosphere 76: 127–133.

Chen Y-X, Huang X-D, Han Z-Y, Huang X, Hu B, Shi D-Z, and Wu W-X (2010). Effects of bamboo charcoal and bamboo vinegar on nitrogen conservation and heavy metals immobility during pig manure composting. Chemosphere 78(9): 1177–1181.

Cheng CH, Lehmann J, Thies JE, and Burton SD (2008). Stability of black carbon in soils across a climatic gradient. J Geophys Res, 113:G02027, doi:10.1029/2007JG000642.

Clough TJ, Condron LM, Kammann C, and Müller C (2013). A review of biochar and soil nitrogen dynamics. Agronomy 3(2): 275–293.

Downie A, Crosky A and Munroe P (2009). Physical properties of biochar. In: Lehmann J, Joseph S (eds) Biochar for Environmental Management: Science and Technology. Earthscan, London, UK. pp. 13–32.

Elad Y, Rav David D, Meller Harel Y, Borenshtein M, Ben Kalifa H, Silber A, and Graber ER (2010). Induction of systemic resistance in plants by biochar, a soil applied carbon sequestering agent. Phytopathology 100: 913–921.

Elmer WH, Pignatello JJ (2011). Effect of biochar amendments on mycorrhizal associations and *Fusarium* crown and root rot of asparagus in replant soils. Plant Disease 95: 960–966.

Eykelbosh AJ, Johnson MS, de Queiroz ES, Dalmagro HJ, and Couto EG (2014). Biochar from sugarcane filtercake reduces soil CO_2 emissions relative to raw residue and improves water retention and nutrient availability in a highly-weathered tropical soil. PLoS ONE 9: 1–9.

Fischer D, and Glaser B (2012). Synergisms between compost and biochar for sustainable soil amelioration. In: Sunil K, Bharti A (eds) Management of Organic Waste. In Tech 167–198.

Gaskin JW, Speir A, Morris LM, Ogden K, Harris K, Lee D, and Das KC (2007). Potential for pyrolysis char to affect soil moisture and nutrient retention status of a loamy sand soil. Proceedings of the 2007 Georgia Water Resources Conference, Athens, GA, 27-29 March 2007, pp. 1–3.

Glaser B, and Amelung W (2003). Pyrogenic carbon in native grassland soils along a climosequence in North America. Global Biogeochem Cy 17(2): 1064.

Glaser B, Lehmann J, and Zech W (2002). Ameliorating physical and chemical properties of highly weathered soils in the tropics with charcoal-a review. Biol Fert Soils 35: 219–230.

Gokila B, and Baskar K (2015). Characterization of *Prosopis juliflora* L. biochar and its influence of soil fertility on maize in alfisols. International Journal of Plant, Animal and Environmental Sciences 5(1): 123–127.

Goldberg E (1985). Black Carbon in the Environment: Properties and Distribution. John Wiley & Sons Inc, New York.

Hely FW, Bergersen FJ, and Brockwell J (1957). Microbial antagonisms in the rhizosphere as a factor in the failure of inoculation of subterranian clover. Aust J Agric Res 8: 24–44.

Huang XD, and Xue D (2014). Effects of bamboo biochar addition on temperature rising, dehydration and nitrogen loss during pig manure composting. J Appl Ecol 25(4): 1057–1062.

Inal A, Gunes A, Sahin O, Taskin MB, and Kaya EC (2015). Impacts of biochar and processed poultry manure, applied to a calcareous soil, on the growth of bean and maize. Soil Use Manage 31: 106–113.

Jin H (2010). Characterization of microbial life colonizing biochar and biochar amended soils. PhD Dissertation, Cornell University, Ithaca, NY.

Jindo K, Suto K, Matsumoto K, García C, Sonoki T, and Sánchez-Monedero MÁ (2012). Chemical and biochemical characterization of biochar-blended composts prepared from poultry manure. Biores Technol 110: 396–404.

Kappler A, Wuestner ML, Ruecker A, Harter J, Halama M, and Behrens S (2014). Biochar as an electron shuttle between bacteria and Fe(III) minerals. Environ Sci Technol Lett 1(8): 339–344.

Karhu K, Mattila T, Bergström I, and Regina K (2011). Biochar addition to agricultural soil increased CH_4 uptake and water holding capacity-results from a short-term pilot field study. Agric Ecosyst Environ 140: 309–313.

Keiluweit M, Nico PS, Johnson MG, and Kleber M (2010). Dynamic molecular structure of plant biomass-derived black carbon (biochar). Environ Sci Technol 44: 1247–1253.

Khavazi K, Rejali F, Seguin P, and Miransari M (2007). Effects of carrier, sterilization method, and incubation on survival of *Bradyrhizobium japonicum* in soybean (*Glycine max* L.) inoculants. Enzyme Microb Tech 41: 780–784.

Kidgell JT, de Nys R, Hu Y, Paul NA, and Roberts DA (2014a). The sequential application of macroalgal biosorbents for the bioremediation of a complex industrial effluent. PLoS One 9: e101309.

Kidgell JT, de Nys R, Hu Y, Paul NA, and Roberts DA (2014b). Bioremediation of a complex industrial effluent by biosorbents derived from freshwater macroalgae. PLoS One 9: e94706.

Klüpfel L, Keiluweit M, Kleber M, and Sander M (2014). Redox properties of plant biomass derived black carbon (Biochar). Environ Sci Technol 48(10): 140505153602003.

Koide RT, Petprakob K, and Peoples M (2011). Quantitative analysis of biochar in field soil. Soil Biol Biochem 43: 1563–1568.

Laird DA, Brown RC, Amonette JE, and Lehmann J (2009). Review of the pyrolysis platform for coproducing bio-oil and biochar. Biofuels, Bioproducts and Biorefining 3: 547–562.

Laird DA, Fleming P, Davis DD, Horton R, Wang B, and Karlen DL (2010). Impact of biochar amendments on the quality of a typical Midwestern agricultural soil. Geoderma 158: 443–449.

Lehmann J, Czimczik C, Laird D, and Sohi S (2009). Stability of biochar in soil. In: Lehmann J, Joseph S (eds) Biochar for Environmental Management: Science and Technology. Earthscan, London, pp. 183–205.

Lehmann J, Rillig MC, Thies J, Masiello CA, Hockaday WC, and Crowley D (2011). Biochar effects on soil biota: A review. Soil Biol Biochem 43: 1812–1836.

Liang B, Lehmann J, Solomon D, Kinyangi J, Grossman J, O'Neill B, Skjemstad JO, Thies J, Luizao FJ, Petersen J, and Neves EG (2006). Black carbon increases cation exchange capacity in soils. Soil Sci Soc Am J 70: 1719–1730.

Liu J, Schulz H, Brandl S, Miehtke H, Huwe B, and Glaser B (2012). Short-term effect of biochar and compost on soil fertility and water status of a Dystric Cambisol in NE Germany under field conditions. J Soil Sci Plant Nutr 175(5): 698–707.

Major J, Lehmann J, Rondon M, and Goodale C (2010). Fate of soil-applied black carbon: downward migration, leaching and soil respiration. Global Change Biol 16: 1366–1379.

Mao JD, Johnson RL, Lehmann J, Olk DC, Neves EG, Thompson ML and Schmidt-Rohr K (2012). Abundant and stable char residues in soils: Implications for soil fertility and carbon sequestration. Environl Sci Technol 46(17): 9571–9576.

Nerome M, Toyota K, Islam TM, Nishijima T, Matsuoka T, Sato K, and Yamaguchi Y (2005). Suppression of bacterial wilt of tomato by incorporation of municipal biowaste charcoal into soil. Soil Microorganisms 59: 9-14.

Nguyen B, Lehmann J, Hockaday WC, Joseph S, and Masiello CA (2010). Temperature sensitivity of black carbon decomposition and oxidation. Environ Sci Technol 44: 3324–3331.

Noble R, and Coventry E (2005). Suppression of soil-borne plant diseases with composts: a review. Biocontrol Sci Techn 15: 3–20.

Novak JM, Busscher WJ, Watts DW, Amonette JE, Ippolito JA, Lima IM, Gaskin J, Das KC, Steiner C, Ahmedna M, Rehrah D, and Schomberg H (2012). Biochars impact on soil-moisture storage in an ultisol and two aridisols. Soil Science 177: 310–320.

Novak JM, Lima I, Xing B, Gaskin JW, Steiner C, Das KC, Ahmedna M, Rehrah D, Watts DW, Busscher WJ, and Schomberg H (2009). Characterization of designer biochar produced at different temperatures and their effects on a loamy sand. Annals of Environmental Science 3: 195–206.

Okimori Y, Ogawa M, and Takahashi F (2003). Potential of CO_2 emission reductions by carbonizing biomass waste from industrial tree plantation in South Sumatra, Indonesia. Mitig Adapt Strat Gl 8(3): 261–280.

Old SM (1969). Microclimate, fire, and plant production in an Illinois prairie. Ecol Monograph 355–384.

Paz-Ferreiro J, Liang C, Fu S, Mendez, and Gasco G (2015). The effect of biochar and Its interaction with the earthworm *Pontoscolex corethrurus* on soil microbial community structure in tropical soils. PLoS ONE 10(4): e0124891.

Rajkovich S, Enders A, Hanley K, Hyland C, Zimmerman AR, and Lehmann J (2012). Corn growth and nitrogen nutrition after additions of biochars with varying properties to a temperate soil. Biol Fertil Soils 48: 271–284.

Rizhiya EY, Buchkina NP, Mukhina IM, Belinets AS, and Balashov EV (2015). Effect of biochar on the properties of loamy sand Spodosol soil samples with different fertility levels: A laboratory experiment. Euras Soil Sci 48(2): 192–200.

Roberts DA, Paul NA, Dworjanyn SA, Bird MI, and de Nys (2015). Biochar from commercially cultivated seaweed for soil amelioration. Sci Rep 5:9665.

Schmidt MWI, and Noack AG (2000). Black carbon in soils and sediments: analysis, distribution, implications, and current challenges. Global Biogeochem Cy 14: 777–793.

Schulz H, Dunst G, and Glaser B (2013). Positive effects of composted biochar on plant growth and soil fertility. Agron Sustain Dev doi:10.1007/s13593-013-0150-0.

Skjemstad JO, Reicosky DC, Wilts AR, and McGowan JA (2002). Charcoal carbon in U.S. agricultural soils. Soil Sci Soc Am J 66: 1249–1255.

Steiner C, Das KC, Garcia M, Förster B, and Zech W (2008a). Charcoal and smoke extract stimulate the soil microbial community in a highly weathered xanthic ferralsol. Pedobiologia-International Journal of Soil Biology 51: 359–366.

Steiner C, Das KC, Melear N, and Lakly D (2010). Reducing nitrogen loss during poultry litter composting using biochar. J Environ Qual 39: 1–7.

Steiner C, Glaser B, Teixeira WG, Lehmann J, Blum WEH, and Zech W (2008b). Nitrogen retention and plant uptake on a highly weathered Central Amazonian Ferralsol amended with Compost and Charcoal. J Soil Sci Plant Nutr 171: 893–899.

Steiner C, Melear N, Harris K, and Das KC (2011). Biochar as bulking agent for poultry litter composting. Carbon Management 2(3): 227–230.

Tilak KVBR, and Subba Rao NS (1978). Carriers for legume (rhizobium) inoculants. Fertilizer News 23: 25–28.

U.S. Department of Agriculture, Economic Research Service (2011). Data sets: ARMS Farm Financial and Crop Production Practices.

U.S. Forest Service (2011). Western Bark Beetle Strategy: Human Safety, Recovery and Resiliency.

Wilson K (2014). How biochar works in soil. The Biochar Journal, Arbaz, Switzerland. www.biochar-journal.org/en/ct/32.

Zhang H, Voroney RP, and Price GW (2015). Effects of temperature and processing conditions on biochar chemical properties and their influence on soil C and N transformations. Soil Biol Biochem 83: 19–28.

Zhang J, Lü F, Shao L, and He P (2014). The use of biochar-amended composting to improve the humification and degradation of sewage sludge. Biores Technol 168: 252–258.

Zimmerman AR, Gao B, and Ahn M-Y (2011). Positive and negative carbon mineralization priming effects among a variety of biochar-amended soils. Soil Biol Biochem 43: 1169–1179.

8

Resource - Conserving Agriculture and Role of Microbes

A.A. Haleem Khan[1], Naseem[2] and B. Vidya Vardhini[1]

[1]Dept. of Botany, University College, Telangana University
Nizamabad-503322, Telangana State, India
[2]Dept. of Pharmaceutical Chemistry, University College
Telangana University, Nizamabad-503322, Telangana State, India
Corresponding authors E-mail: *aahaleemkhan@gmail.com, naseemprof@gmail.com*
drvidyavardhini@rediffmail.com

1. INTRODUCTION

Man of prehistoric times gathered food from wildlife resources with progress in time demanded for cultivation of crops. Forest life was prime source of exchange among the different groups and civilizations. Tools and techniques were improved in agriculture to increase yield and enhance earnings. Agriculture is also known as farming or husbandry that covers the cultivation of animals, plants, fungi, and other life forms for food, fibre, fuel, drugs and products used to sustain and enhance human life. Agriculture was the key development in the rise of sedentary human civilization, whereby farming brought the domesticated species that created food surpluses to nurture the civilization (Bhattacharyya and Jha, 2012). The history of agriculture dates back thousands of years that had been driven and defined by climates, cultures, and technologies. However, the forms of farming generally rely on techniques to expand and maintain the lands that are suitable for raising domesticated species (De-la-Pena and Loyola-Vargas, 2014). The farming of plants suffered lot of obstacles, such as soil erosion, scarcity of water, scant rainfall, lack of seeds supply, industrialization and labour problem, pests attack, etc. The developments in science witnessed use of machines in plant farming, synthetic chemicals as fertilizers, herbicides, weedicides and pesticides (Pereg and McMillan, 2015). The over application of synthetic chemicals in agriculture, increased input and reduced output (Ryan et al., 2009). These chemicals created resistant pest species and physico-chemical characteristics of soil (Matson et al., 1997).

Modern agronomy, plant breeding, agrochemicals, such as pesticides, fertilizers and technological improvements have sharply increased yields from cultivation, but also resulted in widespread ecological damage and negative human health effects (Sessitsch et al., 2013). The total cultivated area of land worldwide is expanding with significant negative impact on global environment (Kumar et al., 2013). Green revolution was introduced with motive to improve the crop yields by use of high yielding seed varieties, improved irrigation practices, use of mechanization and application of agrochemicals resulted in degrading the environment quality with fair increase in yields (Castro-Sowinski et al., 2007). Intense and expanded agriculture witnessed increase in erosion, decrease in soil fertility, reduction in biodiversity, pollution of groundwater and eutrophication of surface water bodies like lakes and rivers (Nadeem et al., 2013). In India, the intensive rice-wheat cultivation in the Punjab region declined in soil quality and enhanced problems of plant health. The cropping systems with high yields by monoculture influenced the abundance of the associated biota, such as the pests, soil invertebrates and microorganisms that, in turn, affect plant and soil processes (Pretty et al., 2011, Uphoff, 2003).

The herbivorous insects (pests), their predators and parasites, and microbial insects from both agricultural and natural ecosystems by attacking crops decreased plant productivity. The crop diversification inhibits pathogen growth such as fungi, viruses and parasitic insect vectors can be lower in polycultures, i.e., intercropping. Pest management was targeted with use of pesticides that resulted in pesticide resistance a ubiquitous problem, spread of pesticide residues through developed human health and environmental threat.

Integrated pest management (IPM) was introduced to overcome the pest and pesticide related problems that suggest the use of host plant resistance, biological and cultural control along with pesticides. The IPM practices have been promoted for long time run with notable success. The agrochemicals used in any of the form pose severe threat to ecosystem as there is contamination of food chain and change in the abiotic components. Pests are the invertebrates belonging to the phyla-arthropoda, adult form of insect with adaptation to flight and the larvae feeding on the green forms causing the decrease of crop quality and yield (Glick, 2012). The pests are one of leading problems in agriculture. There are multifarious ways to reduce, eradicate or kill these pests, but the practice of pest management has limitations. There are commercially available synthetic pesticides with hydrocarbons or organophosphates as essential ingredients. These components possess long half life in environment, toxic to non-targets and remain as recalcitrant (Berestetskiy, 2008). As there are several constraints to apply synthetic pesticides in cropping systems, certainly there is need to replace with compound that is eco-safe, inexpensive, more selective and efficacious against the pest at wide spectrum (Campbell et al., 2014; Styger et al., 2011).

In the integrated crop management behavioural manipulation of insect pests is a useful tool in ecologically sustainable and economical through host plant preferences (Pimentel et al., 2005). There are different types of plants grown together with primary crop for pest management purposes in the crop systems known as secondary crops. These crops are additional plants that increased efficiency and sustainability to biologically control the pests by natural enemies (Saxena and Pandey, 2001). These plants are non-crop plants, added to a given crop area and influence multi-trophic interactions by providing, additional shelter or food to beneficial organisms. They also have direct effects on crop plants like either positive effect by fixing nitrogen or negative influence by competition. These plants are divided into: companion, repellent, barrier, indicator, trap, insectary and banker (Parolin et al., 2012a,b,c; Huang et al., 2011).

Push-pull technology is a novel cropping system developed by the International Centre of Insect Physiology and Ecology (ICIPE), Rothamsted Research (UK), Kenyan Agricultural Research Institute (KARI) for integrated pest, weed and soil management in cereal-livestock based farming systems. In this approach attracting stem borers with Napier grass (*Pennisetum purpureum*), planted on the border of field as trap plant (pull) that drive away from the main crop using a repellent intercrop (push). In the crops best known trap plant is *Desmodium* sp. forage legume, nitrogen fixer as push. The trap cropping is based on the different attractiveness of host plants for pests. The trap cropping reduces dependence on insecticide applications, with one or more host plant species for particular pest, grown to decoy pest species away from the main cash crop. These crops are classified as either conventional or dead-end trap crops. The conventional trap crops attract than main crop for pest feeding or oviposition and the dead-end trap crops attract insects, but do not support their development. The dead-end trap crops are either through intrinsic characteristics of a plant species, such as nutritional inadequacy, through its chemical defence properties or created by introducing of an insecticidal transgene, such as a *Bacillus thuringiensis* gene (Khan et al., 2011).

2. SUSTAINABLE AGRICULTURE

The set of practices of integrated farming with preserving the natural resource base, minimum use of synthetic inputs like agrochemicals (fertilizers and pesticides) and maximum conservative practices (crop rotation, integrated pest management, natural fertilizer supply, biological control) (Sayyed et al., 2012; Singh et al., 2011). **Benbrook** defined '*Sustainable agriculture*' as 'the production of food and fibre by using a system that inherent productive capacity of natural and biological resources in step with demand. At the same time, it must allow farmers to earn adequate profits provide consumers with wholesome, safe food and minimize adverse impacts on the environment' (Godfray and Garnett, 2014; Stoop, 2011).

3. ORGANIC AGRICULTURE

The farming system with utilization of optimum resources, maintaining of soil fertility, biological activity and maximum output in the cropping pattern is commonly termed organic farming. The crop selection, rotation, compost/manures, water management, tillage, increased beneficial insect population and biopesticide application are measures of this farming approach (McGee, 2015). The crop nutrients like C, N, P, K, Ca, Mg are increased significantly by activities of soil microflora and their enzymes (Dimkpa et al., 2009). The compost from agro-residues consists of heterotrophic microorganisms efficient in lignocellulolytic activities (Hayatsu et al., 2008). The lignocellulolytic microbes like

Azotobacter chroococcum, Bacillus subtilis, B. megaterium, B. polymyxa, Burkholderia, Serratia marcescens, Pseudomonas striata, P. fluorescens, Aspergillus awamori, A. fumigatus, Paecilomyces fusisporus, P. lilacinus, Penicillium funiculosum, Phanerochete chrysosporium, Pleurotus sojarcaju, P. florida, Pochonia chlamydosporia, Pycnoporus, Trichurus spiralis, Trichoderma harzianum, T. viride, Verticillium hydrolyse the contents in compost (Sreenivasa, 2012; Gomiero and Paoletti, 2008; Hole et al., 2005).

Climate smart agriculture: Increase in the agricultural productivity that supports enhanced income and food security with decrease in greenhouse gas emissions. Banana-coffee intercropping reduced coffee leaf rust than monocropping system (Campbell et al., 2014).

Resource-conserving agriculture: The technologies that facilitate soil replenishment with integrated pest management, integrated nutrient management, conservation tillage, use of nitrogen-fixing legumes, improved water management, crop diversification and crop rotation without harm to the environment (Hobbs et al., 2008).

Laser land leveling, zero or minimum tillage, direct seeding, permanent/semi-permanent residue cover, use of plant varieties that utilize nutrients more efficiently, furrow irrigated raised bed (FIRB) technique, system of rice intensification (SRI), direct seeded rice (DSR), precision farming, use of leaf colour chart (LCC), optical sensors (GreenSeeker, chlorophyll meter) are resource conserving techniques reported to increase crop yield and nitrogen-use efficiency (NUE) (Dass et al., 2015; Bennett and Franzel, 2013; Uphoff, 2012; Wolfe et al., 2008).

Undersowing and mixed cultivation are two practical approaches where simultaneous cultivation of various plant species with or between plantings of a main crop that conserve resources like water, soil and energy. For example, white clover (*Trifolium repens* L.) and Lucerne/alfalfa (*Medicago sativa* L.) are proven for undersowing that reduces weeds in legumes. The mixed cropping of false flax (*Camelina sativa*) and field pea (*Pisum sativum*) inhibits the weeds and herbicide treatment is not required (Ramseier and Crismaru, 2014).

In wheat cultivars conservation practices, such as zero/minimum tillage adaptation, root health improvement, foliar disease resistance enhanced water-use efficiency (WUE). WUE is production of more grains or biomass per of water used for crop plants. Marker assisted selection and disease bioassays in combination improved root health in wheat cultivars (Trethowan et al., 2005). The water-use efficiency showed gain in crop yield in rainfed crops and sustainability was achieved without adverse effects on environmental goods and services (Pretty et al., 2006).

4. AGRICULTURE AND MICROBES

4.1 Plant Growth Promoting Rhizobacteria (PGPR) and Crop productivity

The discovery of microorganisms was first proved by Anton von Leeuwenhoek (1683), as 'animalcules' i.e., little animals later confirmed as microscopic prokaryotic living forms as bacteria. The utilization of bacteria for stimulating plant growth was there even in ancient times. Theophrastus (372–287 BC) suggested that by mixing of different soil samples could be remedy to recover soil defects. The legume crops increase the fertility of cultivated land and improve yield of successive crops was recorded by Virgil. In grasses and legumes roots were found to get colonized by rhizosphere with atmospheric nitrogen (N_2) converted into plant utilizable form was suggested by Hellriegel and Wilfarth (1888). The area surrounding the healthy plant roots where maximum microbial activity is found compared to distant surrounding soil is known as rhizosphere by Hiltner (1904) (Ahemad and Kibret, 2014; de Weert and Bloemberg, 2006).

The experiments of Kloepper and Schroth (1978) on radishes introduced term rhizosphere to the microbial community that competitively colonized plant roots and stimulate growth and suppress the incidence of plant pathogens. Kloepper and Schroth (1981) termed the beneficial bacteria surrounding the root 'plant growth-promoting rhizobacteria' (PGPR). PGPR are indispensable part of rhizosphere biota that grow in association with the host plants and benefit the host to improve its growth. PGPR get accomplished wide variety of ecosystems, show fast growth rate and metabolize the natural and xenobiotic compounds due to their innate genetic potential (Dobereiner 1988). The enrichment of microorganisms, such as bacteria, algae, fungi, and protozoa increase the nutrients in soil from 10- to 100-fold than bulk soil (Santi et al., 2013). To fulfill the concept of PGPR there are three criteria: aggressive colonization, plant growth stimulation and biocontrol.

As per Whipps (2001), PGPR are divided into three categories based on their interactions as: neutral, negative and positive. Among the PGPR most of them are commensals and establish an innocuous interaction with healthy plant without visible effect (Rao, 2014; Dixon and Kahn, 2004). The phytopathogenic rhizobacteria exert negative effect by producing phytotoxic substances, i.e., hydrogen cyanide (HCN) or ethylene. In response to the deleterious or negative effect exerting bacteria there are positive effects showing bacteria improve the growth by solubilization of nutrients, nitrogen fixation, production of plant growth regulators etc. (Kaur et al., 2013; Figueiredo et al., 2011). The bacteria of this group inhabit the healthy plants in and around the roots (Rhizosphere), root surface (Rhizoplane) or the root itself (Endophytes) (Raaijmakers et al., 2009).

Plants and microorganisms live in intimate associations. These organisms may exist as free-living, attached to root surface or establish symbiotic relation with plants (mutualism/commensalism or parasitism) that exert beneficial, neutral or detrimental effects on plant growth (Prasad et al., 2005, 2015; Rout and Southworth, 2013; Momota et al., 2012). The beneficial association can be illustrated in legumes and *Rhizobium* that show positive interactions through inhabiting of bacterium in the plant tissues (Ludwig-Muller, 2015; Mendes et al., 2013). The bacteria associated inside the plant tissues and organs are called endophytes (Taurian et al., 2012; Gaiero et al., 2013). The endophytic bacteria with nitrogen fixing ability in the Gramineae were Alpha-Proteobacteria (*Rhizobium* sp., *Ochrobacterum* sp., *Azospirillum* sp., and *Gluconacetobacter* sp.), Beta-Proteobacteria (*Herbaspirillum* sp., *Azoarcus* sp., and *Burkholderia* sp.) and Gamma-Proteobacteria (*Pseudomonas stutzeri* and *Klebsiella pneumoniae*) (Rothballer et al., 2009).

The endophytic bacteria from corn (*Zea mays* L.) roots were screened and *Enterobacter* sp. and *Bacillus* sp. showed production of IAA, siderophores and hydrolytic enzymes. The isolates increased root volume and could be biological inoculants for corn (Szilagyi-Zecchin et al., 2014). The leaves of *Sorghum bicolor* L. Moench was examined for N-fixing endophytic bacteria and showed *Herbaspirillum seropedicae,* and *H. rubrisubalbicans*. The endophytic *H. rubrisubalbicans* expressed symptoms of 'red stripe disease' and this proved colonization could be detrimental (James et al., 1997).

The effects of mycelia extracts and polysaccharide fraction from endophytic fungus *Trichoderma atroviride* isolated from plant root of *Salvia miltiorrhiza* showed the biosynthesis of phenolic acids and tanshinones. The hairy root cultures of the host plant showed the role of endophyte in plant survival (Ming et al., 2013). The endophytic fungus-grass (*Neotyphodium*) associations are mutualistic and could be parasitic, pathogenic and saprophytic interactions (Saikkonen et al., 2004). The medicinal plant orchid *Dendrobium candidum* wall. Ex. Lindl., was screened for endophytic fungus with ability for biotransformation of gentiopicroside (GPS). The fungus identified was *Penicillium crustosum* showed multienzyme system with metabolite potency (Zeng et al., 2014).

PGPRs were classified into two groups (Verma et al., 2010):

- Intracellular plant growth promoting rhizobacteria (iPGPR)
- Extracellular plant growth promoting rhizobacteria (ePGPR)

Table 8.1: PGPR: Types and their genera

<table>
<tr><td rowspan="2">PGPR</td><td>Intracellular plant growth promoting rhizobacteria (iPGPR)</td><td>a. Bacteria that reside in the root tissue (endophytes)
b. Bacteria that inhabit the specialized root structures or nodules.
Legumes – Rhizobia,
Woody plants – Frankia</td><td>Genera
Allorhizobium, Azorhizobium, Bradyrhizobium, Mesorhizobium, Sinorhizobium, Rhizobium from family Rhizobiaceae.
Frankia sp.</td></tr>
<tr><td>Extracellular plant growth promoting rhizobacteria (ePGPR)</td><td>These bacteria exist
a. Bacteria near the roots (rhizosphere)
b. Bacteria that colonize the root surface (Rhizoplane)</td><td>Acinetobacter, Achromobacter, Alcaligenes, Agrobacterium, Aereobacter, Arthrobacter, Azotobacter, Azomonas, Azospirillum, Bacillus, Beijerinckia, Burkholderia, Caulobacter, Citrobacter, Chromobacterium, Derxia, Erwinia, Enterobacter, Flavobacterium, Herbaspirillum, Hyphomicrobium, Klebsiella, Micrococcus, Paenibacillus, Pseudomonas, Serratia</td></tr>
</table>

PGPR are categorized into two groups based on the effect on plant growth as direct promoters and indirect promoters (Beneduzi et al., 2012)

Direct PGPR are subdivided into:

- Nitrogen fixers

- Phosphate solubilizers
- Potassium solubilizers
- Sulfur oxidizers
- Iron oxidizers
- Iron chelators
- Plant growth regulator producers
- Indirect PGPR are subdivided into:
- Suppress the plant pathogens
- Resistance against plant pathogens
- Abiotic stress

4.2 Nitrogen Fixers

Among the PGPR nitrogen-fixing bacteria are beneficial plant-interacting bacteria that supply nitrogen in the form of ammonia (NH_3) (Tkacz and Poole, 2015; Pii et al., 2015). Most of the nitrogen fixers are found to be associated with plant root (Carvalho et al., 2014). The rhizosphere soils of plant species *Sabina squamata, Weigela florida* and *Zanthoxylum simulans* was reported for nitrogen-fixing bacterial strains of genus *Paenibacillus*. The phylogenetic analysis based on 16S rRNA gene sequences revealed novel species *Paenibacillus sabinae* sp. nov. along with *Paenibacillus azotofixans* (Ma et al., 2007). The different plant rhizospheres were selected for isolation of nitrogen-fixing *Paenibacillus* by using 16S rRNA gene sequence, G+C content and DNA-DNA hybridization (Seldin 2011). The different plant rhizospheres were prevalent nitrogen-fixing *Paenibacillus* (Jin et al., 2011). The phylogenetic analyses of *nif* operon were reported in *Paenibacillus polymyxa* an endophytic diazotroph through single specific primer-polymerase chain reaction (SSP-PCR) and Southern blot. The arrangement of genes was consistent with other species of the genus *Paenibacillus* (Anand and Chanway, 2013). The nodules of sweet clover (*Melilotus alba*), alfalfa (*Medicago sativa*), *Phaseolus vulgaris, Medicago lupulina, Macroptilium atropurpureum* were reported for species of genus *Ensifer, Phyllobacterium, Rhizobium* with phage-resistant and susceptible bacteria (Bromfield et al., 2010). The amino acids and FAD cofactor under derepressing conditions of NifL/NifA regulatory system in nitrogen fixing *Klebsiella pneumoniae* in response to ammonium and molecular oxygen was reported (Milenkov et al., 2011). The random point mutagenesis by PCR amplification was investigated in nitrogen fixing *K. pneumoniae* and found that GlnK protein binds NifA that result in gene expression (Gloer et al., 2008; Jack et al., 1999). The interaction between silicate minerals (feldspar and biotite) and weathering bacteria was evaluated based on 16S rRNA gene sequence analysis. The cultures with minerals increased the production of organic acids (citric, tartaric and succinic acids) and extracellular polysaccharides by weathering bacterium. The bio-weathering process was found to be associated with synergistic action of acid and polysaccharide production by *Rhizobium tropici* (Wang et al., 2015).

The pond soil surrounded with area contaminated with hexachlorocyclohexane (HCH) was investigated with 16S rRNA gene sequence and chemotaxonomic data. The phylogenetic analysis revealed novel species belonging to family *Hyphomicrobiaceae, Rhizobiales* order and named *Devosia lucknowensis* sp. nov. (Dua et al., 2013). The aquatic legume *Neptunia natans* (L.f.) Druce nodules were studied for isolation of symbiotic nitrogen fixers and found new species *Devosia* (Rivas et al., 2002).

The nodules of alfalfa (*Medicago sativa* L.) from acid soils were used to screen the nitrogen fixers by RAPD fingerprinting. The strains were identified as *Ensifer meliloti* and *Ensifer medicae* based on intergenic spacer (ITS) sequences. Microcosm experiments showed that alfalfa inoculation with acid-tolerant strains increased yields in acid soils (Ramirez-Bahena et al., 2015). The denitrification in *Ensifer meliloti* was studied and reported that *napA, nirK, norC* and *nosZ* genes were involved in nitrate respiration and expression of denitrification enzymes (Torres et al., 2014). The woody legumes root nodules were reported to possess 71 strains of the genus *Ensifer* by using multilocus sequence sequence analysis (MLSA) and phenotypic approaches. The cross-inoculation study showed potential to improve nitrogen fixation by selection of broad-host-range inoculants (Degefu et al., 2012). The fungicidal activity of *Ensifer adhaerens* isolated from soil sample with ability of enzyme alliinase (*Allium* genus-garlic, onion and others) production. The (garlic) enzyme generated a fungicidal activity against *Saccharomyces cerevisiae* (Yutani et al., 2011).

The root inoculation of PGPR (*Pseudomonas putida, P. fluorescens, Bacillus subtilis*) in aromatic plant species *Mentha piperita* was found to affect the biosynthesis of secondary metabolites and accumulation of monoterpenes (Cappellari et al., 2015). The perennial legume species alfalfa and white clover ability to N_2 fix and transfer was compared in field and greenhouse experiments through ^{15}N dilution. The differences were found in fixation, legume proportion, population dynamics or tissue composition (Louarn et al., 2015). The nodules from bean (*Phaseolus vulgaris* L.) were assessed for diversity and distribution of native population of rhizobia with RFLP markers and 16S rDNA by sequencing. *Rhizobium leguminosarum* was common in all soils and acid soils showed *R. etli, R. tropici,* and *R. leucaenae* (Baginsky et al., 2015). The legume bacteria (*Rhizobium* sp.) as PGPR were inoculated in non-nitrogen fixing plant (*Linum usitatissimum*) seeds to investigate the plant protection and plant-stimulating agents.

The PGPR stimulated plant growth by producing the secondary metabolites facilitating the uptake of nutrients and inhibiting plant pathogenic organisms in the rhizosphere (Hossain and Martensson, 2008). The greenhouse experiments were reported to test the effects of biochar (a co-product of the thermal decomposition of organic material in a low-oxygen environment) on common beans (*Phaseolus vulgaris* L.) and biochar proved to contribute enhanced nitrogen fixation (Guerena et al., 2015). The non-cellulose producing bacterial strain *Komagataeibacter medellinensis* (*Gluconacetobacter xylinus*) was isolated through repetitive static culture. The frameshift mutation produced cellulose as exopolysaccharide (Matsutani et al., 2015).

The co-inoculation of maize with *Herbaspirillum seropedicae* was evaluated for plant-bacterial interactions in the rhizosphere through root exudates. The root exudates showed nitrogenous compounds, fatty acids, organic acids, steroids and terpenoid derivatives that improved crop yield (Lima et al., 2014; Chubatsu et al., 2012). The citrate operon in *Herbaspirillum seropedicae* was reported to enhance growth in cereals by organic acid secretion, mineral phosphate solubilization along with nitrogen fixation (Wagh et al., 2014). The mutagenesis using gamma rays of *Paracoccus denitrificans* was reported for improved and cost effective production of Coenzyme Q_{10} (CoQ_{10}) (Tokdar et al., 2015).

The N-fixing filamentous cyanobacterium *Tolypothrix* sp. from seasonal creek was evaluated for production of biomass and bioproducts in outdoor suspension and biofilm prototype cultivation systems. This cyanobacterium proved as an attractive alternative for bioproducts (Velu et al., 2015). The biofilms with *Anabaena torulosa* after co-culturing with *Azotobacter chroococcum, Pseudomonas striata, Serratia marcescens, Mesorhizobium ciceri* were evaluated. These biofilms were effective inoculants for sustainable and eco-friendly environment (Prasanna et al., 2011). The nitrogen-fixation by acetylene reduction in the marine non-heterocystous cyanobacteria *Trichodesmium thiebautii,* and *T. erythraeum* was showed under presence of light and aerobic conditions (Saino and Hattori, 1982).

In boreal habitats nitrogen-fixation by moss (*Sphagnum* sp.) associated bacteria were investigated by $^{15}N_2$ experiment and phylogenetic analysis of *nifH* genes. The gene sequences were found related to the order Rhizobiales, and availability of diazotrophic community structure (Leppanen et al., 2015). The rhizospheric soil of different crops (mustard, barseem, wheat, sugarcane, brinjal, onion, cauliflower, cabbage, and chickpea) was used for isolation of PGPR – a total of 72 bacterial isolates belonging to *Azotobacter, Pseduomonas, Mesorhizobium* and *Bacillus* from soil and plant root nodules (Ahmad et al., 2008). The leaf, root microorganisms were measured by acetylene reduction assay for N_2 fixation. The bacterial species isolated were *Azotobacter, Beijerinckia, Derxia, Clostridium* sp. and *Klebsiella pneumoniae* from roots of plants, i.e., *Araucaria cunninghamii, Pinus ellotti, Avicennia germinans, Laguncularia racemosa, Rhizophora mangle. Azospirillum lipoferum* from roots of Poaceae (Gajendiran and Mahadevan, 1989). The PGPR *Pseudomonas* genus affects the mineral nutrition and plant uptake nutrients from the soil allows the increase the crop yield (Shabayev, 2012).

The six-month-old sugarcane (*Saccharum officinarum* L.) field was evaluated for endophytic diazotrophs and found isolates of *Acetobacter diazotrophicus, Herbaspirillum seropedicae,* and *H. rubrisubalbicans* among the nitrogen-fixers (Asis et al., 2000). The chickpea (*Cicer arietinum* L.) was studied for interactions among *Piriformospora indica* with PGPR, viz., *Pseudomonas* sps. and *Mesorhizobium ciceri*. This study proved tripartite combination as biofertilizer for improved productivity (Mansotra et al., 2015). Two field experiments were conducted in legumes and non-legumes to monitor the biomass production and nitrogen content. The legumes with short growing period showed high N accumulation (Buchi et al., 2015). The random mutagenesis of MsiR in *Mesorhizobium tianshanense* was investigated and showed the functional domain for canavanine upregulation (Zhong et al., 2015).

Table 8.2: Activities of PGPR in rhizospheric soil

Rhizosphere soil	Availability of nutrients—amino acids, fatty acids, nucleotides, organic acids, phenols, plant growth regulators, plant growth promoters, putrescine, sterols, sugars, and vitamins released from the roots by exudation, secretion and deposition

4.3 Phosphate Solubilizers

Phosphorus (P) is a major macronutrient, plant growth-limiting nutrient in soils besides nitrogen. It is required for synthesis of nucleic acids, membranes, proteins and energy molecules, such as ATP, GTP and NADPH. P contributes in several metabolic processes including breakdown of sugar, photosynthesis, energy transfer, cell division, nutrient transport within the plant, expression and maintenance of genetic material, signal transduction, macromolecular synthesis and reactions in respiratory chain. The total P content of arable soil is 0.02–0.5 percent with approximate of 0.05 percent in inorganic and organic forms. The soil table possesses most of the nutrients in millimolar concentration, but P is found about micromolar quantities. The low level of P in soil is due to its high reactivity with Ca, Fe and Al. In the alkaline soils hydroxyapatite and fluoroapatite are calcium phosphates, the acidic soils possess variscite ($AIPO_4.H_2O$), strengite ($FePO_4.2H_2O$) as amorphous iron and aluminum phosphates. Though the soil has high levels of P content, there is deficiency of P due to low levels of soluble PO_4 availability to plants from soil. To meet the P requirement of plants, fertilizers with P content or leaf litter, plant residues or animal remains application in common practice.

In the soil, there are groups of microorganisms efficient in solubilizing the phosphorus from natural ecosystems. These organisms play fundamental role in biogeochemical cycling of P in soil. Phosphate solubilizing microbes (PSM) can transform the insoluble phosphorus into soluble forms HPO^{2-}_4 and $H2PO^-_4$ by chelation, exchange reactions, production of polymeric substances, organic acids (oxalic, citric, butyric, malonic, lactic, gluconic, acetic, glyconic, fumaric, adipic and 2-ketogluconic acid), inorganic acids, CO_2 and enzymes-acid phosphatase, phytase, phosphohydrolases, phosphonoacetate hydrolase, D-α-glycerophosphatase and C-P lyase.

Table 8.3: Phosphate solublizing microorganisms (PSM)

PSM	Bacteria	*Acinetobacter, Achromobacter, Aereobacter, Agrobacterium, Alcaligenes, Arthrobacter, Azotobacter, Bradyrhizobium, Burkholderia, Chromobacterium, Citrobacter, Enterobacter, Erwinia, Escherichia, Flavobacterium, Klebsiella, Micrococcus, Pantoea, Pseudomonas, Salmonella, Serratia, Streptomyces, Thiobacillus*
	Fungi	*Aspergillus, Fusarium, Glomus, Penicillium, Rhizopus, Sclerotium, Trichoderma*

Utilization of PSM is an attractive approach to increase the availability of P content of soil that increases the crop yield and sustains the agriculture system. The heavy metal-polluted paddy soil samples were investigated for PGPR with 16S rRNA gene sequence indicated that the isolate was member of genus *Burkholderia* (*B. tropica, B. bannensis, B. unamae, B. metalliresistens*). On the basis of genotypic, phenotypic and phylogenetic data novel species was reported with phosphate solubilizing ability *Burkholderia metalliresistens* (Guo et al., 2015).

The rhizospheric soil was screened for the mineral phosphate solubilization and found oxalic acid secreting *Klebsiella* sp. was isolated that contributed mineral PO_4 through catabolite repression (Rajput et al., 2013).

4.4 Potassium Solubilizers

Potassium (K) is essential element for plant growth and development. It is important for maintaining cellular osmotic pressure and transport of solutes in plants. The activation of enzymes, the utilization of nitrogen (N) and the syntheses of sugar and protein is promoted by potassium. It enhances photosynthesis in plants; its deficiency causes slow plant growth and root development, yellowing of leaf margins, with burnt appearance. The soil is K rich reservoir; about 90–98 percent of K exists as silicate minerals (K-feldspar, illite and mica), only plant K uptake source. K is released from its mineral form at slow rate; there are varieties of soil microbes release soluble form from K-bearing minerals.

Potassium solubilizing bacteria (KSB) are Gram positive, heterotrophic, aerobic bacteria obtain their energy and carbon source from pre-existing organic matter. These bacteria form humus in soil, cycling of minerals in organic matter, maintain soil structure by formation and stabilization of water-stable soil aggregates and stimulate plant growth and inhibit the root pathogens. Among the KSB, *Bacillus mucilaginosus* and *Bacillus edaphicus* were reported the dissolution rate of silicate and alumino-2-silicate minerals with release K^+ and SiO_2 by organic acids (tartaric acid and citric acid) and extracellular polysaccharide (EPS) formation. The solubilizing activity of KSB will alleviate K^+ deficiency soils.

Tea (*Camellia sinensis*) was studied for efficacy of indigenous KSB in combination with variety doses of potash fertilizers along with recommended quantity of N, P fertilizers. The analysis of soil and leaf samples from these plants improved N, P, K concentration and high chlorophyll, carotenoid in crop shoots. The indigenous KSB improved productivity and nutrient uptake in tea planatations (Bagyalakshmi et al., 2012).

4.5 Sulfur Oxidizers

Soil rich with reduced sulfur compounds are exclusively oxidized by prokaryotes and the sulfate as byproduct obtained from the oxidation reaction. In the sulfur cycle around the globe, hydrogen sulfide (H_2S), sulfur, sulfite, thiosulfate and various polythionates undergoes biological oxidation to form sulfate (SO_4) in acidic, alkaline and neutral conditions. Sulfur is absorbed from soils as divalent sulfate anion (SO_4) by roots and translocated to shoots through xylem. Deficiency of sulfur is rare, symptoms of deficiency is expressed as chlorosis, common in young leaves. The sulfur content is predominant in plant proteins with amino acids-cysteine and methionine, vitamins-thiamine and biotin, co-enzyme-A.

Table 8.4: Sulfur oxidizing bacteria

Sulfur oxidizers	Aerobic-mesophilic	*Acidianus*, *Acidithiobacillus*, *Aquaspirillum*, *Aquifex*, *Bacillus*, *Beggiatoa*, *Methylobacterium*, *Paracoccus*, *Pseudomonas*, *Starkeya*, *Sulfolobus*, *Thermithiobacillus*, *Thiobacillus* and *Xanthobacter*
	Anaerobic-phototrophic (neutrophilic & mesophilic)	*Allochromatium*, *Chlorobium*, *Rhodobacter*, *Rhodopseudomonas*, *Rhodovulum*, *Rhodocyclus*, and *Thiocapsa*

4.6 Iron Oxidizers

Iron (Fe) is essential in certain enzymes and proteins of photosynthesis and respiration. It alternate undergoes oxidation and reduction between the Fe^{2+} to Fe^{3+} states to serve as electron carriers in proteins. Fe is accumulated in older leaves as insoluble oxide or inorganic or organic ferric-phosphate compounds. In leaves one abundant and stable iron-protein complex is called phytoferritin; this protein is store house of iron in chloroplast. Fe deficiency is often in plants of rose family, shrubs and trees, maize and sorghum, high pH, bicarbonates in soil were found to be associated with this deficiency.

In the 1830s, iron oxidizing bacteria (FeOB) were recognized as chemolithoautotrophs; grow on Fe as a sole source of energy. FeOB and iron reducing bacteria are responsible to precipitate iron deposits, such as oxides (magnetite), carbonates (siderite) and sulfides (greigite) through controlled or induced mineralization process.

4.7 Iron Chelators: Siderophores

The term siderophore derives from Greek word for iron carriers. Siderophores are designed to form light and stable ferric iron complexes. These are ferric ion specific chelating agents synthesized by prokaryotic and eukaryotic microbes, such as bacteria, actinomycetes, fungi and certain algae growing in low iron stress (Ahmed and Holmstrom, 2014). Siderophores are iron binding proteins with molecular weight ranging from 400–1500 Da. These compounds play important role in scavenging iron from the environment. Secreted siderophores form extracellular Fe (III) complexes with stabilities. The iron charged siderophore is taken up by ferric-chelate-specific transporters or siderophore-bound Fe (III) undergoes reduction to Fe (II) that is catalyzed by ferric-chelate reductases. Most of the aerobic and facultative anaerobic microorganism synthesizes at least one siderophore. Siderophores have also been found to be associated with virulence mechanisms in pathogenic microorganism of animals and plants (D'Onofrio et al., 2010). These compounds have applications in clinical, agricultural and environmental fields. Presently, 500 siderophores are reported from different microorganisms.

The important groups of siderophores are:

- Hydroxamate (α-hydroxy-) siderophores,
- Catecholate (catecholates and phenolates) siderophores
- Carboxylate siderophores

Hydroxamate siderophores possess three secondary hydroxamate groups and each hydroxamate group provides two oxygen molecules that form a bidentate ligand with iron. The siderophores were reported to be produced by both bacteria and fungi.

Catecholate siderophores are produced by certain bacteria. The catecholate group provides two oxygen atoms for chelation with iron so that a hexadentate octahedral complex is formed as in the hydroxamate siderophores. The finest example for catecholate siderophore is enterobactin produced by *Escherichia coli*.

Carboxylate siderophores are produced by *Rhizobium* and *Staphylococcus* strains and fungi belonging to Mucorales, coordinating iron with carboxyl and hydroxyl groups.

4.8 Microorganisms Producing Plant Growth Regulators

Phytohormones play an important role in regulation of growth and development of plants. The conventional classification made five groups of phytohormones: auxins, gibberellins, cytokinins, ethylene and abscisic acid. The microorganisms PGPR (rhizosphere, epiphytes, and symbionts) associated with plants were reported to synthesize phytohormones, stimulate and facilitate plant growth (Duca et al., 2014; Dodd and Perez-Alfocea, 2012; Ortiz-Castro et al., 2009).

The bacterial strains *Acinetobacter baumannii* and *Escherichia coli* were investigated for lacR, MarR family protein of *iac* locus and found that IAA serves as specific ligand of lacR (Shu et al., 2015).

Table 8.5: Iron oxidizers

Iron oxidizing bacteria	*Acidothiobacillus ferroxidans* (*Thiobacillus ferrooxidans*), *Acidothiobacillus thiooxidans* (*Thiobacillus thiooxidans*), *Acidothiobacillus caldus,* *Acidianus ambivalens,* *Acidianus infernus,* *Acidianus brierleyi,* *Leptospirillum ferrooxidans,* *Leptospirillum ferriphilum,* *Ferroplasma acidiphilum,* *Geobacter metallireducens,* *Geobacter sulfurreducens,* *Shewanella putrefaciens,* *Shewanella denitrificans,* *Gallionella ferruginea,* *Magnetospirillum magnetotacticum,* *Leptothrix ochracea,* *Sulfobacillus thermosulfidooxidans,* *Sulfobacillus montserratensis* *Sulfobacillus metallicus,* *Sulfolobus* sp. *Metallosphaera* sp. *Rhodobacter* sp. *Dechlorosoma* sp.

Table 8.6: Siderophore producing microorganisms

Siderophore producing microorganisms	Bacteria	*Escherichia coli,* *Salmonella,* *Klebsiella pneumoniae,* *Vibrio cholerae,* *Vibrio anguillarum,* *Aeromonas* *Aerobacter aerogens,* *Enterobacter,* *Yersinia and* *Mycobacterium* sp.
	Actinomycetes	*Actinomadura madurae,* *Nocardia asteroides, and* *Streptomyces griseus*
	Cyanobacteria	*Anabaena flos-aquae and* *Anabaena cylindrica*

	Fungi	*Aspergillus nidulans,* *A. versicolor,* *Penicillium chrysogenum,* *P. citrinum,* *Mucor,* *Rhizopus,* *Trametes versicolor.* *Ustilago sphaerogina,* *Saccharomyces cerevisiae,* *Rhodotorula minuta and* *Debaryomyces* sp.

Table 8.7: Microorganisms producing plant growth regulators (Tsavkelova et al., 2006)

Auxins	Prokaryotes	Phototrophs	*Anabaena, Anabaenopsis, Calothrix, Chlorogloeopsis, Cylindrospermum, Gloeothece, Nostoc, Plectonema, Synechocystis*
		Heterotrophs	*Acetobacter, Achromobacter, Acinetobacter, Agrobacterium, Alcaligenes, Aminobacter, Arthrobacter, Azospirillum, Azotobacter, Bacillus, Bradyrhizobium, Corynebacterium, Enterobacter, Erwinia, Flavobacterium, Herbaspirillum, Klebsiella, Methylobacterium, Methylovorus, Microbacterium, Micrococcus, Mycobacterium, Paracoccus, Pseudomonas, Rhizobium, Rhodococcus, Sphingomonas, Streptomyces, Sulfolobus, Xanthomonas*
	Eukaryotes	Algae	*Chlorella, Dunaliella, Fucus*
		Fungi	*Absidia, Actinomucor, Alternaria, Amanita, Aspergillus, Colletotrichum, Fusarium, Hebeloma, Laccaria, Lentinus, Monilia, Paxillus, Penicillium, Phoma, Phymatotrichum, Phytophtora, Pisolithus, Plasmodiophora, Pythium, Rhizoctonia, Rhizopogon, Rhizopus, Saccharomyces, Sclerotium, Taphrina, Ustilago*
Gibberellins	Prokaryotes	Phototrophs	*Anabaenopsis, Cylindrospermum*
		Heterotrophs	*Acinetobacter, Agrobacterium, Arthrobacter, Azospirillum, Azotobacter, Bacillus, Clostridium, Flavobacterium, Micrococcus, Pseudomonas, Rhizobium, Xanthomonas*
	Eukaryotes	Algae	*Chara, Fucus, Porphyra, Tetraselmis*
		Fungi	*Aspergillus, Botryodiplodia, Cercospora, Fusarium, Penicillium, Phaeosphaeria, Rhizopus, Shizophillum, Sphaceloma, Ustilago, Verticillum, as well as certain yeasts*

Contd...

Contd...

Cytokinins	Prokaryotes	Phototrophs	*Anabaena, Calothrix, Chlorogloeopsis, Rhodospirillum*
		Heterotrophs	*Agrobacterium, Aminobacter, Arthrobacter, Azospirillum, Azotobacter, Bacillus, Blastobacter, Escherichia, Erwinia, Hyphomicrobium, Methylarcula, Methylobacillus, Methylobacter, Methylobacterium, Methylomonas, Methylophylus, Methylosinus, Methylovorus, Paracoccus, Pseudomonas, Rhizobium, Rhodococcus, Streptomyces, Xanthobacter*
	Eukaryotes	Algae	Certain red and brown algae
		Fungi	*Dictyostellium, Fusarium, Paxillus, Phoma, Rhizopogon, Saccharomyces, Schizophyllum, Schizosaccharomyces, Suillus, Taphrina, Trichoderma, Uromyces*

4.9 Microorganisms Under Abiotic and Biotic Stresses

There are reports on bacteria that provide tolerance to host plants under different abiotic stress environments, such as temperature (cold and heat), water (drought and flooding), humidity, light intensity, nutrients (lack of ions, over-fertilization and salinity), carbon dioxide (CO_2), oxygen (photosynthesis, respiration/photorespiration, oxidative stress and anaerobiosis) (Yang et al., 2009).

The PGPR are known to enhance plant growth directly or indirectly by various mechanisms, i.e., antibiosis, competition for nutrients, colonization of specific sites by pathogen and inducing resistance to pathogen by bacterial biocontrol agents (BCA) (Narayanasamy, 2013; Choudhary and Johri, 2009). Several PGPR that colonize root systems protect plant against foliar disease include *P. fluorescens, P. putida, B. pumilus,* and *S. marcescens*. There is a network of interconnected signaling pathways to regulate induced defenses of plants against pathogens (Majumder et al., 2014; Kumar et al., 2012). The primary components of the network are plant signal molecules: salicylate (SA), jasmonate (JA), ethylene (ET) and nitric oxide (NO). The signal transduction leading to ISR requires responsiveness to both JA and ethylene. Methyl jasmonate (MeJA) and the ethylene precursor, i.e., 1-aminocyclopropane-1-carboxylate (ACC) are effective in inducing resistance against phytopathogenic microflora (Hartmann et al., 2009; Vessey 2003).

Mediation of ethylene levels by microbial ACC plays a key role in signal transduction, viz., the ISR-inducing *P. fluorescens* bacterium enhances bacterial ACC-converting capacity and leads to potentiated level of ethylene emission in infected plant (Berg, 2009; Saleem et al., 2007). The PGPR contain a vital enzyme, 1-aminocyclopropane-1-carboxylate (ACC) deaminase that regulates ethylene production by metabolizing ACC (an precursor of ethylene biosynthesis in higher plants) into ketobutyrate and ammonia. Inoculation with PGPR containingACC deaminase activity could sustain plant growth and development under stress conditions by reducing stress-induced ethylene production (Choudhary et al., 2007).

Table 8.8: Bacteria that protect from abiotic stress

Abiotic stresses	*Rhizobium, Bacillus, Pseudomonas, Pantoea, Paenibacillus, Burkholderia, Achromobacter, Azospirillum, Microbacterium, Methylobacterium, Variovorax*, and *Enterobacter*

The PGPR with heavy metal resistance to cadmium (Cd) uptake and soil microbial community structure during phytoremediation was investigated by isolation of PGPR from root samples from *Sedum arboreum, Sedum plumbizincicola, Sedum X graptosedum, Crassula multicava* and *Carpobrotus rossii. Actinospica, Bradyrhizobium, Rhizobium, Mesorhizobium,* and *Mycobacterium* were found to enhance Cd phytoremediation (Liu et al., 2015). The metagenomes were reported to contain DNA from both culturable and non-culturable microorganisms and could be a source of novel genes. The metagenomic DNA library from soil sample contaminated with aromatic compounds was studied by using cosmid vector. The genes for enzymes hydroxylase and oxygenases for indigo-forming activity was demonstrated (Nagayama et al., 2015). The acid-tolerance in seven levels of low pH media were used with three plant species, i.e., *Acacia auriculiformis, A. confusa,* and *Melaleuca armillaris* and 50 plant species was investigated. Two tolerant species were found in plants like *Jatropha curcas* and *Acacia auriculiformis* useful for acid remediation and biofuel feedstock (Ma et al., 2015).

The growth response of tomato (*Lycopersicon pimpinellifolium* Mill.) was studied for identification and selection of PGPR with ACC-deaminase and/or nitrogen fixers. The inoculation of PGPR improved physical, chemical (nutrients: N, K, Ca, Mg, Zn, Cu, Fe, Mn and Chlorophyll-a, b and carotenoids), and enzymatic (glutathione S-transferase, peroxidase, ACC-deaminase, and catalase) growth parameters in plants (Hassan et al., 2014). The salinity and drought stress in cucumber (*Cucumis sativus*) was assessed with role of PGPR strains, viz., *Burkholderia cepacia, Promicromonospora* sp. and *Acinetobacter calcoaceticus* to increase crop productivity in marginalized agricultural lands. The PGPR were found to reduce the adverse effects of stress by regulating the phytohormones and antioxidants (Kang et al., 2014). The nitrate-fed and nitrogen-fixing soybeans (*Glycine max* L.) were evaluated for photosynthetic responses under water stress. The nodulated plants were less sensitive to drought than nitrate-fed plants (Kirova et al., 2008). The symbiosis of *Medicago lupulina* under copper stress with *Sinorhizobium meliloti* was evaluated for growth, metal uptake and antioxidant responses. The symbiosis showed improved responses to excess copper stress (Kong et al., 2015). The soil samples treated with two β-triketone herbicides were used for isolation of bacteria and isolates were identified by phylogenetic study based on 16S rRNA gene sequence showed *Bradyrhizobium* sp. (Romdhane et al.,

2016). The PGPR as bio-inoculants were assessed to ameliorate *Ilex paraguariensis* St. Hill growth in nursery with different substrates. The native PGPR increased crop in nursery represented sustainable strategy to improve plant growth in low-fertility soils (Bergottini et al., 2015).

The active association between *in vitro* growing potato (*Solanum tuberosum* L.) microclones and the bacterium *Azospirillum brasilense* was studied and found the inoculation improved minituber yield (Tkachenko et al., 2015).

The post-harvest pathogen *Colletotrichum acutatum* causative agent of glomerella leaf spot (GLS) disease in apple was studied for biological control with bacterial strains (*Bacillus* sp., *B. alcalophilus*) and commercial product Serenade® (*B. subtilis*). There were significant differences between the epidemiological parameters of GLS and proved as new alternative fungicide sprays to control disease in apple (Moreira et al., 2014). The biological control activities of *Streptomyces* sp. were studied against fusarium wilt (*Fusarium oxysporum* f. sp. *ciceri*) of chickpea in sorghum and rice and were found to improve the plant growth due to antagonistic activities (Gopalkrishnan et al., 2013).

The biocontrol formulations of cyanobacteria (*Calothrix elenkinii, Anabaena laxa, A. torulosa*) and *Trichoderma viride* efficacy against root rot disease in two cotton varieties (*Gossypium hirsutum, G. arboreum*) were evaluated. The microbial inoculation elicited high levels of hydrolytic enzymes and available N in soil (Babu et al., 2015). The common beans pathogen white mold *Sclerotinia sclerotiorum* was studied for biocontrol affectivity with 10 *Trichoderma* sp. isolates in field trials. The cell wall-degrading enzymes (lipase, NAGase, β-glucosidase and protease) and sclerotia parasitism were key components to reduce the disease severity (Geraldine et al., 2013). The leaf, root and vascular diseases of dry beans-bacterial wilt, fusarium wilt, charcoal rot and angular leaf spot were studied for biocontrol with isolates of *Bacillus cereus* and *Pseudomonas fluorescens*. The seeds of microbiolized with the test bacteria in single and in combinations, test *Pseudomonas* singly was significantly to control four diseases in beans (Correa et al., 2014). The biocontrol of white mold (*Sclerotinia sclerotiorum*) on processing tomatoes was evaluated with *Trichoderma* sp. treatments by drip irrigation for control of disease incidence. The white mold biological control was reported to be advantageous by drip irrigation with *Trichoderma* sp. delivery close to the plants and to the inoculums source (de Anguiar et al., 2014; Markovich and Kononova, 2003).

5. MICROORGANISMS AS INOCULANTS

Microbial inoculants or soil inoculants are use as beneficial microbes (bacteria and fungi) to promote plant growth (Bhardwaj et al., 2014; Brahmaprakash and Sahu, 2012). These organisms show symbiotic relationship with crop plants where in both partners get benefit (Calvo et al., 2014; Franche et al., 2009). These inoculants improve plant nutrition and promote growth by stimulation of growth regulators (Kuhad, 2012; Malusa et al., 2012; Prasad et al., 2015). Nitrogen-fixers and phosphate-solubilizers are common inoculants that increase availability of macronutrients – nitrogen and phosphorus to plant (Adesemoye and Kloepper, 2009). These bacteria are known as plant growth promoting rhizobacteria (PGPR).

Table 8.9: Bacteria that confer resistance in stress

Induced systemic resistance (ISR)	*Pseudomonas fluorescens, P. putida, Bacillus pumilus, B. amyloliquefaciens, B. subtilis, B. pasteurii, B. cereus, B. pumilus, B. mycoides, and B. sphaericus Paenibacillus alvei , Acinetobacter lwoffii, Chryseobacterium balustinum, Azospirillum brasilense, Curtobacterium sp., Arthrobacter oxidans, Stenotrophomonas and Serratia marcescens*
Bacteria releasing volatiles against phytopathogenic fungi	*Alcaligenes, Bacillus, Ensifer, Lysobacter, Planomicrobium, Sporosarcina, and Stenotrophomonas, Burkholderia, Pseudomonas, Serratia, Xanthomonas, Pectobacterium, and Agrobacterium* sp.
Microbes producing 1-aminocyclopropane-1-carboxylate (ACC) deaminase	*Agrobacterium genomovars and Azospirillum lipoferum, Alcaligenes and Bacillus, Burkholderia, Enterobacter, Methylobacterium fujisawaense, Pseudomonas, Ralstonia solanacearum, Rhizobium, Rhodococcus, Sinorhizobium meliloti and Variovorax paradoxus*

Biofertilizers contain microorganisms in combination with organic fertilizers (manure) that provide high nutrient availability to host plants due to the interactions with microorganisms (Mohammadi and Sohrabi, 2012; Zhao et al., 2005). These fertilizers are of the following types:

1. **Symbiotic nitrogen fixers** (*Rhizobium* spp.) are used for legume crops like groundnut, soybean, increases crop yield from 10 to 35 percent by adding 50–200 kg N/ha. *Rhizobium* sp. is Gram-negative bacterium, enter legume plants through nodules in the roots and fix the atmospheric nitrogen (Gopalakrishnan et al., 2015). This bacterium improves the crop yield of inoculated crop and also benefits the subsequent crop. *Rhizobium leguminosarum, R. melilotii, R. trifoli, R. phaseoli, R. lupinii,* and *R. japonicum* are common rhizobia used as inoculants for legume plants (Willems, 2007; Sawada et al., 2003).
2. **Non-symbiotic nitrogen fixers** (*Azotobacter, Azospirillum*) used for cereal crops like maize, oats, sorghum, sugarcane etc. increase yield from 10 to 15 percent by adding 20–25 kg N/ha. Application of *Azotobacter* to crop plants supply nitrogen by free atmospheric nitrogen fixation, synthesize growth promoters and antibiotics (Babalola, 2010). *Azospirillum* colonizes the roots and aerial parts of host plants by associative symbiosis and it is effective with farmyard manure in sorghum and millets (Hui-Qin, 1997).
3. **Algal biofertilizers** (Cyanobacteria, *Azolla*) used for paddy crop, add 10 to 35 kg N/ha, Azolla supplies 30–100 kg N/ha. *Anabaena, Cylindrospermum, Nostoc, Plectonema, Aulosira, Tolypothrix, Oscillatoria,* etc., are used in combination as dried flakes at the rate of 10 kg/ha after tenth day of paddy seedlings transplantation (Johansen et al., 2014). *Azolla* is an aquatic fern that harbors *Anabaena* in leaf as symbiont that is common as green manure in paddy crop (Dalton and Kramer, 2006).
4. **Phosphate solubilizers** can be used for a variety of crops to improve 5 to 30 per cent yield. *Pseudomonas, Bacillus* spp. are phosphate solubilizers that convert insoluble form of inorganic phosphates into soluble form that can be easily consumed by applied crop plants

Mycorrhizae used for crops and ornamental plants, increases intake of phosphorus, zinc, sulphur and water in the host plants and thereby enhances 10 to 50 percent yield. The symbiotic association of fungi in plant roots is called mycorrhiza. The colonization showed greater growth response in taproots than plant species with fibrous root (Raudaskoski and Kothe, 2015). The fungal (family: *Glomeraceae, Gigasporaceae* and *Claroideoglomeraceae*) association increases nutrient uptake in host plant from soil (Kawaguchi and Minamisawa, 2010). Ectomycorrhiza and endomycorrhiza are two types of mycorrhiza (Manchanda and Garg, 2007). Ectomycorrhiza are common on roots of forest trees like eucalyptus, oak, pine, etc. This association increases phosphorus, potassium and protects from pathogens (Rodriguez and Redman, 2008). Endomycorrhiza are found in the roots of horticultural crops like coffee, betel vine, cardamom, etc., and promote phosphorus nutrition (Yang et al., 2015).

The arbuscular mycorrhizal fungi (AMF) in phosphorus-deficient agricultural soils of semi-arid tropics showed six species of *Glomus*, and *Gigaspora* sp., *Scutellospora* sp., *Acaulospora* sp. were found low abundance. The infectivity and spore density was affected by phosphorus (P), soil organic carbon (SOC), dehydrogenase and alkaline phosphatase activities (Cibichakravarthy et al., 2015). The trifoliate oranges (*Poncirus trifoliata*) grown under different P levels were evaluated in pot study for AMF (*Funneliformis mosseae*) on production of mycorrhizal induced glomalin-related soil protein (GRSP) and soil enzymes (β-glucosidase, catalase, peroxidase, polyphenol oxidase and phosphatase). The AMF mediated GRSP and soil enzymes activities were not found dependent on external P concentrations (Wu et al., 2015). The influence of plant functional groups and seasonality on AMF was reported and found root colonization and spore density was decreased in grassland, and scrub vegetation; intermediary in secondary forest and lowest in mature forest (Zangaro et al., 2013). The mycorrhizal fungal community in primary lacustrine dune succession was investigated and found that AMF performance was not depending directly on sequence of plant community succession (Sikes et al., 2014). The roots of the grass *Deschampsia flexuosa* and AMF communities in different sand area was studied and found that the early succession was higher in the AMF than late succession (Francini et al., 2014). The salinity-induced oxidative stress was studied in *Cicer arietinum* to evaluate the role of AMF (*Funneliformis mosseae*) for antioxidant potential. The AMF was found to be an important protective mechanism by shifting the redox system that could impart the salinity tolerance in *C. arietinum* (Garg and Singla, 2015). The drought stress on pomegranate (*Punica granatum*) after incoluation AMF was investigated and found that photosynthesis was promoted in stress by protection against oxidative stress with accumulation of anthocyanins (Bompadre et al., 2015).The co-inoculation of AMF and PGPR *Pseudomonas* in strawberry grown at low nitrogen and phosphorus levels, and found increase in flowering, fruit production and vitamin content (Bona et al., 2015).

6. MICROORGANISMS IN INSECT CONTROL

Pathogenic microorganisms with ability to invade and kill the insect pests were commonly called entomopathogens. Metchnikoff (1879) and Krassilnikow (1888) were the first to introduce entomopathogens that could be used to control the pests. *Metarhizium anisopliae*, a muscardine fungus was first documented as pathogenic microorganism that was mass produced and applied as microbial pesticide for crops like sugar beets and grains (Compant et al., 2010; Lugtenberg et al., 2002). In 1940, White and Dutky introduced bacterium *Bacillus popilliae* spores to control Japanese beetle. *Bacillus thuringiensis* was introduced in 1960–63 as effective bacterial insecticide and today nearly 40 types of products are popular as insecticides. Studies of Balch and Bird (1944) and Thompson and Steinhaus (1949) succeeded in use of viruses against the insect pests. Baculoviruses are the common choice among the 1600 viruses to control insect species. The selected viruses were targeted against sawflies in forestry during 1950. Oil based formulations of entomopathogenic fungi were used to target the pests like aphids and white flies, but these fungi were effective in highly humid locations. Protozoal species of microsporidia were found disease-causing agents in insects.

6.1 Bacteria as Insecticides

Bacillus popilliae: It causes milky disease in Japanese beetle *Popillio japonica*. After ingestion, spores germinate and penetrate the alimentary canal and blood appears milky due to spores and then the larva dies. This bacterium is produced commercially by feeding or injecting in the host that multiplies in the body of host. Then it is recovered by grinding the host with talc. This dust is expected to contain 100 million spores/g. when it is ingested by host; the mortality is due to septicemia.

Bacillus thuringiensis: It was detected in 1902 in dying larva of *Bombyx mori* by Ishiwata and Berliner in 1915 from Germany in larva of *Ephestia kuehniella*. In 1938, it was first used as microbial insecticide against Lepidopterous larvae with name 'sporeine' and this resulted in the development of several pesticide companies all over the world. It is pathogenic in Lepidopteran larvae and 4 insect orders. Bacterium is transmitted orally, after sporulation bacteria form a toxic crystal and ingestion of these spores causes cessation of feeding. The crystal activity depends on the pH of larval gut (pH 9–10.5) and the action of proteolytic enzymes. The crystal is a protoxin activated by enzymatic hydrolysis. The crystal protein is ecologically safe and successfully tested for controlling insects affecting cotton, cabbage, maize, sunflower, sunflower and pigeon pea.

It is formulated under trade names such as - Thuricide (Sandoz, AG), Bakthane, Dipel (Abbott labs), Biotrol, Karab, Agritol, Bactospeine (Rhone-Poulene) in dust, liquid concentrate, spray concentrate, granules and baits formulations.

6.2 Viruses as Insecticides

Nuclear polyhedrosis viruses (NPV): Virus is rod-like in shape and the outer envelope is with protein crystals called polyhedra. NPV is transmitted by oral ingestion of polyhedra. Ingested polyhedra dissolve, releasing virus rods into the lumen of the insect host midgut, then larval skin darkens, hemolymph becomes turbid and prior to death, infected larva usually climbs to the highest point available and then dies.

The disease of *Heliothis armigera* is visible after 2–3 days of ingestion of virus. The cuticle of infected larva in the first instar become milky white in appearance, but the later infected instars turn pinkish. The diseased larva discharges a yellowish grey fluid from their mouth. The death occurs in 3–7 days; the integument became fragile and burst liberating the liquefied body contents.

***Granulosis viruses* (GV):** GV particles are surrounded by an envelope similar to NPV envelope. The virion is rod shaped with DNA as genetic material with inclusion body, i.e., capsule or granule hence the virus is named Granulosis virus. This virus is active against rice leaf roller *Cnaphalocrocis medinalis*, castor semi looper *Achaea janata*, hairy cater pillar *Pericallia ricini*, cut worm *Agrotis ipsilon*, internode borer *Chilo* spp. and sugarcane shoot borer *C. infuscatellus*. The fat body of Lepidopteran larvae is the primary site of infection. GV are transmitted orally and via the egg. GV infections involving the epidermis cause liquefaction of larvae similar to NPV infections.

***Cytoplasmic polyhedrosis viruses* (CPV):** Particles are not enclosed in membranes as of NPV, but contain protein crystals similar to those of NPV. CPV infect the cytoplasm of the midgut epithelium of Lepidopterous larvae. Infected larvae appear to have small bodies and large heads.

6.3 Fungi as Insecticides

More than 36 different genera of fungi contain species, which cause insect diseases. Most fungi are transmitted from one host to another by a conidium. Conidia germinate and form a special structure, which penetrates the insect cuticle. Fungus then grows in insect's body until the insect is filled with mycelia. Fungal infections depend on environmental conditions, such as high humidity and temperature, and high population densities. Examples: *Metarhizium anisopliae, Beauveria bassiana, Entomophthora* sp. (Gibson et al., 2014).

Beauveria bassiana belongs to class fungi imperfecti and it is actively biological control product (2.3 × 10^7 conidia/ml of product). Upon contact with target insect, the fungal spores attach themselves to the cuticle. The spores then secrete enzymes that dissolve the cuticle. This helps the fungal spore to penetrate hyphae into the body of insect. The production and proliferation of the fungal hyphae cause the release of toxicants from the fungus into the insects. Insect death occurs due to water loss and effects of nutrient loss. *Beauveria bassiana* is popular with trade name Dispel and used at the rate of 2-4 ml/L of water is effective against coleopterous, hemipterous and lepidopterous insects.

7. CONCLUSION

Multiple forms of agriculture are available that can be improved for increase in yields by employing the microorganisms. PGPR are known to be mutualistic/synergistic that supply the nutrients in sustainable mode. These organisms are beneficial in nutrient supply and enhance the plant physiological responses in both biotic and abiotic stress conditions. The farming systems extensively use the synthetic products that can be reduced with PGPR. The alternative methods of farming with application of combinations of PGPR will remediate the agroecosytem constraints. Resource conserving farming along with PGPR could be a better approach to improve the ecosafe agrosystems.

References

Adesemoye AO, and Kloepper JW (2009). Plant-microbes interactions in enhanced fertilizer-use efficiency. Appl Microbiol Biotechnol 85: 1–12.

Ahemad M, and Kibret M (2014). Mechanisms and applications of plant growth promoting rhizobacteria: current perspective. J King Saud Univ Sci 26: 1–20.

Ahmed E, and Holmstrom SJM (2014). Siderophores in environmental research: roles and applications. Microbial Biotechnol 7(3): 196–208.

Ahmad F, Ahmad I, and Khan MS (2008). Screening of free-living rhizospheric bacteria for their multiple plant growth promoting activities. Microbiol Res 163: 173–181.

Anand R, and Chanway CP (2013). nif gene sequence and arrangement in the endophytic diazotroph *Paenibacillus polymyxa* strain P2b-2R. Biol Fertil Soils 49: 965–970.

Asis Jr CA, Kubota M, Ohta H, Arima Y, Chebotar VK, Tsuchiya K, and Akao S (2000). Isolation and partial characterization of endophytic diazotrophs associated with Japanese Sugarcane cultivar. Soil Sci Plant Nutr 46(3): 759–765.

Baginsky C, Brito B, Scherson R, Pertuze R, Seguel O, Canete A, Araneda C, and Johnson WE (2015). Genetic diversity of *Rhizobium* from nodulating beans grown in a variety of Mediterranean climate soils of Chile. Arch Microbiol 197: 419–429s.

Babalola OO (2010). Beneficial bacteria of agricultural importance. Biotechnol Lett 32: 1559–1570.

Babu S, Bidyarani N, Chopra P, Monga D, Kumar R, Prasanna R, Kranthi S, and Saxena AK (2015). Evaluating microbe-plant interactions and varietal differences for enhancing biocontrol efficacy in root rot disease challenged cotton crop. Eur J Plant Pathol 142: 345-362.

Bagyalakshmi B, Ponmurugan P, and Marimuthu S (2012). Influence of potassium solubilizing bacteria on crop productivity and quality of tea (*Camellia sinensis*). Afr J Agric Res 7(30): 4250–4259.

Bompadre MJ, Bidondo LF, Silvani VA, Colombo RP, Pergola AG, and Godcas AM (2015). Combined effects of arbuscular mycorrhizal fungi and exogenous cytokinins on pomegranate (*Punica granatum*) under two contrasting water availability conditions. Symbiosis 65: 55–63.

Bergottini VM, Otegui MB, Sosa DA, Zapata PD, Mulot M, Revord M, Zopfi J, Wiss F, Benrey B, and Junier P (2015). Bio-inoculation of yerba mate seedlings (*Ilex paraguariensis* St. Hill.) with native plant growth-promoting rhizobacteria: a sustainable alternative to improve crop yield. Biol Fert Soils 51(6): 749–755.

Berg G (2009). Plant-microbe interactions promoting plant growth and health: perspectives for control use of microorganisms in agriculture. Appl Microbiol Biotechnol 84(1): 11–18.

Berestetskiy AO (2008). A review of fungal phytotoxins: from basic studies to practical use. Appl Biochem Microbiol 44(5): 453–465.

Beneduzi A, Ambrosini A, and Passaglia LMP (2012). Plant growth-promoting rhizobacteria (PGPR): their potential as antagonists and biocontrol agents. Genet Mol Bio 35(4): 1044–1051.

Bhardwaj D, Ansari MW, Sahoo RK, and Tuteja N (2014). Biofertilizers function as key player in sustainable agriculture by improving soil fertility, plant tolerance and crop productivity. Microb Cell Fact 13: 66.

Bhattacharyya PN, and Jha DK (2012). Plant growth-promoting rhizobacteria (PGPR): emergence in agriculture. World J Microbiol Biotechnol 28: 1327–1350.

Bennett M, and Franzel S (2013). Can organic and resource-conserving agriculture improve livelihoods? A synthesis. Int J Agr Sustain 11(3): 193–215.

Bona E, Lingua G, Manassero P, Cantamessa S, Marsano F, Todeschini V, Copetta A, D'Agostino G, Massa N, Avidano L, Gamalero E, and Berta G (2015). AM fungi and PGP pseudomonads increase flowering, fruit production, and vitamin content in strawberry grown at low nitrogen and phosphorus levels. Mycorrhiza 25: 181–193.

Brahmaprakash GP, and Sahu PK (2012). Biofertilizers for sustainability. J Indian Inst Sci 92(1): 37–62.

Bromfield ESP, Tambong JT, Cloutier S, Prevost D, Laguerre G, van Berkum P, Thi TVT, Assabgui R, and Barran LR (2010). *Ensifer, Phyllobacterium* and *Rhizobium* species *Medicago sativa* (alfalfa) and *Melilotus alba* (sweet clover) grown at a Canadian site without a history of cultivation. Microbiology 156: 505–520.

Buchi L, Gebhard CA, Liebisch F, Sinaj S, Ramseier H, and Charles R (2015). Accumulation of biologically fixed nitrogen by legumes cultivated as cover crops in Switzerland. Plant and Soil 393(1-2): 163–175.

Cappellari LR, Santoro MV, Reinoso H, Travaglia C, Giordano W, and Banchio E (2015). Anatomical, morphological, and phytochemical effects of inoculation with plant growth promoting rhizobacteria on peppermint (*Mentha piperita*). J Chem Ecol 41: 149–158.

Campbell BM, Thornton P, Zougmore R, Asten PV, and Lipper L (2014). Sustainable intensification: what is its role in climate smart agriculture? Curr Opin Environ Sustain 8: 39–43.

Calvo P, Nelson L, and Kloepper JW (2014). Agricultural uses of plant biostimulants. Plant Soil 383(1-2): 3–41.

Carvalho TLG, Pires EB, Saraiva RM, Ferreira PCG, and Hemerly AS (2014). Nitrogen signaling in plant interactions with associative and endophytic diazotrophic bacteria. J Exp Bot 65(19): 5631–5642.

Castro-Sowinski S, Herschkovitz Y, Okon Y, and Jurkevitch E (2007). Effects of inoculation with plant growth-promoting rhizobacteria on resident rhizosphere microorganisms. FEMS Microbiol Lett 276: 1–11.

Chubatsu LS, Monteiro RA, de Souza EM, de Oliveira MAS, Yates MG, Wassem R, Bonatto AC, Huergo LF, Steffens MBR, Rigo LU, and Pedrosa FO (2012). Nitrogen fixation control in *Herbaspirillum seropedicae*. Plant Soil 356(1-2): 197–207.

Choudhary DK, and Johri BN (2009). Interactions of *Bacillus* spp. and plants-with special reference to induced systemic resistance (ISR). Microbiol Res 164: 493–513.

Choudhary DK, Prakash A, and Johri BN (2007). Induced systemic resistance (ISR) in plants: mechanism of action. Indian J Microbiol 47: 289–297.

Compant S, Clement C, and Sessitsch A (2010). Plant growth-promoting bacteria in the rhizo- and endosphere of plants: their role, colonization, mechanisms involved and prospects for utilization. Soil Biol Biochem 42: 669–678.

Correa BO, Schafer JT, and Moura AB (2014). Spectrum of biocontrol bacteria to control leaf, root and vascular diseases of dry bean. Biol Control 72: 71–75.

Cibichakravarthy B, Kumutha K, and Balachandar D (2015). Arbuscular mycorrhizal fungal diversity in phosphorus-deficient Alfisols of a dry North-western agro-ecosystem of Tamil Nadu, India. Ann Microbiol 65: 143–153.

de Weert S, and Bloemberg GV (2006). Rhizosphere competence and the role of root colonization in biocontrol. In: Gnanamanickam SS(ed.) Plant-associated bacteria. Springer, Dordrecht, pp. 317–333.

de Aguiar RA, da Cunha MG, and Junior ML (2014). Management of white mold in processing tomatoes by *Trichoderma* spp. and chemical fungicides applied by drip irrigation. Biol Control 74:1–5.

De-la-pena C, and Loyola-Vargas VM (2014) Biotic interactions in the rhizosphere: A diverse cooperative enterprise for plant productivity. Plant Physiol 166: 701–719

Dass A, Jat SL, and Rana KS (2015). Resource conserving techniques for improving nitrogen-use efficiency. In: Rakshit A et al., (eds.), Nutrient use efficiency: from basics to advances 45–58.

Dalton DA, and Kramer S (2006). Nitrogen-fixing bacteria in non-legumes. In: Gnanamanickam SS (ed.), Plant-associated bacteria. Springer, Dordrecht, pp. 105–130 .

Degefu T, Wolde-meskel E, and Frostegard A (2012). Phylogenetic multilocus sequence analysis identifies seven novel *Ensifergeno* species isolated from a less-well-explored biogeographical region in East Africa. Int J Syst Evol Microbiol 62: 2286–2295.

D'Onofrio A, Crawford JM, Stewart EJ, Witt K, Gavrish E, Epstein S, Clardy J, and Lewis K (2010). Siderophores from neighboring organisms promote the growth of uncultured bacteria. Chem Biol 17: 254–264.

Dixon R, Kahn D (2004). Genetic regulation of biological nitrogen fixation. Nat Rev Microbiol 2: 621–631.

Dimkpa C, Weinand T, and Asch F (2009). Plant-rhizobacteria interactions alleviate abiotic stress conditions. Plant Cell Environ 32: 1682–1694.

Dobereiner J (1988). Isolation and identification of root associated diazotrophs. Plant and Soil 110: 207–212.

Dodd IC, and Perez-Alfocea F (2012). Microbial amelioration of crop salinity stress. J Exp Bot 63(9): 3415–3428.

Dua A, Malhotra J, Saxena A, Khan F, and Lal R (2013). *Devosia lucknowensis* sp. nov., a bacterium isolated from hexachlorocyclohexana (HCH) contaminated pond soil. J Microbiol 51(5): 689–694.

Duca D, Lorv J, Patten CL, Rose D, and Glick BR (2014). Indole-3-acetic acid in plant-microbe interactions. Antonie van Leeuwenhoek 106: 85–125.

Figueiredo MB, Seldin L, de Araujo FF, and Mariano RR (2011). Plant growth promoting rhizobacteria: fundamentals and applications. In: Maheshwari DK (ed), Plant growth and helath promoting bacteria. Microbiology Monographs pp. 21–43.

Francini G, Mannisto M, Alaoja V, and Kytoviita MM (2014). Arbuscular mycorrhizal fungal community divergence within a common host plant in two different soils in a subarctic Aeolian sand area. Mycorrhiza 24: 539–550.

Franche C, Lindstrom K, and Elmerich C (2009). Nitrogen-fixing bacteria associated with leguminous and non-leguminous plants. Plant Soil 321: 35–59.

Gaiero JR, McCall CA, Thompson KA, Day NJ, Best AS, and Dunfield KE (2013). Inside the root microbiome: Bacterial root endophytes and plant growth promotion. Am J Bot 100(9): 1738–1750.

Gajendiran N, and Mahadevan A (1989). Nitrogenase activity associated with leaf, root, and rhizosphere soils in an Indian tropical forest. Biol Fertil Soils 8: 71–74.

Godfray HCJ, and Garnett T (2014). Food security and sustainable intensification. Phil. Trans. R. Soc. B: Biol Sci 369: 1–10.

Garg N, and Singla P (2015). Naringenin- and *Funneliformis mosseae*-mediated alterations in redox state synchronize antioxidant network to alleviate oxidative stress in *Cicer arietinum* L. genotypes under salt stress. J Plant Growth Reguldoi: 10.1007/s00344-015-9494-9 [Epub ahead of print].

Geraldine AM, Lopes FAC, Carvalho DDC, Barbosa ET, Rodrigues AR, Brandao RS, Ulhoa CJ, and Junior ML (2013). Cell wall-degrading enzymes and parasitism of sclerotia are key factors on filed biocontrol of white mold by *Trichoderma* spp. Biol Control 67: 308–316.

Glick BR (2012). Plant growth-promoting bacteria: mechanisms and applications. Scientifica 2002 Article ID 963401, 15 pages doi:10.6064/2012/963401.

Gibson DM, Donzelli BGG, Krasnoff SB, and Keyhani NO (2014). Discovering the secondary metabolite potential encoded within entomopathogenic fungi. Nat Prod Rep 31(10): 1287–1305.

Gopalakrishnan S, Sathya A, Vijayabharathi R, Varshney RK, Gowda CLL, and Krishnamurthy L (2015). Plant growth promoting rhizobia: challenges and opportunities. 3 Biotech 5(4): 355–377.

Gopalakrishnan S, Srinivas V, Vidya MS, and Rathore A (2013). Plant growth-promoting activities of *Streptomyces* spp. in sorghum and rice. Springer Plus 2: 574.

Guerena DT, Lehmann J, Thies JE, Enders A, Karanja N, and Neufeldt H (2015). Partitioning the contributions of biochar properties to enhanced biological nitrogen fixation in common bean (*Phaseolus vulgaris*). Biol Fertil Soils 51: 479–491.

Guo JK, Ding YZ, Feng RW, Wang RG, Xu YM, Chen C, Wei XL, and Chen WM (2015). *Burkholderia metalliresistens* sp. nov., a multiple metal-resistant and phosphate-solubilising species isolated from heavy metal-polluted soil in Southeast China. Antonie van Leeuwenhoek 107(6):1591–1598.

Gloer J, Thummer R, Ullrich H, and Schmitz RA (2008). Towards understanding the nitrogen signal transduction for nif gene expression in *Klebsiella pneumoniae*. FEBS Journal 275: 6281–6294.

Gomiero T, and Paoletti MG (2008). Organic and sustainable agriculture and energy conservation. In: Pimentel D (ed) Biofuels, Solar and Wind as Renewable Energy Systems. Springer 425–464.

Hayatsu M, Tago K, and Saito M (2008). Various players in the nitrogen cycle: Diversity and functions of the microorganisms involved in nitrification and denitrification. Soil Sci Plant Nutr 54: 33–45.

Hassan W, David J, and Bashir F (2014). ACC-deaminase and/or nitrogen-fixing rhizobacteria and growth response of tomato (*Lycopersicon pimpinellfolium* Mill.) J Plant interact 9(1): 869–882.

Hartmann A, Schmid M, van Tuinen D, and Berg G (2009). Plant-driven selection of microbes. Plant Soil 321: 235–257.

Huang N, Enkegaard A, Osborne LS, Ramkers PMJ, Messelink GJ, Pijnakker J, and Murphy G (2011). The Banker plant method in biological control. Cri Rev Plant Sci 30(3): 259–278.

Hossain MS, and Martensson A (2008). Potential use of *Rhizobium* spp. to improve fitness of non-nitrogen-fixing plants. Acta Agric Scand, Sect B 58: 352–358.

Hole DG, Perkins AJ, Wilson JD, Alexander IH, Grice PV, and Evans AD (2005). Does organic farming benefit biodiversity? Biological Conservation 122: 113–130.

Hobbs PR, Sayre K, and Gupta R (2008). The role of conservation agriculture in sustainable agriculture. Phil Trans R Soc B 363: 543–555.

Hui-Qin Y (1997). Biological nitrogen fixation.In: Zhu ZL, Wen QX, Freney JR (eds) Nitrogen in Soils of China. 136–158.

Jack R, de Zamaroczy M, and Merrick M (1999). The signal transduction protein GlnK is required for NifL-dependent nitrogen control of *nif* gene expression in *Klebsiella pneumoniae*. J Bacteriol 181(4): 1156–1162.

James EK, Olivares FL, Baldani JI, and Dobereiner J (1997). Herbaspirillum, an endophytic diazotroph colonizing vascular tissue in leaves of *Sorghum bicolor* L. Moench. J Exp Bot 48(308): 785–797.

Jin HJ, Tu R, Xu F, and Chen SF (2011). Identification of nitrogen-fixing Paenibacillus from different plant rhizospheres and a novel *nifH* gene detected in the *P. stellifer*. Microbiol 80(1): 117–124.

Johansen JR, Bohunicka M, Lukesova A, Hrckova K, Vaccarino MA, and Chesarino NM (2014). Morphological and molecular characertization within 26 strains of the genus *Cylindrospermum* (Nostocaceae, Cyanobacteria), with descriptions of three new species. J Phycol 50: 187–202.

Kaur S, Dhillon GS, Brar SK, Chauhan VB, Chand R, and Verma M (2013). Potential eco-friendly soil microorganisms: Road towards green and sustainable agriculture. In:

Malik A, Grohmann E, Alves M (eds) Management of microbial resources in the environment. Springer 249–287.

Kang SM, Khan AL, Waqas M, You YH, Kim JH, Kim JG, Hamayun M, and Lee IJ (2014). Plant growth-promoting rhizobacteria reduce adverse effects of salinity and osmotic stress by regulating phytohormones and antioxidants in *Cucumis sativus*. J Plant Interact 9(1): 673–682.

Kawaguchi M, and Minamisawa K (2010). Plant-microbe communications for symbiosis. Plant Cell Physiol 51(9): 1377–1380.

Khan Z, Midega C, Pittchar J, Pickett J, and Bruce T (2011). Push-pull technology: a conservation agriculture approach for integrated management of insect pests, weeds and soil health in Africa. Int J Agr Sustain 9(1): 162–170.

Kong Z, Mohamad OA, Deng Z, Liu X, Glick BR, and Wei G (2015). Rhizobial symbiosis effect on the growth, metal uptake, and antioxidant responses of *Medicago lupulina* under copper stress. Environ Sci Pollut Res 22(16):12479–12489.

Kloepper JW, and Schroth MN (1981). Relationship of *in vitro* antibiosis of plant growth promoting rhizobacteria to plant growth and the displacement of root microflora. Phytopathology 71: 1020–1024.

Kirova E, Tzvetkova N, Vaseva I, and Ignatov G (2008). Photosynthetic responses of nitrate-fed and nitrogen-fixing soybeans to progressive water stress. J Plant Nutrition 31: 445–458.

Kumar P, Dubey RC, and Maheshwari DK (2012). *Bacillus* strains isolated from rhizosphere showed plant growth promoting and antagonistic activity against phytopathogens. Microbiol Res 167: 493–499.

Kumar D, Shivay YS, Dhar S, Kumar C, and Prasad R (2013). Rhizospheric flora and the influence of agronomic practices on them: A review. Proc Natl Acad Sci India Sect B. Biol Sci 83(1): 1–14.

Kuhad RC (2012). Microbes and their role in sustainable development. Indian J Microbiol 52(2): 309–313.

Leppanen SM, Rissanen AJ, and Tiirola M (2015). Nitrogen fixation in *Sphagnum* mosses is affected by moss species and water table level. Plant Soil 389: 185–196.

Louarn G, Pereira-Lopes E, Fustec J, Mary B, Voisin AS, Carvalho PCF, and Gastal F (2015). The amounts and dynamics of nitrogen transfer to grasses differ in alfalfa and white clover-based grass-legume mixtures as a result of rooting strategies and rhizodeposit quality. Plant Soil 389: 289–305.

Lima L da S, Olivares FL, de Oliveira RR, Vega MRG, Aguiar NO, and Canellas LP (2014). Root exudates profiling of maize seedlings inoculated with *Herbaspirillum seropedicae* and humic acids. Chemical and Biological Technologies in Agriculture 1: 23.

Liu W, Wang Q, Wang B, Hou J, Luo Y, Tang C, and Franks AE (2015). Plant growth-promoting rhizobacteria enhance the growth and Cd uptake of *Sedum plumbizincicola* in a Cd-contaminated soil. J Soils Sediments 15: 1191–1199.

Ludwig-Muller J (2015). Bacteria and fungi controlling plant growth by manipulating auxin: balance between development and defense. J. Plant Physiol 172: 4-12.

Lugtenberg BJJ, Weong TFCCA, and Bloemberg GV (2002). Microbe-plant interactions: principles and mechanisms. Antonie van Leeuwenhoek 81: 373–383.

Ma Y, Xia Z, Liu X, and Chen S (2007). *Paenibacillus sabinae* sp. nov., a nitrogen-fixing species isolated from the rhizosphere soils of shrubs. Int J Syst Evol Microbio l57: 6–11.

Ma L, Rao X, Lu P, Huang S, Chen X, Xu Z, and Xie J (2015). Acid-tolerant plant species screened for rehabilitating acid mine drainage sites. J Soils Sediments 15: 1104–1112.

Majumder D, Kongbrailatpam JD, Suting EG, Kangjam B, and Lyngdoh D (2014). *Pseudomonas fluorescens*: A potential biocontrol agent for management of fungal diseases of crop plants. Future challenges in crop protection against fungal pathogens. Fungal Biology 2014: 317–342.

Malusa E, Sas-Paszt L, and Ciesielska J (2012). Technologies for beneficial microorganisms inocula used as biofertilizers. The Scientific World Journal 2012: Article ID 491206, 12 pages doi:10.1100/2012/491206.

Manchanda G, and Garg N (2007). Endomycorrhizal and rhizobial symbiosis: how much do they share? J Plant Interact 2(2): 79–88.

Matsutani M, Ito K, Azuma Y, Ogino H, Shirai M, Yakushi T, and Matsushita K (2015). Adaptive mutation related to cellulose producibility in *Komagataeibacter medellinensis* (*Gluconacetobacter xylinus*) NBRC 3288. Appl Microbiol Biotechnol doi: 10.1007/s00253-015-6598-x [Epub ahead of print].

Matson PA, Parton WJ, Power AG, and Swift MJ (1997). Agricultural intensification and ecosystem properties. Science 277: 504–509.

Mansotra P, Sharma P, and Sharma S (2015). Bioaugmentation of *Mesorhizobium cicer*, *Pseudomonas* spp. and *Piriformospora indica* for sustainable Chickpea production. Physiology Molecular Biology of Plants 21(3): 385–393.

McGee JA (2015) Does certified organic farming reduce greenhouse gas emissions from agricultural production? Agriculture and Human Values 32: 255–263

Markovich NA, and Kononova GL (2003). Lytic enzymes of *Trichoderma* and their role in plant defense from fungal diseases: A review. Appl Biochem Microbiol 39(4): 341–351.

Mendes R, Garbeva P, and Raaijmakers JM (2013). The rhizosphere microbiome: significance of plant beneficial, plant pathogenic, and human pathogenic microorganisms. FEMS Microbiol Rev 37: 634–663.

Moreira RR, Nesi CN, and Mio LLMD (2014). *Bacillus* spp., and *Pseudomonas putida* as inhibitors of the *Colletotrichum acutatum* group and potential to control Glomerella leaf spot. Biol Control 72: 30–37.

Mohammadi K, and Sohrabi Y (2012). Bacterial biofertilizers for sustainable crop production: A review. ARPN Journal of Agricultural and Biological Science 7(5): 307–316.

Ming Q, Su C, Zheng C, Jia M, Zhang Q, Zhang H, Rahman K, Han T, and Qin L (2013). Elicitors from the endophytic fungus *Trichoderma atroviride* promote *Salvia miltiorrhiza* hairy root growth and tanshinone biosynthesis. J Exp Bot 64(18): 5687–5694.

Milenkov M, Thummer R, Gloer J, Grotzinger J, Jung S, and Schmitz RA (2011). Insights into membrane association of *Klebsiella pneumoniae* NifL under nitrogen-fixing conditions from mutational analysis. J Bacteriol 193(3): 695–705.

Momota P, Singh BK, and Devi SI (2012). Role of endophytic microorganisms in sustainable agriculture. NeBIo 3(2): 69–77.

Muller JL (2015). Plants and endophytes: equal partners in secondary metabolite production? Biotechnol Lett 37(7): 1325–1334.

Nadeem SM, Naveed M, Zahir ZA, and Asghar HN (2013). Plant-microbe interactions for sustainable agriculture: fundamentals and recent advances. In: Arora NK (ed) Plant Microbe Symbiosis: Fundamentals and Advances Publisher, Springer 51–103.

Nagayama H, Sugawara T, Endo R, Ono A, Kato H, Ohtsubo Y, Nagata Y, and Tsuda M (2015). Isolation of oxygenase genes for indigo-forming activity from an artificially polluted soil metagenome by functional screening using *Psuedomonas putida* strains as hosts. Appl Microbiol Biotechnol 99(10):4453–4470.

Narayansamy P (2013). Mechanisms of action of bacterial biological control agents. Biological management of diseases of crops, Progress in Biological Control 15: 295–429.

Ortiz-castro R, Contreras-cornejo HA, Macias-Rodriguez L, and Lopez-Bucio J (2009). The role of microbial signals in plant growth and development. Plant Signal Behav 4(8): 701–712.

Parolin P, Bresch C, Poncet C, and Desneux N (2012a). Functional characteristics of secondary plants for increased pest management. Int J Pest Manage 58(4): 369–377.

Parolin P, Bresch C, Desneux N, Brun R, Bout A, Boll R, and Poncet C (2012b). Secondary plants used in biological control: A review. Int J Pest Manage 58(2): 91–100.

Parolin P, Bresch C, Poncet C, and Desneux N (2012c). Introducing the term 'Biocontrol plants' for integrated pest management. Scientia Agricola 71(1): 77–80.

Prasad R, Garg AP, and Varma A (2005). Interaction of medicinal plants with plant growth promoting rhizobacteria and symbiotic fungi. In: Podila GK, Varma A (eds), Basic Research and Applications of Mycorrhizae. IK International Pvt. Ltd., Delhi, India, 1: 363–407.

Prasad R, Kumar M, and Varma A (2015). Role of PGPR in soil fertility and plant health. In: Egamberdieva D, Shrivastava S, Varma A (eds)Plant Growth-Promoting

Rhizobacteria and Medicinal Plants, Springer International Publishing Switzerland 247–260.

Pretty JN, Noble AD, Dixon J, Hine RE, de Vries FWTP, and Morison JIL (2006). Resource-conserving agriculture increases fields in Developing Countries. Environ Sci Technol 40(4): 1114–1119.

Pretty J, Toulmin C, and Williams S (2011). Susatinable intensification in African agriculture. Inter J Agr Sustain 9(1): 5–24.

Pereg L, and McMillan M (2015). Scoping the potential uses of beneficial microorganisms for increasing productivity in cotton cropping systems. Soil Biol Biochem 80: 349–358.

Pii Y, Mimmo T, Tomasi N, Terzano R, Cesco S, and Crecchio C (2015). Microbial interactions in the rhizosphere: beneficial influences of plant growth-promoting rhizobacteria on nutrient acquisition process. A review. Biol Fertil Soils 51:403–415.

Pimentel D, Hepperly P, Hanson J, Douds D, and Seidel R (2005). Environmental, energetic, and economic comparisons of organic and conventional farming systems. BioScience 55(7): 573–582.

Prasanna R, Pattnaik S, Sugitha TCK, Nain L, and Saxena AK (2011). Development of cyanobacterium-based biofilms and their *in vitro* evaluation for agriculturally useful traits. Folia Microbiol 56: 49–58.

Ramirez-Bahena MH, Vargas M, Martin M, Tejedor C, Velazquez E, and Peix A (2015). Alfalfa microsymbionts from different ITS and nodC lineages of *Ensifer meliloti* and *Ensifer medicae* symbiovar *meliloti* establish efficient symbiosis with alfala in Spanish acid soils. Appl Microbiol Biotechnol 99(11):4855–4865.

Ramseier H, and Crismaru V (2014). Resource-conserving agriculture: undersowing and mixed crops as stepping stones towards a solution. Soil as World Heritage 353–363.

Rao DLN (2014). Recent advances in biological nitrogen fixation in agricultural systems. Proc Natl Acad Sci India Sect B. Biol Sci 80(2): 359–378.

Raaijmakers JM, Paulitz TC, Steinberg C, Alabouvette C, and Moenne-Loccoz Y (2009). The rhizosphere: a playground and battlefield for soilborne pathogens and beneficial microorganisms. Plant Soil 321: 341–361.

Rajput MS, Kumar GN, and Rajkumar S (2013). Repression of oxalic acid-mediated mineral phosphate solubilization in rhizospheric solates of *Klebsiella pneumonaie* by succinate. Arch Microbiol 195: 81–88.

Raudaskoski M, and Kothe E (2015). Novel findings on the role of signal exchange in arbuscular and ectomycorrhizal symbioses. Mycorrhiza 25: 243–252.

Rivas R, Velazquez E, Willems A, Vizcaino N, Rao NSS, Mateos PF, Gillis M, and Molina EM (2002). A new species of *Devosia* that forms a unique nitrogen-fixing root-nodule symbiosis with the aquatic legume *Neptunia natans* (L.f.) Druce. Appl Environ Microbiol 68(11): 5217–5222.

Rodriguez R, and Redman R (2008). More than 400 million years of evolution and some plants still can't make it on their own: plant stress tolerance via fungal symbiosis. J Exp Bot 59(5): 1109–1114.

Rothballer M, Schmid M, and Hartmann A (2009). Diazotrophic bacterial endophytes in Gramineae and other plants. K. Pawlowski: Prokaryotic symbionts in plants. Microbiol Monograph 8: 273–302.

Romdhane S, Devers-Lamrani M, Martin-Laurent F, Calvayrac C Rocaboy-Faquet E, Riboul D, Cooper J-F, and Barthelmebs L (2015). Isolation and characterization of Bradyrhizobium sp. SR1 degrading two β-triketone herbicides. Environ Sci Pollut Res 23(5): 4138-4148.

Rout ME, and Southworth D (2013). The root microbiome influences scales from molecules to ecosystems: the unseen majority. Am J Bot 100(9): 1689–1691.

Ryan PR, Dessaux Y, Thomashow LS, and Weller DM (2009). Rhizosphere engineering and management for sustainable agriculture. Plant Soil 321: 363–383.

Saleem M, Arshad M, Hussain S, and Bhatti AS (2007). Perspective of plant growth promoting rhizobacteria (PGPR) containing ACC deaminase in stress agriculture. J Ind Microbiol Biotechnol 34: 635–648.

Santi C, Bogusz D, and Franche C (2013). Biological nitrogen fixation in non-legume plants. Annals of Botany 111(5): 743–767.

Saikkonen K, Wali P, Helander M, and Faeth SH (2004). Evolution of endophyte-plant symbioses. Trends in Plant Sci 9(6): 275–279.

Saxena S, and Pandey AK (2001). Microbial metabolites as eco-friendly agrochemicals for the next millennium. Appl Microbiol Biotechnol 55: 395–403.

Sayyed RZ, Reddy MS, Kumar KV, Yellareddygari SKR, Deshmukh AM, Patel PR, and Gangurde NS (2012). Potential of plant growth-promoting rhizobacteria for sustainable agriculture. In: Maheshwari DK (ed.). Bacteria in Agrobiology: Plant Probiotics, Springer-Verlag Berlin Heidelberg, pp. 287–313.

Sawada H, Kuykendall LD, and Young JM (2003) Changing concepts in the systematics of bacterial nitrogen-fixing legume symbionts. J Gen Appl Microbiol 49: 155–179

Saino T, and Hattori A (1982). Aerobic nitrogen fixation by the marine non-heterocystous Cyanobacterium *Trichodesmium* (*Oscillatoria*) spp.: its protective mechanism against oxygen. Marine Biology 70: 251–254.

Seldin L (2011). *Paenibacillus*, nitrogen fixation and soil fertility. In: Logan NA and De Vos P. (eds.), Endospore-forming soil bacteria. Soil Biology 27, Springer-Verlag Berlin Heidelberg, pp. 287-307.

Sessitsch A, Kuffner M, Kidd P, Vagronsveld J, Wenzel WW, Fallmann K, and Puschenreiter M (2013). The role of plant-associated bacteria in the mobilization and phytoextraction of trace elements in contaminated soils. Soil Biol Biochem 60: 182–194.

Shabayev VP (2012). Mineral nutrition of plants inoculated with plant growth-promoting rhizobacteria of *Pseudomonas* genus. Biol Bull Rev 2(6): 487–499.

Szilagyi-Zecchin VJ, Ikeda AC, Hungria M, Adamoski D, Kava-Cordeiro V, Glienke C, and Galli-Terasawa LV (2014). Identification and characterization of endophytic bacteria from corn (*Zea mays* L.) roots with biotechnological potential in agriculture. AMB Express 4: 26.

Sikes BA, Maherali H, and Klironomos JN (2014). Mycorrhizal fungal growth responds to soil characteristics, but not host plant identity, during a primary lacustrine dune succession. Mycorrhiza 24: 219–226.

Singh JS, Pandey VC, and Singh DP (2011). Efficient soil microroganisms: A new dimension for sustainable agriculture and environmental development. Agric Ecosys Environ 140: 339–353.

Shu HY, Lin LC, Lin TK, Chen HP, Yang HH, Peng KC, and Lin GH (2015). Transcriptional regulation of the iac locus from *Acinetobacter baumannii* by the phytohormone indole-3-acetic acid. Antonie van Leeuwenhoek 107: 1237–1247.

Sreenivasa MN (2012). Organic farming: for sustainable production and environmental protection. In: Satyanarayana T, Johri BN, Prakash A (eds) Microorganisms in Sustainable Agriculture and Biotechnology, Springer, Netherlands. 55–76.

Stoop WA (2011). The scientific case for system of rice intensification and its relevance for sustainable crop intensification. Int J Agr Sustain 9(3): 445–455.

Styger E, Aboubacrine G, Attaher MA, and Uphoff N (2011). The system of rice intensification as sustainable agricultural innovation: introducing, adapting and scaling up a system of rice intensification practices in the Timbuktu region of Mali. Int J Agr Sustain 9(1): 67–75.

Taurian T, Ibanez F, Angelini J, Tonelli ML, and Fabra A (2012). Endophytic bacteria and their role in legumes growth promotion. Maheshwari DK (ed.) Bacteria in Agrobiology: Plant Probiotics Springer-Verlag Berlin Heidelberg,141–168.

Tokdar P, Sanakal A, Ranadive P, Khora SS, George S, and Deshmukh SK (2015). Molecular, physiological and phenotypic characterization of Paracoccus denitrificans ATCC 19367 mutant strain P-87 producing improved coenzyme Q_{10}(CoQ_{10}). Indian J Microbiol 55(2): 184–193.

Torres MJ, Rubia MI, de la Pena TC, Pueyo JJ, Bedmar EJ, and Delgado MJ (2014). Genetic basis for denitrification in *Ensifer meliloti*. BMC Microbiol 14: 142.

Tkacz A, and Poole P (2015). Role of root microbiota in plant productivity. J Exp Bot 66(8): 2167–2175.

Tkachenko OV, Evseeva NV, Boikova NV, Matora LY, Burygin GL, Lobachev YV, and Shchygoiev SY (2015). Improved potato microclonal reproduction with the plant growth-promoting rhizobacteria *Azospirillum*. Agron Sustainable Dev 35(3): 1167–1174.

Trethowan RM, Reynolds M, Sayre K, and Monasterio IO (2005). Adapting wheat cultivars to resource conserving farming practices and human nutritional needs. Annals Appl Biol 146: 405–413.

Tsavkelova EA, Klimova SY, Cherdyntseva TA, and Netrusov AI (2006). Microbial producers of plant growth stimulators and their practical use: a review. Appl Biochem Microbiol 42(2): 117–126.

Uphoff N (2003). Higher yields with fewer external inputs? The system of rice intensification and potential contributions to agricultural sustainability. Int J Agr Sustain 1(1): 38–50.

Uphoff N (2012). Supporting food security in the 21st century through resource-conserving increases in agricultural production. Agriculture and Food Security, 1: 18, DOI: 10.1186/20487010-1-18. BioMed Central Ltd. 2012.

Velu C, Cires S, Alvarez-Roa C, and Heimann K (2015). First outdoor cultivation of the N2-fixing cyanobacterium *Tolypothrix* sp. in low-cost suspension and biofilm systems in tropical Australia. J Appl Phycoldoi 10.1007/s10811-014-0509-x[Epub ahead of print].

Vessey JK (2003). Plant growth promoting rhizobacteria as biofertilizers. Plant and Soil 255: 571–586.

Verma JP, Yadav J, Tiwari KN, Lavakush, and Singh V (2010). Impact of plant growth promoting rhizobacteria on crop production. Int J AgricRes 5(11): 954–983.

Wagh J, Bhandari P, Shah S, Archana G, and Kumar GN (2014). Overexpression of citrate operon in *Herbaspirillum seropedicae* Z67 enhances organic acid secretion, mineral phosphate solubilization and growth promotion of *Oryza sativa*. Plant Soil 383: 73–86.

Wang RR, Wang Q, He LY, Qiu G, and Sheng XF (2015). Isolation and the interaction between a mineral-weathering *Rhizobium tropici* Q34 and silicate minerals. World J Microbiol Biotechnol 31: 747–753.

Whipps JM (2001). Microbial interactions and biocontrol in the rhizosphere.J Exp Bot 52: 487–511.

Willems A (2007). The taxonomy of rhizobia: an overview. In: Velazquez E, Rodriguez-Barruecco C (eds) First International Meeting on Microbial Phosphate Solubilization, 3–14.

Wolfe MS, Baresel JP, Desclaux D, Goldringer I, Hoad S, Kovacs G, Loschenberger F, Miedaner T, Ostergard H, and van Bueren ETL (2008). Developments in breeding cereals for organic agriculture. Euphytica 163(3): 323–346.

Wu QS, Li Y, Zou YN, and He XH (2015). Arbuscular mycorrhiza mediates glomalin-related soil protein production and soil enzyme activities in the rhizosphere of trifoliate orange grown under different P levels. Mycorrhiza 25: 121–130.

Yang J, Kloepper JW, and Ryu CM (2009). Rhizosphere bacteria help plants tolerate abiotic stress. Trends Plant Sci 14(1): 1–4.

Yang H, Zhang Q, Dai Y, Liu Q, Tang J, Bian X, and Chen X (2015). Effects of arbuscular fungi on plant growth depend on root system: a meta-analysis. Plant Soil 389: 361–374.

Yutani M, Taniguchi H, Borjihan H, Ogita A, Fujita K, and Tanaka T (2011). Alliinase from *Ensifer adhaerens* and its use for generation of fungicidal activity. AMB Express 1: 2.

Zangaro W, Rostirola LV, de Souza PB, Alves RA, Lescano LEAM, Rondina ABL, Nogueira MA, and Carrenho R (2013). Root Colonization and spore abundance of arbuscular mycorrhizal fungi in distinct successional stages from an Atlantic rainforest biome in southern Brazil. Mycorrhiza 23: 221–233.

Zhong Z, Wang Y, Ping W, Ling J, Zheng H, Wang H, and Zhu J (2015). Function of MsiR on canavanine-mediated repression in Mesorhizobium tianshanense. Arch Microbiol 197(5): 729–735.

ZhaoY, Li W, Zhou Z, Wang L, Pan Y, and Zhao L (2005). Dynamics of microbial community structure and cellulolytic activity in agricultural soil amended with two biofertilizers. Eur J Soil Biol 41: 21–29.

Zeng WL, Li WK, Han H, Tao YY, Yang L, Wang ZT, and Chen KX (2014). Microbial biotransformation of gentiopicroside by the endophytic fungus Penicillium crustosum 2T01Y01. Appl Environ Microbiol 80(1): 184–192.

9

Secondary Metabolites: A Gift of Natural Weapon for Defense Mechanism in Plants

RAJESH KUMAR PANDEY[1*] AND RAM PRASAD[2,3]
[1]Department of Botany, Bundelkhand University, Jhansi–284128, UP, India
[2]Amity Institute of Microbial Technology, Amity University
Uttar Pradesh, Noida–201303, India
[3]Department of Mechanical Engineering
Johns Hopkins University, Baltimore–21218, USA
*Corresponding author e-mail: *rkp_vam@rediffmail.com*

1. INTRODUCTION

In plants, secondary metabolites are usually considered non-essential byproducts of metabolism, because they are not involved in the growth and development. This does not mean that they are not important, as they relate plants with the different components of their ecosystem, i.e., the physical environment (biotope) and the living community (biocenosis). This indicates that secondary metabolites are indispensable for the survival of the species. Moreover, it must be pointed out that the widely accepted division of metabolism in primary and secondary is somewhat unsatisfactory and shows mosaicism in statement of their characters. In fact, some major chemical groups still classified as secondary metabolites, such as lipids, are required by all living organisms, as are some metabolites arising from a secondary pathway (i.e., gibberellins A1 from isoprenoid pathway) that can have a determinant endogenous regulatory role. Additionally, stress metabolism could be regarded as a particular expression of secondary metabolism, when stressful conditions, of both biotic and abiotic nature, change the dynamic equilibrium of the ecosystem.

From the literature it is clear that chemical diversity in plants and microbes have determined their evolutionary success. Due to their sessile habits, plants are unable to avoid worsening environmental conditions and they cannot escape from the effect of biotic stresses. Consequently, unlike animals, plants have evolved an enormous secondary metabolites to overcome any danger, though most of them are species-specific. This implies that the actual number of secondary metabolites within a given species is much more limited. The functional role of these plant chemicals range from some ecological aspects to defense mechanism, improving protection against both biotic and abiotic stresses. The secondary metabolites have also found involved in ecological roles as attractants or repellents for pollinators and phytophagy, respectively as well as colors and scents of flowers and fruits.

Generally, precursors of secondary metabolic pathways are products of the primary metabolism. Therefore, a severe or long lasting stress factor could induce an excessive shift between primary and secondary metabolism and consequently, a diversion of essentially available resources from growth to defense. To a large extent, secondary metabolites derive from three biosynthetic routes, namely the phenylpropanoid, isoprenoid and alkaloid pathways. Phytochemicals arising from these pathways include not only compounds with a broad-spectrum antibiotic activity, but also powerful antioxidants that are able to counteract oxidative stress (Iriti and Faoro, 2004; Shirsat et al., 2012).

Secondary metabolites are organic molecules that are not involved in the normal growth and development of an organism. While primary metabolites have a key role in the survival of the species, playing an active function in the photosynthesis and respiration, absence of secondary metabolites does not result in immediate death, but rather in long-term impairment of the organism's survivability, often playing an important role in plant defense. These compounds belong to an extremely diverse group of natural products synthesized by plants, fungi, bacteria, algae, and animals. Most of secondary metabolites, such as terpenes, phenolic compounds and alkaloids are classified based on their biosynthetic origin. Different classes of these compounds are often associated to a narrow set of species within a phylogenetic group and constitute the bioactive compound in several medicinal, aromatic, colorant, and spice plants and/or functional foods. Secondary metabolites are frequently produced at highest levels during a transition from active growth to stationary phase. The producer organism can grow in the absence of their synthesis, suggesting that secondary metabolism is not essential, at least for short-term survival. A second view proposes that the genes involved in secondary metabolism provide a "genetic playing field" that allows mutation and natural selection to fix new beneficial traits via evolution. A third view characterizes secondary metabolism as an integral part of cellular metabolism and biology; it relies on primary metabolism to supply the required enzymes, energy, substrates and cellular machinery and contributes to the long-term survival of the producer (Roze et al., 2011).

A simple classification of secondary metabolites includes trees and main groups of secondary metabolites are: **terpenes** (such as plant volatiles, cardiac glycosides, carotenoids and sterols), **phenolics** (such as phenolic acids, coumarins, lignans, stilbenes, flavonoids, tannins and lignin) and **nitrogen** containing compounds (such as alkaloids and glucosinolates).

A number of traditional separation techniques with various solvent systems and spray reagents, have been described as having the ability to separate and identify secondary metabolites. This chapter proposes to discuss major secondary metabolites classes (terpenoids, phenolic compounds and alkaloids) with different chemical structures and functions being screened, separated, fractionated, purified or analyzed using various adsorbents and eluents through column chromatography (CC) and thin layer chromatography (TLC).

2. TERPENOIDS

Terpenoids are the largest and most diverse family of natural products, ranging in structure from linear to polycyclic molecules and in size from the five-carbon hemiterpenes to natural rubber, comprising thousands of isoprene units. All terpenoids are synthesized through the condensation of isoprene units (C5) and are classified by the number of five-carbon units present in the core structure (Mahmoud and Croteau, 2002). Many flavor and aromatic molecules, such as menthol, linalool, geraniol and caryophyllene are formed by monoterpenes (C10), with two isoprene units, and sesquiterpenes (C15), with three isoprene units. Other bioactive compounds, such as diterpenes (C20), triterpenes (C30) and tetraterpenes (C40) show some special properties.

2.1 Monoterpenes and Sesquiterpenes (Plant Volatiles)

Plant volatiles are typical lipophilic liquids with high vapor pressures. Non-conjugated plant volatiles can cross membranes freely and evaporate into the atmosphere when there are no barriers to diffusion. The number of identified volatile chemicals synthesized by various plants exceeds 1000 and is likely to grow as more plants are examined with new methods for detecting and analyzing quantities of volatiles that are often minute (Pichersky et al., 2006; Dudareva et al., 2004).

The gas chromatography (GC and GC-MS) is a powerful analytical tool for identifying of essential oil components. However, GC-MS has its limitations. Isomers usually give similar mass spectra. This is particularly true for terpenes and even more for sesquiterpenes. Therefore, a favorable match factor between mass spectra is not sufficient for identification (Zellner et al., 2010). Retention indices have been used, together with mass spectrometry, for the proper identification of essential oils composition.

Misidentification is not rare, however, either if a non-authentic mass spectra library is used, which means a database built with data from the literature, not from the analysis of real standards, or by misuse of retention indices (Joulain and König, 1998). CC has been applied to solve this problem. After isolation, a NMR analysis can be performed for the unknown or suspect compound, so that the correct mass spectrum and retention index can be recorded.

Despite the advances in analytical methods to evaluate the composition of essential oil, CC remains a powerful technology for separation and characterization of specific compounds of interest. Column chromatography has been largely used and reported for searching and identification of new molecules, sometimes associated with their antimicrobial, antibacterial and antifungal activities. Also, this technique has been successfully used to obtain sufficient amounts of a substance for the investigation of its biological properties and allowing the detection of its olfactory properties. Isolation is also applied to and, very important for volatiles, to evaluate its odor.

2.1.1 Compound Identification

The major constituents present in the volatile oils of different aromatic species may also be isolated or fractionated by silica gel on preparative TLC or CC. For example, the essential oil of *Tanacetum chiliophyllum* was subjected to silica gel CC, using a solvent mixture of *n*-hexane and ethyl acetate, to isolate one of its major components, dihydro-α-cyclogeranyl hexanoate (Salamci et al., 2007).

Djabou et al. (2010) have characterized the essential oils of *Teucrium massiliense* L. from Corsican and Sardinian islands, using a combination of capillary GC/retention indices, GC/MS and 13C NMR spectroscopy after fractionation on CC. A mixture of all Corsican oil samples was submitted to flash chromatography [FC; silica gel, elution with *n*-pentane, then with diethyl ether]. The polar fraction was separated on silica gel and 14 fractions were eluted with a mixture of *n*-pentane and diethyl ether of increasing polarity to give the major components: 6-methyl-3-heptyl acetate, 3-octyl acetate, isobutyl isovalerate, germacrene D and linalool. Successive CC revealed the unknown compound 6-methyl-3-heptyl acetate.

Chemical investigations on the essential oil of *Lippia integrifolia* performed by CC (silica gel using *n*-hexane with increasing amounts of diethyl ether), followed by HPLC, GC–MS, 1H and 13C-NMR spectroscopy led to the identification of 78 components. A new sesquiterpenic alcohol, trans-africanan-1α-ol, was identified (Coronel et al., 2006).

Sutour et al. (2008) studied the essential oil of *Mentha suaveolens* ssp. *insularis* (Req.) Greuter, finding pulegone to be the major constituent. However, the second most abundant component of the essential oil remained unknown, being isolated by repeated chromatography on silica gel and submitted to a full set of NMR experiments, and identified as *cis-cis-p*-menthenolide. Fractions were eluted with a gradient of solvents of increasing polarity (pentane:diethyl ether 100:0–0:100).

2.1.2 Biological Activities

Human enteropathogens (*Campylobacter jejuni*, *Campylobacter coli*, and *Campylobacter lari* strains) were inhibited by the essential oil of wild *Daucus carota* L. DCEO. Major components of the essential oil responsible for the antibacterial activity were isolated on CC and identified as (*E*)-methyl-isoeugenol and elemicin (Rossi et al., 2007).

Confertifolin (6,6,9a-trimethyl-4,5,5a,6,7,8,9,9a-octahydronaphtho[1,2-c] furan-3 (1*H*)-one), isolated from the essential oil of *Polygonum hydropiper* L. (Polygonaceae) leaves using CC, showed activity both against bacteria and fungi. The oil was chromatographed on a column packed with silica gel and eluted with hexane-ethyl acetate gradient solvent system with increasing polarity, yielding 117 fractions (Duraipandiyan et al., 2010).

The essential oil of *Daucus crinitus* Desf. was submitted to flash chromatography (silica gel, using petroleum ether with increasing amounts of diethyl ether as eluent), and fractions were analyzed by GC-MS. Those containing the isochavicol esters were gathered for the preparation of isochavicol. The antibacterial and antifungal activities of the whole essential oil, of these two esters, and of isochavicol itself, were investigated against a wide range of bacteria and fungi.

Additionally, their antimalarial and antiradical properties were also evaluated, showing an interesting antiplasmodial activity of isochavicol (Lanfranchi et al., 2010).

Patchouli oil obtained from *Pogostemon cablin* (Blanco) Benth and its main constituent, patchouli alcohol, were tested for their repellency and toxicity against Formosan subterranean termites (*Coptotermes formosanus* Shiraki). Both were found to be toxic and repellent. The fraction containing patchouli alcohol was purified by silica CC, and its purity was determined by GC-MS. Unusual tissue destruction was noted inside the exoskeleton of the termite after patchouli alcohol was topically applied to the dorsum (Zhu et al., 2003).

A successful breeding program of *Cannabis sativa* cultivars with a low content of the hallucinogenous D-9-tetrahydrocannabinol (THC) is a prerequisite to avoid drug abuse of hemp. Four TLC systems for analyzing cannabinoids were compared, being able to identify the absence of THC in the hemp essential oils from five commercial fibre varieties. The system using HPTLC, RP-18 and acetonitrile-water (9:1) achieved the best separation of hallucinogenous and non-hallucinogenous cannabinoids (Novak et al., 2001).

Nine fractions of essential oil extracted from *Thymus praecox* were separated by silica gel CC and the fractions were demonstrated to have antioxidant activity (Ozen et al., 2011). The essential oil of *Coriandrum sativum* was submitted to silica gel on dry-CC, resulting in six fractions. Essential oil and its fractions presented antimicrobial potential against Candida yeast infections (Begnami et al., 2010). Essential oils, distilled from seeds of *Coriander sativum* and *Carum carvii* and from leaves of *Ocimum basilicum,* were fractionated by silica gel over CC and tested in the laboratory for volatile toxicity against three stored rice pests. Fractions, where combinations of products occurred with or without other minor compounds, were often more toxic than any one compound alone (Lopez et al., 2008).

The main component of the essential oil of *Anemia tomentosa* var. *anthriscifolia,* (−)-9-epipresilphiperfolan-1-ol was isolated by CC over silica gel, eluted with hexane and hexane-ethyl acetate (95:5), then tested against *Mycobacterium tuberculosis* (H37Rv) and *M. smegmatis*. A minimum inhibitory concentration of 120 µg/ml was recorded for *M. tuberculosis* (Pinto et al., 2009).

2.2 Diterpenes and Sesterterpenes

The diterpene compounds arise from geranyl geranyl diphosphate, and present 20 carbon units in their basic skeletal type. One of the simplest and most important of the diterpenes is phytol, a reduced form of geranylgeraniol, which forms the lipophilic side-chain of the chlorophylls (Vetter and Schröder, 2011).

Cyclization reactions of geranyl geranyl diphosphate led to many structural types of diterpenoids, presenting a large range of polarity nature, from a polar hydrocarbons, such as cembrene (Villanueva and Setzer, 2010), a 14-membered ring, to fully oxidized skeleton of virescenoside, isolated from marine fungus *Acremonium striatisporum* (Ebel, 2010).

Sesterterpenes (C25) may be the least common group of terpenoids. This class of compounds arises from geranylfarnesyl diphosphate, which by cyclization can give rise to various skeletal types, presenting different oxidation levels and several biological activities.

Although many examples of these natural terpenoids are known, they are primarily isolated from fungi and marine organisms.

Considering the large range of polarity nature presented by both diterpene and sesterterpene, the isolation and purification techniques vary and can be classic TLC, preparative thin-layer chromatography (PTLC), CC, flash chromatography (FC), or modern high performance liquid chromatography (HPLC), multiflash chromatography, vacuum liquid chromatography (VLC), solid-phase extraction and others (Lanças, 2008). However, the most important factor that has to be considered before designing an isolation protocol is the nature of the target compound, such as solubility (hydrophobicity or hydrophilicity), acid-base properties, charge, stability, molecular size. Taking these factors into account, the choice of chromatographic methods and the stationary phases to be used are important for the design of a purification system.

Silica gel-based material is the most usual stationary phase. However, polyamide, Sephadex® LH-20, alumina (Al_2O_3), florisil and others are also used, mainly for isolation of more polar compounds (Ajit-Simh et al., 2003). The air-dried and ground roots of Peltodon longipes were extracted with hexane. The methanol soluble fraction of the hexane crude extract was subjected to open CC on silica gel 60 eluted with hexane/ethyl acetate step gradient. After subsequent purification, several abietane diterpenes were isolated, including royleanone, inuroylleanol, deoxyneocryptotanshinone and 7α-ethoxyroyleanone. The sesterterpenes 24-O-methylmanaolide and 24-O-ethylmanaolide, besides others, were isolated from the sponge Luffariella cf. variabilis. Briefly, the frozen sponge tissue was extracted by methanol:chloroform (1:2). The crude extract was successively partitioned between equal volumes of aqueous methanol and a solvent series of hexane, carbon tetrachloride and chloroform. The hexane fraction was subjected to silica gel columns, using eluents of increasing polarity from hexane:ethyl acetate (Gauvin-Bialecki et al., 2008).

Olefins can form co-ordination products with silver ions, and this ability has been used for chromatographic separation of several terpenes, mainly those with a carbon skeleton higher than C_{20}. The procedure to prepare the modified silica was described as follows: Silver nitrate ($AgNO_3$) in water (1:2 w:v) is added to silica. After water evaporation, the residue is dried in oven at 130°C for 15 h, resulting in an almost white powder that should be stored in a dark bottle. Column packing is carried out in the same way as ordinary silica gel column; however, should be wrapped in dark paper or built with amber glass (Norin and Westfelt, 1963).

Methyl ester from kaurenoic acid, iso-kaurenoic acid and grandiflorenic acid were purified by using argentum silica (Batista et al., 2010). The diterpene acids were isolated from aerial parts of Wedelia paludosa. The plant material was extracted by percolation first using hexane, followed by ethanol. The crude ethanol extract was chromatographed over silica gel column. A mixture of these three compounds was obtained and further submitted to isocratic CC on silica gel impregnated with 20 percent of $AgNO_3$, eluting with hexane:diethyl ether 97:3 (Batista et al., 2010). Polyamide 6 (e-aminopolycaprolactam) has been used in our laboratory to isolate polar natural products due to its low cost and reusability. The pretreatment of polyamide is carried out as follows: before column package, polyamide should be washed with methanol, rinsed with distilled water, and poured into the glass column, using an approximate proportion of 1 kg of polyamide to 10 L of each solvent, in order to eliminate oligomers and other impurities (Ye et al., 2011). After elution with H_2O until column stabilization, elution is performed with H_2O and a miscible polar solvent (usually methanol) gradient.

A polyamide column was used in the first step to obtain diterpenes from aerial parts of Euphorbia pannonica. The plant material was extracted by methanol at room temperature. The obtained crude extract was partitioned by H_2O and dichloromethane and the organic fraction was chromatographed on a polyamide column with mixtures of methanol and H_2O (2:3, 3:2 and 4:1) as eluent. The fraction obtained with methanol-H_2O (3:2) was subjected to silica gel VLC using a gradient system of cyclohexane, chloroform and acetone (3:2:0, 1:1:0, 2:3:0, 3:7:0, 1:4:0, 10:40:1, 4:16:1 and 1:4:1). After purification, two tigliane-type diterpenes, 4,12-dideoxy(4β)phorbol-20-benzoate-13-isovalerianate and 4,12-dideoxy(4β) phorbol-20- benzoate-13-isobutyrate, were isolated (Sulyok et al., 2009). Sephadex LH-20 column was used to isolate halisulfate 1 (16) from dark brown sponge. The freeze-dried sponge was extracted by methanol. After the solvent had been filtered and removed, the residue was partitioned between dichloromethane and water. The organic extracts were dried over sodium sulfate and evaporated to obtain dark brown oil that was chromatographed on Sephadex LH-20 with 1:1 methanol-dichloromethane as eluant to obtain two antimicrobial fractions. Fraction B gave crystals of halisulfate 1 from dichloromethane (Kernan and Faulkner, 1988).

2.3 Triterpenes

Triterpenes (C30) are a large class of compounds presenting a number of important biological activities; they arise from squalene, a coupling of two farnesil diphosphate units (Abe, 2007). Cyclization of squalene, or squalene oxide leads to a large number of diverse structural triterpene skeletal types (with 30 carbons), such as lupane, oleane, ursane types (Tantillo, 2011).

Steroids are modified triterpenoids, lacking the three-methyl groups at C-4 and C-14. Sterols are characteristic of eukaryotes. In bacteria they are sparsely distributed with limited array of products (Summons et al., 2006). Skeletal modifications, especially to the side-chain, originate a wide range of biologically important natural products, e.g., sterols, steroidal saponins, cardioactive glycosides, bile acids, corticosteroids, and mammalian sex hormones (Abe, 2007). Saponins are a group of natural compounds presenting triterpenoidal or steroidal aglycone, designated genin or sapogenin, covalently linked to one or more sugar moieties (Augustin et al., 2011). The most usual technique to isolate triterpenes and steroids is by silica open column. As an example, six triterpenes were isolated from petroleum ether extract of Azorella trifurcate whole plant (Areche et al., 2009). The crude extract was chromatographed over silica gel using petroleum ether-ethyl acetate gradient. The fraction eluted with 20 percent ethyl acetate was purified by CC on silica gel impregnated with 10 percent $AgNO_3$, furnishing four triterpenes: lanost-7-en-3β-ol, lanost-9(11)-en-3 β-ol, lanosta-7,24-dien-3 β-ol and cycloartenol.

The acetylation product from the fraction eluted with 40 percent ethyl acetate, after purification by argentum silica gel CC, yielded the triterpenes lanosta-7,24-dien-3 β-yl acetate and 28- acetoxycycloartenyl acetate (Areche et al., 2009).

Several cycloartane-type saponins were isolated from *Astragalus wiedemannianus* methanol crude extract (Polat et al., 2010). Plant material was extracted under reflux. After filtration and solvent removal, the methanol crude extract was solved in water and successively partitioned with hexane, dichloromethane and butanol saturated with water. The butanol fraction was submitted to VLC on reversed-phase material (Lichropep RP-18), employing a H_2O:methanol gradient. Fractions developed with H_2O: methanol 2:8 were rich in saponins.

The first fraction was chromatographed over Lichropep RP-18 (H_2O: methanol gradient). The obtained sub-fractions were submitted to new chromatographic procedures, depending on the nature of target compounds furnishing the saponins.

2.4 Tetraterpenes (Carotenoids)

More than 650 carotenoids (C40) are found in nature, constituting the largest group of natural dyes. The carotenoids are substances with exclusive properties possessed by no other group of substances; these form the basis of their many varied functions and actions in all kinds of living organisms (Britton, 1995). Carotenoids are biosynthesized by plants, algae, fungi, yeasts and bacteria.

The carotenoids are isoprenoid compounds, biosynthesized by tail-to-tail linkage of two geranylgeranyl diphosphate molecules. This produces the parent C40 carbon skeleton from which all the individual variations are derived. This skeleton can be modified: (a) by cyclization at one end or both ends of the molecule to give the seven different end groups; (b) by changes in hydrogenation level, and (c) by addition of oxygen-containing functional groups. Carotenoids that contain one or more oxygen functions are known as xanthophylls, the parent hydrocarbons as carotenes (Britton, 1995). After being absorbed through human diet, some carotenes, among them β-carotene, are pro-vitamin A; other such as lycopene, are important due to their antioxidant properties.

Carotenoid extracts have been screened by TLC and separated by CC, involving liquid-solid chromatography (adsorption). Various adsorbents have been applied in carotenoid analysis, including Al_2O_3, silica, magnesium oxide (MgO), calcium hydroxide [$Ca(OH)_2$], calcium carbonate [$CaCO_3$], siliceous earth as hyflosupercell and others. In normal phase CC, the adsorption affinity depends on the number of conjugated double bonds, cyclization and the presence of oxygen substituents (Rodriguez-Amaya, 1999). CC has been used for separations of mixtures of carotenes and xanthophylls, aiming for mainly analytical determinations, standard purifications, biological evaluations of carotenoids and the purification of synthesized carotenoids, especially by flash chromatography.

Separations on basic adsorbents, such as MgO and $Ca(OH)_2$ are mainly determined by the number and type of double bonds in the carotenoid molecules (Bernhard, 1995). The procedure for isolating and purifying carotenoid standards was established because of the difficulty in obtaining standards commercially. The procedure consists of carotenoid extraction with cold acetone, partition to petroleum ether in a separatory funnel with addition of water, concentration in a rotatory evaporator and chromatographic separation of carotenoids on CC developed with petroleum ether containing increasing percentages of ethyl ether and acetone (Rodriguez-Amaya, 1999). A variety of carotenoid standards have been isolated and purified using MgO:Hyflosupercel (1:1) CC developed with 2–8 percent ethyl ether in petroleum ether and 2–95 percent acetone in petroleum ether: 98 percent β-carotene (isolated from carrot), 94 percent lycopene, 99 percent β-cryptoxanthin, 91 percent β-carotene and 91 percent rubixanthin (from pitanga) (Porcue Rodriguez Amaya 2008), 91–97 percent neoxanthin, 95–98 percent violaxanthin, 97–100 percent lactucaxanthin, 92–96 percent lutein, 93 percent β-cryptoxanthin, 96 percent zeaxanthin and 90–99 percent β-carotene (from lettuce, papaya or green corn) (Kimura and Rodriguez-Amaya, 2002; Oliveira and Rodriguez-Amaya, 2007) and 97 percent trans-ζ-carotene, 99 percent cis-violaxanthin, 97 percent *trans*-violaxanthin, 92 percent prolycopene and 94 percent lutein (from passion fruit extracts) (Wondracek et al., 2011). The β-cryptoxanthin elutes before lycopene in MgO: Hyflosupercel and after lycopene in Al_2O_3 column, indicating that the influence of cyclization is greater than that of the presence of hydroxyl substituents in the MgO:Hyflosupercel column (Rodriguez-Amaya 1999). MgO: Hyflosupercel on CC has been also widely used to analyse the main carotenoids of tropical fruits such as Pouteria campechiana (Agostini-Costa et al., 2010) and to evaluate the variation in the carotenoid composition of Malpighia glabra pulp (Agostini-Costa et al., 2003) and Eugenia uniflora fruit (Porcu and Rodriguez-Amaya, 2008). Hydrocarbon carotenes, such as β-carotene, lycopene, β-carotene, ζ-carotene and xanthophylls such as β-cryptoxanthin, violaxanthin and neoxanthin, were extracted from these tropical fruits, separated by MgO: Hyflosupercel on CC developed with ethyl ether and acetone gradient in petroleum ether, identified by chemical tests and quantified by visible spectroscopy. The chromatographic behavior of carotenoids bears a definite relationship with their structure.

Although these data cannot be used as the sole criteria for identifying carotenoids, they serve as useful complementary information (Rodriguez-Amaya, 1999). Carotenoids of *Lycium barbarum* L. fruits, a traditional Chinese herb that possesses vital biological properties, were isolated by a column containing MgO: diatomaceous earth. The β-carotene was eluted with n-hexane, β-cryptoxanthin and neoxanthin with ethyl acetate and zeaxanthin with ethyl acetate-ethanol (80:20 v/v). The zeaxanthin fraction was most effective in scavenging hydroxyl-free radicals (Wang, 2010).

Other basic materials, especially $Ca(OH_2)$, $ZnCO_3$ and $CaCO_3$, are particularly useful for separating geometrical isomers (Bernhard, 1995). Five lutein derivatives (four isomers and two epoxides) were isolated and separated from inflorescences of *Solidago canadensis* L. and from flowers of *Chelidonium majus* L., using $CaCO_3$ CC developed with different compositions of toluene-hexane and acetone-hexane. They were identified as lutein-5,6-epoxide, (9Z)-lutein-5,6 epoxide, (9Z,9′Z)-lutein (neolutein C), (9Z)- and/or (9′Z)-lutein (neolutein B) and (13Z)- and/or (13′Z)-lutein (neolutein A), in addition to (all-E)-lutein, violaxanthin, (9Z)-violaxanthin, flavoxanthin and/or chrysanthemaxanthin. Xanthophylls, such as 5,6-diepikarpoxanthin, mutatoxanthin, anteraxanthin, zeaxanthin, β-cryptoxanthin and apo-carotenoids, such as capsorubin, capsochrome, capsanthin 5,6-epoxide and capsanthone, were also isolated from *Asparagus falcatus* using $CaCO_3$ as adsorbent on CC (Deli et al., 2000). The isomers of β-carotene present in acetone extracts from Malpighia glabra pulp (Agostini-Costa et al., 2003) and from *Annona coriaceae* fruit (Agostini et al., 1996) were separated on $Ca(OH)_2$ CC eluted with 2 percent ethyl ether in petroleum ether.

Separations on silica and Al_2O_3 are mainly determined by carotenoid polarity. The hydrocarbon carotenes are weakly adsorbed, whereas xanthophylls, containing polar substituents, especially hydroxyl groups, are adsorbed more strongly (Bernhard, 1995). A novel carotenoid (xanthophyll) derivative, lutein-3-acetate, extracted from senescing leaves of rice, was isolated and purified using silica gel TLC (Kusaba, 2009). Phytofluene, β-carotene, ζ-carotene, lycopene, zeinoxanthin, cryptoxanthin and lutein were separated from ripe and partly ripe *Momordica charantia* seeds and tomatoes through Al_2O_3 and MgO CC (Rodriguez et al., 1975). The main carotenoid (xanthophyll) in the orange muscle of Yesso scallop was isolated and purified by acetone extraction and silica gel CC. Its structure was identified as that of pectenolone (3,3′-dihydroxy-β,β-caroten-4-one) (Li et al., 2010). Carotenoid extracts of eight species of lichens were separated by Al_2O_3 column and divided into fractions on silica gel TLC and analyzed by ion-pairing in reverse-phase HPLC.

Fourteen carotenoids, such as β-carotene, β-cryptoxanthin, lutein, zeaxanthin, canthaxanthin, astaxanthin, violaxanthin and neoxanthin were separated and identified (Czeczuga et al., 2010). Stereoisomers and epoxy/carbonyl derivatives of β-carotene were separated through Al_2O_3 CC developed with 10-60% diethyl ether in n-hexane (Marty and Berset, 1990).

CC on silica gel 60 was also used for partial separation of carotenoids and DEAE-Toyopearl 650 M to remove B Chlorophyll and polar lipids during carotenoid isolation from a thermophilic filamentous photosynthetic bacterium *Roseiflexus castenholzii*. Purified fractions provided keto-myxocoxanthin and methoxy-keto-mycoxanthin, besides ketomyxocoxanthin glucoside, keto-myxocoxanthin glucoside ester, myxocoxanthin, myxocoxanthin glucoside ester and others (Takaichi et al., 2001). The major polar carotenoids, myxoxanthophyll and ketomyxoxanthophyll, produced by two cyanobacterias (*Anabaena* sp. and *Nostoc punctiforme*), were also isolated and purified first on a column of silica gel 60. The β-carotene was eluted with hexane, non-polar carotenoids were eluted with acetone/hexane (2:8) and polar carotenoids were eluted with acetone methanol (9:1, v/v).

The polar carotenoids (last fraction) were then loaded on a column of DEAE-Toyopearl 650M and the carotenoids were eluted with acetone/hexane (1:1 v/v) and finally purified on silica gel TLC developed with dichloromethane/ethyl acetate/acetone/methanol (2:4:2:1, v/v) to give myxoxanthophyll and ketomyxoxanthophyll (Takaichi et al., 2005).

3. PHENOLIC COMPOUNDS

Phenolic compounds are widely distributed in nature. Their chemical structures may vary greatly, including simple phenols (C6), such as hydrobenzoic acid derivatives and catechols, as well as long chain polymers with high molecular weight, such as catechol melanins (C6)6, lignins (C6-C3)n and condensed tannins (C6-C3-C6)n. Stilbenes (C6-C2-C6) and flavonoids (C6-C3-C6) are phenolic compounds with intermediate molecular weight that present many pharmacological and biological activities. Flavonoids, including anthocyanins, flavonols (such as quercetin and myricetin), isoflavones (such as daidzein and genistein) and others are formed by multiple biosynthetic branches that originate from chalcone. Phenolic compounds have been widely fractionated in medicinal, aromatic and food plants using CC. Repeated silica gel, sephadex-LH20, RP-18, RP-8, MCI-gel, diaion and toyopearl chromatography columns have been used to fractionate simple phenolics, flavonoids and tannins from kernels and nuts (Zhang et al., 2009; Karamac, 2009), fruits such as apples, *Morus nigra*, *Punica granatum* (Lee et al., 2010; Pawlowska et al., 2008); olive oil (Khanal et al., 2011), tea (Gao et al., 2010; Liu et al., 2009), seeds such as lentils (Amarowicz and Karamac, 2003), medicinal species, including *Ulmus davidiana* and *Tridax procumbens* (Jung et al., 2008; Agrawal, 2011); and aromatic plants, including mint and sage (She et al., 2010; Wang et al., 1998).

Separations on silica are mainly determined by polarity, where phenolic compounds containing more hydroxyl groups are adsorbed more strongly. Separations on Sephadex LH-20, a crosslinked dextran-based resin for gel permeation, are mainly determined by molecular sizing of the phenolic compound, outside of adsorption and partition mode.

Phenolic acids (such as ferulic acid and gallic acid), flavonoids (such as flavonol, catechins and anthocyanidins derivatives) and procyanidins from fruits of wild black berry *Aristotelia chilensis* were obtained using flash and open CC on silica gel and Sephadex LH-20 eluted with hexane, hexane-ethyl acetate (1:1); ethyl acetate-methanol (1:1) and methanol (Cespedes et al., 2010).

Pyrogallol and phenolic acids were purified from defatted extract of *Juglans regia* kernels by repeated CC on Sephadex LH-20 eluted with methanol, followed by silica gel CC developed with chloroform–acetone 40:1, v/v. Ethyl gallate was obtained by recrystallization from methanol. Protocatechuic acid was obtained after CC on Sephadex LH-20 eluted with CH_2Cl_2–EtOH, 1:1, v/v and further purification with silica gel CC developed with chloroform–methanol, 20:1. Gallic acid and 3,4,8,9,10-pentahydroxydibenzo[b,d]pyran-6-one were separated by silica gel CC and eluted with a mixture of chloroform-methanol (Zhang, 2009). High-molecular-weight tannins of defatted walnut, hazelnut and almond kernels were also isolated by CC on Sephadex LH-20 gel eluted with 50 percent (v/v) acetone (Karamac, 2009). The ethanol fraction of *Dipteryx lacunifera* kernels was found to exhibit high radical scavenging activity and was subjected to further fractionation.

CC over silica gel with gradient from chloroform to methanol and Sephadex LH-20 eluted with methanol afforded (-)-eriodictyol, (-)-butin, luteolin, 3′,4′,7-trihydroxyflavone, butein and sulfuretin (Junior et al., 2008).

The total polyphenols from red wine isolated during vinification and storage were isolated by CC and their contribution to wine sensory properties and antioxidant activity was evaluated. Wine samples were evaporated and loaded onto an open CC packed with LiChroprep RP-18; the column was washed with distilled water followed by methanol to recover total polyphenols extract, further separated by HPLC (Sun et al., 2011).

The alkyl esters of protocatechuic acid were synthesized and the crude product was then purified by CC using petroleum ether/ethyl ether (7:3 to 5:5) as eluent. The increase in the length of ester alkyl chain attached to the catecholic ring had influenced the stabilization of the radicals formed in the oxidation process. Alkyl protocatechuate compounds demonstrated a higher radical-scavenging activity than the natural antioxidant protocatechuic acid. Moreover, the introduction of alkyl groups in the carboxylic acid led to a significant increase in lipophilicity influenced by the antioxidant activity of protocatechuic acid derivatives (Reis et al., 2010).

The virgin olive oil phenol was partitioned successively with hexane and ethyl acetate and fractionated by CC on silica gel and Sephadex LH-20 to give a dialdehydic form of decarboxymethyl ligstroside aglycone (p-HPEA-EDA), a phenolic compound that activates AMP-activated protein kinase to inhibit carcinogenesis (Khanal et al., 2011). Phenols and tocopherols were also removed from olive oil by CC over Al_2O_3 during to evaluate changes in the phenolic composition of virgin oil during frying. The concentration of hydroxytyrosol (3,4- DHPEA) and its secoiridoid derivatives (3,4-DHPEA-EDA and 3,4-DHPEA-EA) in virgin olive oil decreased rapidly when the oil was repeatedly used for preparing fried food. However, tyrosol (p-HPEA) and its derivatives (p-HPEA-EDA and p-HPEA-EA) in the oil were much more stable during frying operations (Gomez-Alonso et al., 2003).

Phytoalexins are secondary metabolites that plants synthesize for self-defense, and they have shown great promise in chronic disease prevention. The best known example is resveratrol, an induced phytoalexin found in yeast-infected grape skin (Wu et al., 2011). Peanuts also contain several active components including flavonoids, phenolic acids, phytosterols, alkaloids and stilbenes. The latter are characterized by a 1,2-diphenylethylene backbone usually derived from the basic unit of trans-resveratrol (3,5,4′-trihydroxy-stilbene) (Lopes et al., 2011). Peanut seeds, when affected by injuries, pathogenic infections, fungal contamination, insect damage and UV light, could produce phytoalexins such as stilbenoid derivatives (resveratrol, arachidins, 3′-isopentadienyl-3,5,4′-trihydroxystilbene, SB-1, chiricanine A, and arahypins), and pterocarpanoid derivatives (e.g., aracarpene-1 and aracarpene-2). Silica gel column eluted with hexane and hexane ethyl acetate (7:3) was used for fractionation the phytoalexin extract from peanut before analysis by HPLC (Wu et al., 2011). Neutral Al_2O_3-silica gel 60 C18 (1:1 w/w) eluted with 80 percent ethanol was used as a clean-up column to separate interfering co-eluting with resveratrol from peanut extracts (Potrebko and Resurreccion, 2009).

A highly efficient column chromatographic extraction of curcumin from *Curcuma longa* was proposed by Zhan et al., (2011). Curcumin was extracted with minimum use of solvent, minimum volume and high concentration of extraction solution by CC. Turmeric material was loaded into a column with 2-fold of 80 percent ethanol. After dissolving target compounds, the column was eluted with 80 percent ethanol. For non-cyclic CCE procedure, 8-fold of eluent was collected, while for cyclic CCE procedure, only the first 2-fold of eluent was collected as extraction solution. A more than 99 percent extraction rate for curcumin was obtained in both procedures, compared to a 59 percent extraction rate through the ultrasonic-assisted extraction with 10-fold of 80 percent ethanol.

A lipid- and essential oil-free infusion of *Cymbopogon citratus* leaves was fractionated on Lichroprep RP-18 column eluted with water and with aqueous methanol solutions. Dry residue was recovered in 50 percent aqueous ethanol and was fractionated by gel chromatography on a Sephadex LH-20 using ethanol as mobile phase. All the fractionation process provided three major fractions: a tannin rich fraction; a flavonoid rich fraction and two phenolic acid rich fractions. Tannin and phenolic acid were the fractions responsible for the anti-inflammatory effect through inhibition of transcription factor NF-kB, inducible nitric oxide synthase expression and nitric oxide production. These fractions probably had a synergistic effect (Francisco et al., 2011).

Oligomeric procyanidins in apples were extracted with boiled water and purified on an ADS-17 macroporous resin column to obtain a procyanidin extract. The extract was fractionated according to its degree of polymerization on a Toyopearl TSK HW- 40 column eluted with methanol to give procyanidins B2 (epicatechin-(4β-8)-epicatechin) and C1 (epicatechin-(4β-8)-epicatechin-(4β-8)-epicatechin). This method was suitable for preparing of procyanidin oligomers (from dimers through tetramers in one run) for laboratory research, and is potentially applicable to large-scale production in industry (Xiao et al., 2008).

Thea (Theaceae) is traditionally used for producing tea, one of the most popular beverages consumed in the world due to its polyphenol-rich content, which are reported to have various bioactivities, such as antioxidative, antimicrobial, antitumor and antimutagenesis (Liu et al., 2009). *Camellia sinensis* and some other species from the genus *Camellia* have also been used for making tea and are consumed widely (Gao et al., 2010). More than 96 phenolic compounds were identified in *C. sinensis* tea. The hydrolyzable tannin epigallocatechin gallate was the major phenolic component of green tea and partially fermented teas, while fully fermented black teas had traces of epigallocatechin gallate, but contained theaflavins.

Glycosylated flavonoids, catechins, proanthocyanidins (condensed tannins) and phenolic acid derivatives were found too (Lin et al., 2008). Aqueous acetonic extract of green tea (*C. crassicolumna*) further partitioned with ethyl acetate and purified by CC led to the identification of various fractions obtained by CC over Sephadex LH-20 eluted with ethanol.

Further repeated CC on MCI-gel CHP20P, Sephadex LH-20, and Toyopearl HW-40F, eluted with methanol/H_2O (0:1–1:0) gave five flavan-3-ols, five flavonol glycosides, three hydrolyzable tannins, two chlorogenic acid derivatives and three simple phenolic compounds (Liu et al., 2009). Phenolic compounds were also isolated from the leaves of *C. pachysandra.* The ethyl acetate and aqueous fractions were separately subjected to repeated CC over Diaion HP-20SS, Sephadex LH-20, MCI-gel CHP20P, and Toyopearl HW-40F to give 22 phenolic compounds, including nine hydrolyzable tannins, 11 flavonol glycosides and two simple phenols, without caffein or catechin (Gao et al., 2010). A new phenolic compound was isolated and purified from butanol fraction of Chinese olive (*Canarium album* L.) fruit through AB-8 adsorption resin CC washed with water to remove impurities and eluted with 90 percent (v/v) aqueous ethanol to get the phenolic eluents.

The phenolics were further separated on a polyamide CC eluted with aqueous ethanol to give several fractions, the ethanol concentration being increased from 0 to 100 percent in increments of 20 percent. Fraction obtained from 20 percent aqueous ethanol was further purified on TSK Toyopearl HW-40 (S) CC developed with aqueous ethanol 0 to 20% to get the purified new compound established as 3-o-galloyl quinic acid butyl ester (He et al., 2009). Polyamide CC eluted with water and 50 percent, 70 percent and 100 percent aqueous methanol further facilitated GC-MS and HPLC separation of phenolic compounds in *Euphrasia rostkoviana* (Blazics et al., 2008). *Lycium barbarum* L., a traditional Chinese herb, possesses vital biological properties, such as prevention of cancer and age related macular degeneration. Flavonoids and phenolic acids extracted from fruits of this plant were separated using a Cosmosil 140 C18 OPN column, with phenolic acids being eluted with deionized water and neutral flavonoids with methanol. The flavonoid fraction showed the most pronounced effect in scavenging free radicals, chelating metal ions and reducing power (Wang et al., 2010). Eight polyphenolic acids were isolated from the aerial part of *Mentha haplocalix* extracted with aqueous acetone 70 percent at room temperature. Repeated CC on silica gel (developed with chloroform–methanol–H_2O, 9:1:0.1 to 7:3:0.5), Sephadex LH-20 and MCI-gel CHP20P and ODS-A, eluted with H_2O–Methanol (1:0 to 0:1) afforded rosmarinic acid, cis-salvianolic acid, lithospermic acid, propanoic acid, sodium lithospermate B, magnesium lithospermate B, and lithospermic acid B. Lithospermic acid B, sodium lithospermate B and magnesium lithospermate B displayed stronger activities than the other compounds (She et al., 2010). Ten phenolic compounds were also isolated from a butanol fraction of sage extracts, using repeated CC on silica gel, Lichroprep RP-18 and Sephadex LH20. Among them, the most active antioxidants were found to be rosmarinic acid and luteolin-7-*O*-*β*- glucopyranoside (Wang et al., 1998).

4. ALKALOIDS

Alkaloids are defined as basic compounds synthesized by living organisms containing one or more heterocyclic nitrogen atoms, derived from amino acids (with some exceptions) and pharmacologically active. The class name is directly related to the fact that nearly all alkaloids are basic (alkaline) compounds. Alkaloids constitute a large group of secondary metabolites, with more than 12,000 substances isolated. A huge variety of structural formulas, coming from different biosynthetic pathways and presenting highly diverse pharmacological activities are characteristic of the group (Brielmann et al., 2006). Archeological evidence has demonstrated the use of alkaloids (plant parts or extracts) since 4000 B.C., and they are important even today (Roberts and Wink, 1998). Poppy (*Papaver somniferum*) and opium have been known and used since antiquity by Sumerians, Arabs, Persians, Egyptians and Greeks. Morphine, obtained from the poppy latex, was the first crude drug isolated. It was named after Morpheus, the god of dreams, one of the sons of Hypnos, the god of sleep, in Greek mythology (Wink, 1998). Morphine is legally used nowadays as an analgesic for severe pain.

4.1 Alkaloids from Plants

CC is an important and much used technique in alkaloid isolation and analysis. Usually, after CC, the sample is re-chromatographed either by preparative TLC or HPLC, in order to obtain a pure sample for spectroscopic studies.

Phytochemical investigation in plants from the genus *Kopsia* (Apocynaceae) led to the discovery of different classes of alkaloids, presenting antileishmanial, antimitotic and antitumor activities. From *Kopsia hainanensis*, a native medicinal plant from Hainan, China, two new pentacyclic indole alkaloids have been isolated (Chen et al., 2011). The acidic partition of the methanol extract was neutralized and extracted with chloroform. Alkaloids, as basic substances, can be protonated in acidic media and therefore remain in the aqueous phase. This strategy has been useful for the isolation of this class of compounds. The pH of the aqueous phase is further adjusted and alkaloids are extracted with a solvent. The concentrated extract was repeatedly submitted to CC over silica gel and eluted with a gradient of chloroform-methanol, alkalinized with a small quantity of triethylamine to give kopsahainanine B. One fraction was re-chromatographed over reversed-phase silica gel and Sephadex LH-20, using methanol-water and pure methanol as eluents, respectively, to give kopsahainanine A. The butanolic fraction of *Lobelia chinensis* Lour was chromatographed on reversed phase Diaion HP 20 gel, eluted with water-methanol into six fractions. Fraction 6 was rechromatographed on Sephadex LH-20 followed by preparative silica gel TLC and 7.3 mg of lobachine, a new alkaloid, were isolated (Kuo et al., 2011).

From *Campylospermum flavum*, a medicinal plant from Cameroon, flavonoids, flavones, chalcones and alkaloids have been isolated. Among them, a new indole alkaloid, flavumindole, was obtained from the methanol extract after partition with ethyl acetate-water and purification on silica gel, Sephadex LH-20 and RP18 silica gel, eluted with dichloromethane-methanol mixtures, pure methanol and methanol-water mixtures, respectively (Ndongo et al., 2011). The isolated alkaloid presented a cytotoxicity of 90 percent according to the brine shrimp (*Artemia salina*) test.

Nine indole alkaloids were isolated from the roots, bark and leaves of *Tabernaemontana salzmannii* (Apocynaceae). The alkaloids were fully identified by spectroscopic methods and mass spectra by comparison with literature data (Figueiredo et al., 2010). The dichloromethane extract was repeatedly chromatographed over silica gel eluted with dichloromethane-methanol mixtures. Coronaridine, 3-oxo-coronaridine, (19*S*)-heyneanine, voacangine, isovoacangine, hydroxyisovoacangine, isovoachristine and olivacine were isolated. For voachalotine, a further chromatographic step was necessary on preparative TLC (silica gel) eluted with dichloromethane-methanol (99:1). After an *in vitro* screening, it was observed that isovoacangine and voacangine alkaloids were able to induce apoptosis cell death in human leukemic cells, line THP-1.

4.2 Alkaloids from Fungi

Probably the most famous (and infamous) alkaloid producing fungi are the ergots (family Hypocreaceae). Ergotism (holy fire) has killed thousands due to contaminated rye, and even in the last century many cases were reported (Wink, 1998). However, there is a plethora of alkaloid producing-fungi, with different structures and biological activities. Three new alkaloids, lyconadin D, lyconadin E and complanadine E were isolated from *Lycopodium complanatum* (Lycopodiaceae) by Ishiuchi et al. (2011). The acidified methanolic extract was partitioned with ethyl acetate. The pH of the aqueous phase was adjusted and extracted with chloroform. After solvent evaporation, the chloroform-soluble materials were chromatographed on an amino silica gel column and eluted with hexane-ethyl acetate mixtures, then with a chloroform-methanol-water system. The fractions eluting with pure chloroform were re-chromatographed over silica gel, and lyconadin D and lyconadin E were isolated. Part of the chloroform soluble fraction was eluted with chloroform-methanol over a Sephadex LH-20 column. The fractions were further purified by HPLC (C18 column) to yield complanadine E, a dimeric unsymmetrical alkaloid.

The club mosses of the genus *Lycopodium* in particular are prone to produce complex polycyclic alkaloids, with important biological activities, such as inhibition of acetylcholine esterase. From three different species of *Lycopodium*, ten new alkaloids were isolated and characterized (Katakawa et al., 2011). Dihydrolycopoclavamine was isolated from the extract of *L. serratum* after chromatography on amino silica gel and elution with hexane-chloroform. For *L. clavatum*, after neutralization of the acidic fraction from the methanolic extract, it was submitted to flash chromatography over silica gel and eluted with a chloroform-methanol gradient, then chloroform-methanol-aqueous ammonium hydroxide and pure methanol. One fraction was re-chromatographed on amino silica gel with hexane chloroform mixtures leading to licopoclavamine A. Another fraction was rechromatographed on silica gel to give lycopoclavamine B. A similar procedure was applied to *L. squarrosum*, from which lycoposquarrosamine A, acetyllycoposerramine-U, 8-β-hydroxyfawcettimine, 8-β-hydroxyfawcettimine, 8-β-acetoxyfawcettimine, 8-β-acetoxyfawcettimine and lycoflexine *N*-oxide were isolated. It is quite remarkable that all these compounds were separated solely by means of CC using silica gel, amino silica gel or Al_2O_3, and solvent mixtures of hexane, chloroform, methanol and ethyl acetate.

4.3 Marine Alkaloids

The marine environment has emerged as a promising source of new substances with medicinal application. Considering only sponges, for example, more than 15,000 compounds have been isolated and characterized, including terpenoids, nucleosides, alkaloids and steroids. Guanidine alkaloids were isolated from *Monanchora arbuscula* after CC over Sephadex LH-20 with methanol then repeatedly over silica gel and eluted with chloroform-ethyl acetatemethanol mixtures. The compounds were identified by 1H and 13C NMR, IR and MS (Ferreira et al., 2011).

Two new brominated arginine-derived alkaloids were isolated from the Red Sea sponge *Suberea mollis* (Aplysinellidae), namely, subereamine A and subereamine B.

Halogenated compounds are common in marine natural products. The subereamines were isolated after CC on Sephadex LH-20 of the ethyl acetate extract, followed by chromatography on silica gel with dichloromethane-methanol mixtures and preparative HPLC on C18 column (Shaala et al., 2011).

The first natural products containing a 1H-oxazolo [40,50:4,5] benzo[1,2,3-de] [1,6] naphthyridine ring system were isolated from a *Suberites* sp. sponge collected in Okinawa, Japan. Named respectively, nakijinamines C and E, they constitute a group of heteroaromatic alkaloids, hybrids of aaptamine-type and bromoindole alkaloids (Takahashi et al., 2011). Other 23 aaptamine-type alkaloids have been isolated so far, but none contains the hybrid ring system. Another new alkaloid was isolated, namely, nakijinamine D, which does not contain the oxazolo ring. The sponges were extracted with methanol and the residue partitioned with ethyl ether and water. The aqueous phase was extracted with butanol and subjected to reversed phase CC (C18) followed by repeated HPLC also on a C18 column. The alkaloids were isolated initially as racemates. Antifungal activity was recorded for nakijinamines C and E against *Aspergillus niger*.

Marine sponges of the genus *Asteropus* have been reported to contain a variety of compounds, such as saponines, sterols and pteridine derivatives. The concentrated methanolic extract was partitioned between water and dichloromethane. The water phase showed toxicity to brine shrimp larvae and was partitioned with butanol. The butanolic extract was submitted to medium pressure liquid chromatography over ODS-A with a methanol gradient. A fraction selected according to the toxicity test result was chromatographed over Sephadex LH-20. The sub-fractions were purified by reversed phase HPLC. The pyroglutamyl alkaloids were new compounds, and therefore fully characterized by spectroscopic methods (Li et al., 2011).

Carboline alkaloids were also found in tunicates of the genus Eudistoma. After purification on Sephadex LH-20, the methanolic extract of Eudistoma glaucus was chromatographed on a silica gel column with hexane-ethyl acetate mixtures, and then on another column with amino silica gel, also eluted with the same solvents. Normal phase HPLC led to the isolation of eudistomidin H and eudistomidin I. Eudistomidins H and I were new compounds containing a unique fused-tetracyclic ring system consisting of a tetrahydro β-carboline ring and a hexahydropyrimidine ring. Using the same stationary phases, but changing the solvent system to chloroform-methanol, eudistomidin K was obtained. For eudistomidin J, the last solvent system was also applied, but without the need for a HPLC separation step (Suzuki et al., 2011).

Pyridoacridine alkaloids are responsible for the bright colors observed in sponges and ascidians. Two new pyridoacridine alkaloids, namely, 13-didemethylaminocy closhermilamine D and dimethyl-deoxyamphimedine, were isolated from the purple chromotype of the Western Mediterranean morph *Cystodytes dellechiajei* together with other six known alkaloids of the same kind (Bry et al., 2011). The ascidians were extracted with methanol-dichloromethane-trifluoroacetic acid mixture followed by reverse phase CC. The purple fraction obtained was chromatographed three successive times on reverse phase columns eluted with a methanol-water gradient. After reverse phase semi-preparative HPLC, the two new alkaloids were isolated.

In this way, the phytochemical investigation of alkaloids has been deeply dependent on CC techniques. Careful and patient repeated CC has led to the isolation of some structurally varied and fascinating compounds, even without the use of modern HPLC methods. To be more efficient and to enhance productivity, of course, combined approaches are the common practice.

It is worth emphasizing the great opportunity that marine organisms offer to researchers looking for biologically active compounds and synthetic models. The real treasures at the bottom of the sea may not come from shipwrecks, but are those which have been lying (or, better, living) there since long before human beings set foot on the earth.

5. CONCLUSION

Plants have evolved multiple defense mechanisms against microbial pathogens and various types of environmental stress. Plants synthesize a number of antimicrobial secondary metabolite, some of which are performed and some of which are induced after infection. These metabolites synthesized using various pathways, but ultimate role is to protect the plant from stress conditions. One of the important innovative ideas is to investigate the correct correlation between N and S fertilizers in the resistance management of cultivated crops. In a number of previous research articles and review papers, it has been shown that the N and S containing secondary metabolites are influenced by optimum supply of N and S and their good nutrition can enhance the capability of a plant to cope with biotic and abiotic stress. The identification of the mechanisms SAR will be an important milestone for sustainable agricultural production, as the use of functicides could then be minimized or eliminated. Thus, SAR may become an important strategy for efficiently combating the pathogens in cultivated crops and improve the agricultural production in future. Even in laboratories with good HPLC and GC-MS facilities, the traditional and classical methods of TLC and CC are still widely used, for rapid preliminary screening of extracts, for isolating and purifying bioactive compounds of secondary metabolism for further study, for comparison of samples with standards and for monitoring chemical synthesis or the course of reactions.

References

Abe I (2007). Enzymatic synthesis of cyclic triterpenes. Nat Prod Rep 24: 1311–1331.

Agostini TS, Cecchi HM, and Godoy HT (1996). Composição de carotenóides no marolo eem produtos de preparo caseiro. Ciência e Tecnologia de Alimentos 16: 67–71.

Agostini-Costa TS, Abreu LN, and Rossetti AG (2003). Efeito do congelamento e do tempo de estocagem da polpa de acerola sobre o teor de carotenóides. Revista Brasileira de Fruticultura 25: 56–58.

Agostini-Costa TS, Wondracek DC, Lopes RM, Vieira RF, and Ferreira FR (2010). Composição de carotenoides em canistel (*pouteria campechiana* (kunth) baehni). Revista Brasileira de Fruticultura 32: 903–906.

Agrawal SS, and Talele GS (2011). Bioactivity guided isolation and characterization of the phytoconstituents from the *Tridax procumbens*. Braz J Pharmacog 21: 58–62.

Ajit-Simh KA, Bemberis I, Brown S, De-Mill B, Deorkar N, Grund E, Hays E, Levine HL, Levy R, Mazzeo J, Natarajan V, Nevill J, Noyes A, O'Neil P, Seely R, Sofer G, and Wheelwright SM (2003). Column packing for process-scale chromatography: guidelines for reproductibility. Biopharmaceutical International 23–30.

Amarowicz R, and Karamac M (2003). Antioxidant activity of phenolic fractions of lentil *(Lens culznaris)*. Journal of Food Lipids 10:1–10.

Areche C, Cejas P, Thomas P, San-Martin A, Astudillo L, Gutierrez M, and Loyola LA (2009). Triterpenoids from *Azorella trifurcata* (Gaertn.) Pers and their effect against the enzyme acetylcholinesterase. Quimica Nova 32:2023–2025.

Augustin JM, Kuzina V, Andersen SB, and Bak S (2011). Molecular activities, biosynthesis and evolution of triterpenoid saponins. Phytochemistry 72:435–457.

Batista R, García PA, Castro MA, Del Corral JMM, Feliciano AS, and Oliveira AB (2010). Iso-Kaurenoic acid from *Wedelia paludosa* DC. Anais da Academia Brasileira de Ciências, 82: 823–831.

Begnami AF, Duarte MCT, Furletti V, and Rehder VLG (2010). Antimicrobial potential of *Coriandrum sativum* L. against different *Candida* species *in vitro*. Food Chem 118: 74–77.

Bernhard K (1995). Chromatography: Part II Column Chromatography. In: Britton G, Liaaen-Jensen, Pfander H (eds) Carotenoids isolation and analysis. Basel: Birkhauser Verlag, pp. 117–130.

Blazics B, Ludanyi K, Szarka S, and Kery A (2008). Investigation of *Euphrasia rostkoviana* Hayne using GC-MS and LC-MS. Chromatographia Supplement 68: S119–S124.

Brielmann HR, Setzer WN, Kaufman, PB, Kirakosyan A, and Cseke LJ (2006). Phytochemicals: The chemical components of plants. In: Cseke LJ, Kirakosyan A, Kaufman PB, Warber SL, Dule JA, Brielmann HR (eds) Natural Products from Plants (2nd edn.), CRC Press, Boca Raton. Pp.1–50.

Britton G (1995). Structure and properties of carotenoids in relation to function. The FASEB Journal 9: 1551–1558.

Bry D, Banaigs B, Long C, and Bontemps N (2011). New pyridoacridine alkaloids from the purple morph of the ascidian *Cystodytes dellechiajei*. Tetrahedron Lett 52: 3041–3044.

Cespedes CL, Valdez-Morales MV, Avila JG, El-Hafidi M, Alarcon J, and Paredes-Lopez O (2010). Phytochemical profile and the antioxidant activity of Chilean wild black-berry fruits, *Aristotelia chilensis* (Mol) Stuntz (Elaeocarpaceae). Food Chem 119: 886–895.

Chen J, Chen JJ, Yao X, and Gao K (2011). Kopsihainanines A and B, two unusual alkaloids from *Kopsia hainanensis*. Org Biomol Chem 9: 5334–5336.

Coronel AC, Cerda-García-Rojas CM, Joseph-Nathan P, and Catalán CAN (2006). Chemical composition, seasonal variation and a new sesquiterpene alcohol from the essential oil of *Lippia integrifolia*. Flavour Frag J 21: 839–847.

Czeczuga B, Czeczuga-Semeniuk, E, and Semeniuk A (2010). Chromatic adaptation in lichen phyco- and photobionts. Biologia 65:587–594.

Deli J, Molnfir E, Osz E, and Tdth G (2000). Analysis of carotenoids in the fruits *of Asparagus falcatus:* isolation of 5,6-diepikarpoxanthin. Chromatographia Supplement 51:S-183-S-187.

Djabou N, Paolini J, Desjobert JM, Allali H, Baldovini N, Costa J, and Muselli A (2010). Qualitative and quantitative analysis of volatile components of *Teucrium massiliense* L.- identification of 6-methyl-3-heptyl acetate as a new natural product. Flavour Frag J 25:475–487.

Dudareva N, Pichersky E, and Gershenzon J (2004). Biochemistry of plant volatiles. Plant Physiol. 135:1893–1902.

Duraipandiyan V, Indwar F, and Ignacimuthu S (2010). Antimicrobial activity of confertifolin from *Polygonum hydropiper*. Pharmaceutical Biology 48:187–190.

Ebel R (2010). Terpenes from marine-derived fungi. Marine Drugs 8: 2340–2368.

Ferreira EG, Wilke DV, Jimenez PC, Oliveira JR, Pessoa OD, Silveira ER, Viana FA, Pessoa C, Moraes MO, Hadju E, and Costa-Lotufo LV (2011). Guanidine alkaloids from *Monanchora arbuscula*: chemistry and antitumor potential. Chemistry and Biodiversity 8:1433–1445.

Figueiredo ER, Vieira IJC, Souza J, Braz-Filho R, Mathias L, Kanashiro MM, and Côrtes FH (2010). Isolamento, identificação e avaliação da atividade antileucêmica de alcaloides indólicos monoterpênicos de *Tabernaemontana salzmannii* (A. DC.), Apocynaceae. Braz J Pharmacog 20:675–681.

Francisco V, Figueirinha A, Neves BM, Rodríguez CG, Lopes MC, Cruz MT, and Batista MT (2011). *Cymbopogon citratus* as source of new and safe anti-inflammatory drugs: Bio-guided assay using lipopolysaccharide-stimulated macrophages. J Ethnopharmacol 133: 818–827.

Gao DF, Xu M, Yang CR, Xu M, and Zhang YJ (2010). Phenolic Antioxidants from the leaves of *Camellia pachyandra* Hu. J Agric Food Chem 58: 8820–8824.

Gauvin-Bialecki A, Aknin M, and Smadja J (2008). 24-O-Ethylmanoalide, a manoaliderelated sesterterpene from the marine sponge *Luffariella* cf. variabilis. Molecules 13:3184–3191.

Gomez-Alonso S, Giuseppe Fregapane G, Salvador MD, and Gordon MH (2003). Changes in phenolic composition and antioxidant activity of virgin olive oil during frying. J Agric Food Chem 51:667–672.

He Z, Xia W, Liu Q, and Chen J (2009). Identification of a new phenolic compound from Chinese olive (*Canarium album* L.) fruit. Eur Food Res Tech 228:339–343.

Iriti M, and Faoro F (2004). Plant defense and human nutrition: Phenylpropanoids on the menu. Curr Topics Nutr Res. 2: 47–65.

Ishiuchi K, Kubota T, Ishiyama H, Hayashi S, Shibata T, Mori K, Obara Y, Nakanhata N, and Kobayashi J (2011). Lyconadins D and E, and complanadine E, new Lycopodium alkaloids from *Lycopodium complanatum*. Bioorg Med Chem 19:749–753.

Joulain D, and König W (1998). The Atlas of Spectral Data of Sesquiterpenes Hydrocarbons (1st ed.), EB-Verlag, Hamburg.

Jung MJ, Heo S, and Wang MH (2008). Food Chemistry. 108: 482–487.

Junior GMV, Sousa CMM, Cavalheiro AJ, Lago JHG, and Chaves MH (2008). Phenolic Derivatives from fruits of *Dipteryx lacunifera* Ducke and evaluation of their antiradical activities. Helvetica Chimica Acta 91: 2159–2167.

Karamac M (2009). *In vitro* study on the efficacy of tannin fractions of edible nuts as antioxidants. Eur J Lipid Sci Tech 111:1063–1071.

Katakawa K, Mito H, Kogure N, Kitajima M, Wongseripipatana S, Arisawa M, and Takayama H (2011). Ten new fawcettimine-related alkaloids from three species of Lycopodium. Tetrahedron 67: 6561–6567.

Kernan MR, and Faulkner DJ (1988). Sesterterpene sulfates from a sponge of the family Halichondriidae. The Journal of Organic Chemistry 53: 4574–4578.

Khanal P, Oh WK, Yun HJ, Namgoong GM, Ahn SG, Kwon SM, Choi HK, and Choi HS (2011). *p*-HPEA-EDA, a phenolic compound of virgin olive oil, activates AMP-activated protein kinase to inhibit carcinogenesis. Carcinogenesis 32: 545–553.

Kimura M, and Rodriguez-Amaya DB (2002). A scheme for obtaining standards and HPLC quantification of leafy vegetable carotenoids. Food Chem 78: 389–398.

Kuo PC, Hwang TL, Lin YT, Kuo YC, and Leu YL (2011). Chemical constituents from *Lobelia chinensis* and their anti-virus and anti-inflammatory bioactivities. Arch Pharm Res 34: 715–722.

Kusaba M, Maoka T, Morita R, and Takaichi S (2009). A novel carotenoid derivative, lutein 3-acetate, accumulates in senescent leaves of rice. Plant Cell Physiol 50: 1573–1577.

Lanças FM (2008). Avanços recentes e tendências futuras das técnicas de separação: uma visão pessoal. Scientia Chromatographica 0:17–44.

Lanfranchi DA, Laouer H, Kolli ME, Prado S, Maulay-Bailly C, and Baldovini N (2010). Bioactive phenylpropanoids from *Daucus crinitus* Desf. from Algeria. J Agric Food Chem 58: 2174–2179.

Lee CJ, Chen LG, Liang W, and Wang ACC (2010). Anti-inflammatory effects of *Punica granatum* Linne *in vitro* and *in vivo*. Food Chem 118:315–322.

Li H, Dant HT, Li J, Sim CJ, Hong J, Kim DK, and Jung JH (2011). Pyroglutamyl dipeptides and tetrahydro-β-carboline alkaloids from a marine sponge *Asteropus* sp. Biochem Sys Ecol 38:1049–1051.

Li N, Hua J, Wanga S, Cheng J, Hu X, Lu Z, Lin Z, Zhu W, and Bao Z (2010). Isolation and identification of the main carotenoid pigment from the rare orange muscle of the *Yesso scallop*. Food Chem 118: 616–619.

Lin L, Chen P, and Harnly JM (2008). New phenolic components and chromatographic profiles of green and fermented teas. J Agric Food Chem 56: 8130–8140.

Liu Q, Zhang YJ, Yang CR, and Xu M (2009). Phenolic antioxidants from green tea produced from *Camellia crassicolumna* var. multiplex. J Agric Food Chem 57: 586–590.

Lopes RM, Agostini-Costa TS, Gimenes MA, and Silveira D (2011). Chemical composition and biological activities of *Arachis* species. J Agric Food Chem 59: 4321–4330.

Lopez MD, Jordan MJ, and Pascual-Villalobos MJ (2008). Toxic compounds in essential oils of coriander, caraway and basil active against stored rice pests. J Stored Prod Res 44: 273–278.

Mahmoud SS, and Croteau RB (2002). Strategies for transgenic manipulation of monoterpene biosynthesis in plants. Trends Plant Sci 7:366–373.

Marty C, and Berset C (1990). Factors affecting the thermal degradation of all-trans-β-Carotene. J Agric Food Chem 38:1063–1067.

Ndongo JT, Shaaban M, Mbing JN, Bikobo DN, Atchadé AT, Pengyemb DE, and Laatsch H (2010). Phenolic dimers and an indole alkaloid from *Campylospermum flavum* (Ochnaceae). Phytochemistry 71:1872–1878.

Norin T, and Westfelt L (1963). Thin layer, column, and gas liquid chromatography of resin acid esters and related terpenes. Acta Chemica Scandinavia 17:1828–1830.

Novak J, Zitterl-Eglseer K, Deans SG, and Franz CM (2001). Essential oils of different cultivars of *Cannabis sativa* L. and their antimicrobial activity. Flavour Frag J 16: 259–262.

Oliveira GPR, and Rodriguez-Amaya DB (2007). Processed and prepared corn products as sources of lutein and zeaxanthin: compositional variation in the food chain. J Food Sci 72: S79.

Ozen T, Demirtas I, and Aksit H (2011). Determination of antioxidant activities of various extracts and essential oil compositions of *Thymus praecox* subsp. *skorpilii* var. *skorpilii*. Food Chem 124:58–64.

Pawlowska AM, Oleszek W, and Braca A (2008). Quali-quantitative analyses of flavonoids of *Morus nigra* L. and *Morus alba* L. (Moraceae) Fruits. J Agric Food Chem 56: 3377–3380.

Pichersky E, Noel JP, and Dudareva N (2006). Biosynthesis of plant volatiles: nature's diversity and ingenuity. Science 311: 808.

Pinto SC, Leitão GG, Oliveira DR, Bizzo HR, Ramos DF, Coelho TS, Silva PEA, Lourenço MCS, and Leitão SG (2009). Chemical composition and antimycobacterial activity from the essential oil from *Anemia tomentosa* var. *anthriscifolia*. Nat Prod Comm 4: 1675–1678.

Polat E, Bedir E, Perrone A, Piacente S, and Alankus-Caliskan O (2010). Triterpenoid saponins from Astragalus wiedemannianus Fischer. Phytochemistry 71:658–662.

Porcu OM, and Rodriguez-Amaya DB (2008). Variation in the carotenoid composition of the lycopene-rich Brazilian fruit *Eugenia uniflora* L. Plant Food Hum Nutr 63: 195–199.

Potrebko I, and Resurreccion AVA (2009). Effect of ultraviolet doses in combined ultraviolet- ultrasound treatments on trans-Resveratrol and trans-Piceid contents in sliced peanut Kernels. J Agric Food Chem 57:7750–7756.

Reis B, Martins M, Barreto B, Milhazes N, Garrido EM, Silva P, Garrido J, and Borges F (2010). Structure-property-activity relationship of phenolic acids and derivatives. Protocatechuic acid alkyl esters. J Agric Food Chem 58: 6986–6993.

Roberts MF, and Wink M (1998). Alkaloids: Biochemistry, Ecology and Medicinal Applications. Plenum Press, New York.

Rodriguez DB, Lee TC, and Chichester CO (1975). Comparative study of the carotenoid Composition of the seeds of ripening *Momordica charantia* and tomatoes. Plant Physiol 56: 626–629.

Rodriguez-Amaya DB (1999). A guide to carotenoid analysis. Washington: OMINI-ILSI Press.

Rossi PG, Bao L, Luciani A, Panighi J, Desjobert JM, Costa J, Casanova J, Bolla JM, and Berti L (2007). (E)-methylisoeugenol and elemicin: antibacterial components of *Daucus carota* L. essential oil against *Campylobacter jejuni*. J Agric Food Chem 55:7332–7336.

Roze LV, Chanda A, and Linz JE (2011). Compartmentalization and molecular traffic in secondary metabolism: a new understanding of established cellular processes. Fungal Genet Biol 48: 35–48.

Salamci E, Kordali S, Kotan R, Cakir A, and Kaya Y (2007). Chemical compositions, antimicrobial and herbicidal effects of essential oils isolated from Turkish *Tanacetum aucheranum* and *Tanacetum chiliophyllum* var. *chiliophyllum*. Biochem System Ecol 35: 569–581.

Shaala LA, Bamane FH, Badr JM, and Youssef DTA (2011). Brominated arginine- derived alkaloids from the red sea sponge *Suberea mollis*. J Nat Prod 74:1517–1520.

She GM, Xu C, Liu B, and Shi RB (2010). Polyphenolic acids from mint (the aerial of *Mentha haplocalyx* Briq.) with DPPH radical scavenging activity. J Food Sci 75: C359–C362.

Shirsath S, Sonawane S and Gogate P (2012). Intensification of extraction of natural products using ultrasonic irradiation-a review of current status. Eng. Proc. Inten. 53: 10–23.

Sulyok E, Vasas A, Rédei D, Dombi G, and Hohmann J (2009). Isolation and structure determination of new 4, 12-dideoxyphorbol esters from *Euphorbia pannonica* Host. Tetrahedron 65: 4013–4016.

Summons RE, Bradley AS, Jahnke LL, and Waldbauer JR (2006). Steroids, triterpenoids and molecular oxygen. Phil Trans R Soc B: Biol Sci 361:951–968.

Sun B, Neves AC, Fernandes T, Fernandes A, Mateus M, Freitas V, Leandro MC, and Spranger MI (2011). Evolution of phenolic composition of red wine during vinification and storage and its contribution to wine sensory properties and antioxidant activity. J Agric Food Chem 59: 6550–6557.

Sutour S, Bradesi P, Rocca-Serra D, Casanova J, and Tomi F (2008). Chemical composition and antibacterial activity of the essential oil from Mentha suaveolens ssp. insularis (Req.) Greuter. Flavour Frag J 23:107–114.

Suzuki T, Kubota T, and Kobayashi J (2011). Eudistomidins H-K, new β-carboline alkaloids from the Okinawan marine tunicate *Eudistoma glaucus*. Bioorganic and Medicinal Chemistry Letters 21: 4220–4223.

Takahashi Y, Kubota T, Shibazaki A, Gonoi T, Fromont J, and Kobaiashi J (2011). Nakijinamines C-E, New Heteroaromatic Alkaloids from the Sponge *Suberites* Species. Organic Letters 13:3016–3019.

Takaichi S, Maoka T, Yamada M, Matsuura K, Haikawa Y, and Hanada S (2001). Absence of carotenes and a presence of a tertiary methoxy group in a carotenoid from a thermophilic phylamentous photosynthetic bacterium *Roseiflexus castanholzii*. Plant Cell Physiol 42:1355–1362.

Takaichi S, Mochimaru M, Maoka T, and Katoh H (2005). 4,6 Myxol and 4-Ketomyxol 2′- Fucosides, not rhamnosides, from *Anabaena* sp. PCC 7120 and *Nostoc punctiforme* PCC 73102, and proposal for the biosynthetic pathway of carotenoids. Plant Cell Physiol 46:497–504.

Tantillo DJ (2011). Biosynthesis via carbocations: Theoretical studies on terpene formation. Nat Prod Rep 28:1035–1053.

Vetter W, and Schröder M (2011). Phytanic acid–a tetramethyl branched fatty acid in food. Lipid Technology 23: 175–178.

Villanueva HE, and Setzer WN (2010). Cembrene diterpenoids: conformational studies and molecular docking to tubulin. Rec Nat Prod 4: 115–123.

Wang CC, Chang SC, Inbaraj SB, and Chen BH (2010). Isolation of carotenoids, flavonoids and polysaccharides from *Lycium barbarum* L. and evaluation of antioxidant activity. Food Chem 120:184–192.

Wang M, Li J, Rangarajan M, Shao Y, Voie EJ, Huang TC, and Ho CT (1998). Antioxidative phenolic compounds from Sage (*Salvia officinalis*). J Agric Food Chem 46: 4869-4873.

Wink M (1998). A short history of alkaloids. In: Roberts MF and Wink M (eds) Alkaloids: Biochemistry, Ecology and Medicinal Applications, Plenum Press, New York. pp. 11–44.

Wondracek DC, Vieira RF, Silva DB, and Agostini-Costa TS (2012). Saponification influence in carotenoid determination in cerrado passion fruit. Química Nova, in prelo.

Wu Z, Song L, and Huang D (2011). Food grade fungal stress on germinating peanut seeds induced phytoalexins and enhanced polyphenolic antioxidants. J Agric Food Chem 59: 5993–6003.

Xiao JS, Liu L, Wu H, Xie BJ, Yang EN, and Sun ZD (2008). Rapid preparation of procyanidins B2 and C1 from granny smith apples by using low pressure CC and identification of their oligomeric procyanidins. J Agric Food Chem 56: 2096–2101.

Ye JH, Wang LX, Chen H, Dong JJ, Lu JL, Zheng XQ, Wu MY, and Liang YR (2011). Preparation of tea catechins using polyamide. J Biosci Bioeng 111:232–236.

Zellner BD, Dugo P, Dugo G, and Mondello L (2010). Analysis of essential oils. In: Baser KHC, Buchbauer G (eds) Handbook of Essential Oils. CRC Press, Boca Raton. pp. 151–184.

Zhan PY, Zeng XH, Zhang HM, and Li HH (2011). High-efficient column chromatographic extraction of curcumin from *Curcuma longa*. Food Chem 129:700–703.

Zhang Z, Liao L, Moore J, Wua T, and Wang Z (2009). Antioxidant phenolic compounds from walnut kernels (*Juglans regia* L.). Food Chem 113: 160–165.

Zhu BCR, Henderson G, Yu Y, and Laine RA (2003). Toxicity and repellency of patchouli oil and patchouli alcohol against Formosan Subterranean Termites *Coptotermes formosanus* Shiraki (Isoptera: Rhinotermitidae). J Agric Food Chem 51: 4585–4588.

10

Role of Arbuscular Mycorrhizas in Sustainable Management of Agricultural Ecosystem

MUNAWAR FAZAL[1*], MOHD YAQUB BHAT[2] AND MOHD ASHAQ[3]

[1]Department of Botany and Biotechnology
College of Commerce, Arts & Science (A Constituent Unit of Magadh University, Bodh Gaya, Bihar), Patna - 800020, Bihar, India

[2]Department of Botany, Universiry of Kashmir, Srinagar
Jammu and Kashmir, India

[3]Department of Botany, Government College for Women
Gandhi Nagar, Jammu–180001, Jammu and Kashmir, India

*Email: *munawar_fazal@rediffmail.com*

1. INTRODUCTION

An agricultural ecosystem consists of plant roots, the soil microflora and fauna, and the abiotic geochemical soil environment. Plant shoots, as its source of energy, also form an important part of the system. In keeping with the role of plants as source of food, cloth, and shelter, functions of fundamental public importance, the ecosystem has been traditionally used, treated and evaluated from phytocentric and edaphic point of view in agriculture. This view has undergone and undergoing sweeping changes. The importance of soil is now being recognised not only as an agricultural resource base (Stewart et al., 1991), but as a complex living fragile system that must be protected (Reganold et al., 1990) and managed for its own sake (Pierie and Lal, 1991) to guarantee its long term stability and productivity. Expressed scientifically in early 1980s (Bezdicek and Power, 1984; Jackson, 1980; Rodale, 1983), sustainability has a long history in agriculture (Harwood, 1990) and deeply rooted in community awareness.

Increased environmental awareness has progressively led to a shift from conventional intensive management to low input sustainable crop production agro-ecosystem. In low input cropping systems natural, activities of microbes contribute to the biocontrol of plant pathogens and improved supply of nutrients, thus maintaining plant health and production (Garg and Chandel, 2010; Pandey et al., 2010; Pandey and Prasad, 2015).

Microbes are key component of soil plant system where they are immense in a network of interactions affecting plant development. Several microbial groups, phosphorus solubilizers, plant growth promoters and other such beneficial microorganisms are reported from different soils. Balanced microbial system contributes to the sustainability in agriculture, horticulture, forestry and range management.

One vehicle for sustainable agriculture has been awaited in the rhizosphere for decades, in the form of beneficial microbes. Efforts to incorporate microbial management into mainstream agricultural practices have not (meat desired result) been successful. In fact, the same practices, namely, fertilizers, herbicides and pesticides that increased yields in the last several decades have largely eradicated natural rhizosphere microbes in agroecosystems (Matson et al., 1997) and may adversely affect introduced beneficial microbes from establishing in agricultural soils.

Arbuscular mycorrhizal fungi (AMF), belonging to the phylum Glomeromycota, are a main component of beneficial soil microbiota, and probably represent the most important terrestrial symbiosis (Jeffries et al., 2003; Fitter 2005; Finlay 2008). This widespread and advantageous rhizosphere microorganism despite its substantial influence on plant productivity is yet to be fully utilized in agricultural production system. Research suggests that AMF have manifold ecosystem functions and are perfect tool in any field where plants and their communities are manipulated, including sustainable farming (Goltapeh et al., 2008).

2. TAXONOMY OF ARBUSCULAR MYCORRHIZAL (GLOMEROMYCOTAN) FUNGI (TAXONOMIC DIVERSITY)

A perusal of history of research on arbuscular mycorrhizal fungi (Koide and Mosse, 2004) reveals that the first description of AM Fungus was made as early as 1842 by Nageli. Before 1974 most of the AM fungi were placed in the genus Endogyne until Gerdermann and Trappe (1974) placed them under four distinct genera (*Glomus, Sclerocystis, Gigaspora & Acaulospora)* in the order Endogonales under Zygomycetes. In 1990 Morton and Benny established a new order "Glomales" in Phylum Zygomycota comprising two suborders (Glomineae and Gigasoprineae) three families (Acaulosporaceae, Gigasporaceae and Glomaceae) and six genera (Acualospora, Entrophospora, Gogaspora, Glomus, Sclerocystis and Scutellospora) later, Morton and Binny (2001) recognised two other families, the Archaeosporaceae and Paraglomaceae, with two new genera, *Archaeospora* and *Paraglomus.* Since then evidences have been accumulated supporting the view that AM fungi are not related to Zygomycota. They do not seem to form characteristic zygospores and in all cases when nutritional mode has been elucidated they form mutualistic symbiosis. Based on their DNA phylogeny, AM fungi are the sister group of Basidiomycota and Ascomycota. Therefore, Schüβler et al. (2001) raised "Glomales" to the rank of "Glomeromycota" and corrected grammatically incorrect order name "Glomales" to Glomerales. Today Phylum Glomeromycota has been divided into 3 classes (Archaeosporomycetes, Glomeromycetes and Paraglomeromycetes), 5 orders (Archaeosporales, Diversisporales, Gigasporales, Glomerales and Paraglomerales), 14 families, 29 genera and approximately 230 species (Schüβler et al., 2001; Koide amd Mosse, 2004; Oehl and Sieverding, 2004; Walker and Schüβler, 2004; Sieverding and Oehl, 2006; Spain et al., 2006; Oehl et al., 2003, 2006, 2011 (a-g), 2012, 2014; Palenzuela et al., 2008).

3. HOST SPECIFICITY OF ARBUSCULAR MYCORRHIZAL (GLOMEROMCOTAN) FUNGI (HOST DIVERSITY)

Most plant phyla including Bryophytes, Pteridophytes, Gymnosperms and Angiosperms are able to form mycorrhizae with AM Fungi. Families, such as Cyperaceae, Brassicaceae, Caryphyllaceae, Junaceae, Chenopodiaceae and Amaranthaceae rarely forms AM association (Chen et al., 2001; Yang et al., 2002; Liu et al., 2002; Wang and Liu, 2002; Muthukumar et al., 2003; Bao and Yan, 2004). Families, such as Betulaceae, Orchidaceae, Fumariaceae, Commenlinaceae and Ericaceae are assumed never to form AM association. Although AM is confined to the roots, they have been reported from leaves of water fern, *Salvinia, Cucullata,* fruiting peg of peanut (*Arachis hypogaea*) and modified scale leaves and rhizomes of ginger (*Zingiber officinale*) and Canna (Sumbali and Mehrotra, 2009).

Individual fungal strains exhibit little host specificity when grown with different plants under laboratory conditions (Smith and Read, 2008). Likewise a single plant can be colonised by different AM fungal species within the same root (Fitter, 2005; Santos-Gonzalez et al., 2007). Therefore, on one hand, the AM symbiosis is thought to show little host specificity at the level of colonization. On the other hand, the biodiversity of fungal and plant communities are positively correlated with each other (Maherali and Klironomos, 2007) and host preferences seems to play an important role in the natural ecosystem (Vandenkoornhuyse et al., 2003). It is likely that these host preferences reflect different fungal strategies and levels of functional compatibility (Sykorova et al., 2007; Munkvold et al., 2004).

4. BIOLOGY OF ARBUSCULAR MYCORRHIZAL (GLOMEROMYCOTAN) FUNGI

Glomeromycotan fungi are obligate symbionts. This means that each of their life cycle steps requires the association of living plant. It comprises about 230 described morphospecies that are traditionally distinguished by spore characters, spore wall architecture and the morphology of subtending hyphae. Some molecular tools have been developed to differentiate between AM species and strains, but these technologies remain to be tested for a variety of AM species and strains (Millner et al., 2001; Yokoyama et al., 2002; Sèguin et al., 2003). There is no evidence that Glomeromycotan reproduce sexually (Redecker and Raab 2006; Harwani et al., 2008). A study reporting formation of sexual zygospores by *Gigaspora* (Tommerup and Sivasithamparam, 1990) was not confirmed. Studies using molecular marker genes have detected no genetic recombination or only low levels (Kuhn et al., 2001). Therefore, in all Glomeromycotan fungi described spores are formed asexually (Schüβler et al., 2001; Redecker and Raab, 2006). Glomeromycotan fungi are believed to reproduce via spores, vesicles and hyphae. Spores are differentiated by budding intercalary or apical on hyphae.

The fungal development starts with the germination of hyphae from resting spores. In the absence of a host plant, AM fungi shows only limited hyphal growth, whereas in the presence of root exudates growth and branching of hyphae is strongly increased (Tamasloukht et al., 2003). This presymbiotic fungal reaction is characterized by the activation of specific genes followed by subsequent physiological and morphological changes. In return, germinating spores produce diffusible factors which are perceived by plant roots leading to the expression of specific genes even in the absence of direct physical contact (Kosuta et al., 2003).

After the first physical contact between hyphae and plant roots epidermis, the fungus forms appressoria and subsequently penetrates the root colonizing the intercellular space of the root cortex. The plant actively mediates at least two steps allowing the fungus to penetrate the rhizodermis (Demchenko et al., 2004)- anticlinal cell walls of the two adjacent epidermal cells separate from each other in the vicinity of fungal hyphae; and fungal hyphae are allowed to pass intracellularly through an exodermal cell and an adjacent cell from the outermost cortical layer. After penetration of the innermost cortical layers, tree-like fungal structures called arbuscules (from Latin word arbusculum, meaning bush or little tree) are formed within individual root cortical cells by repeated dichotomous branching of the fungal hyphae. Arbuscules are the key features of AM and are responsible for nutrient exchange. Their formation induces activation and reorganization of the host cell contents with the proliferation of organelles and membrane systems (Smith, 1990). The arbuscule represent a dead end in the growth of AM fungi (Bonfante and Perotto, 1995), because they finally senesce and collapse after 4–10 days of symbiosis (Sanders et al., 1977). The fungal structures are then degraded completely by plant cell and cell recovers its original morphology (Jacquelinet-Jeanmougin et al., 1997). This way, cortical cells are able to allow a second fungal penetration and arbuscule formation. Except for the species from the genera *Scutellospora* and *Gigaspora*, all AM fungi form intra- or intercellular storage organs, lipid rich vesicles. Following the development of hyphae, arbuscules and vesicles, the fungus also extends its extraradical hyphae and a network develops in the soil. This hyphal network can be in excess of 100 metres of hyphae per cubic centimetre of soil (Miller et al., 1995). The external mycelium has a range of functions including nutrient acquisition (predominantly phosphate), water uptake, invasion of other roots and the formation of spores (Finlay, 2008). In return for supplying plants with nutrients and water, AM fungi obtain carbohydrates from plants (Solaiman and Saito, 1997; Bago et al., 2003). Up to 20% of the photosynthesis products of terrestrial plants (approx 5 billion tonnes of carbon/year) are estimated to be consumed by AM fungi (Bago et al., 2000).

Hyphal networks of AM fungi are usually aseptate and coenocytic, with hundreds of nuclei having the same cytoplasm. Likewise, individual spores contain hundreds of nuclei. Although there is no confirmed report of sexual stage in the life cycle of AM fungi, it is possible that genetic material is exchanged and recombined; anastomosis between hyphae (Giovannetti et al., 2004, Gandolfi et al., 2003) allows the exchange of of nuclei (Giovannetti et al., 2004), but so far only been observed between hyphae of closely related fungal strains.

Although spores of AM fungi can germinate in the absence of host plants, they are obligate biotrophs, and therefore depends on a living photoautotrophic partner to complete their life cycle and produce the next generation spores. In reported case, however, co-culture with *Paenibacillus validus* led to the production of secondary and infective spores in the absence of a host (Hildebrandt et al., 2006).

5. ECOLOGICAL DIVERSITY OF ARBUSCULAR MYCORRHIZAL (GLOMEROMYCOTAN) FUNGI

Arbuscular mycorrhizal fungi (AMF) are distributed among a variety of environments as symbionts with plants from sub-polar to tropical latitudes and from swamps in low lands to mountainous area of high elevations (Read, 1991). AMF like other soil fungi are present in the top 15–30 cm of soil and their number decreases with increasing depth (Redhead, 1977). Climatic and edaphic factors have significant influence on distribution of AM fungi, e.g., *Glomus intraradices, G. clarum,* are most adapted to the hot and arid environment (Li and Zhao, 2005).

6. ARBUSCULAR MYCORRHIZA FUNCTION

6.1 Improvement of Salt Stress

Salinization of soil is a serious global problem and is increasing steadily, particularly in arid and semi-arid regions (Giri et al., 2003; Al-Karaki, 2006). Increased salinization of Arabic land is expected to have destructive universal effects, resulting in 30 percent land loss within the next 25 years and up to 50 percent by middle of the 21st century (Porcel et al., 2012; Kapoor et al., 2013; Abdel Latef and Chaoxing, 2014). The term salinity refers to the accumulation of excessive salts, in sodic (or alkaline) and saline soils. High levels of salinity in soils is mainly due to the soluble salts in irrigation water and fertilizers used in agriculture production system (Abrol, 1986; Copeland et al., 1996; Al-Karaki 2000), low precipitation and high temperature in these regions and over exploitation of available water resources (e.g., groundwater) (Cantrell and Lindermann, 2001; Al-Karaki, 2006; Mouk and Ishi 2006).

The significance of salinity in agricultural production system is enormous (Tester and Davenport, 2003) as it affects the establishment, growth and development of plants leading to huge losses in productivity (Giri et al., 2003; Mathur et al., 2007). Arbuscular mycorrhizal fungi have been shown to promote plant growth and salinity tolerance. They promote salinity tolerance by utilizing various mechanisms, such as (a) enhancing nutrient uptake (Evelin et al., 2012), (b) promoting plant growth hormones, (c) improving rhizospheric and soil conditions (Asghari et al., 2005), (d) improvement in photosynthetic activity and water use efficiency (Hajiboland et al., 2010), (e) accumulation of compatible solutes (Evelin et al., 2012), (f) production of higher antioxidant enzymes (Manchanda and Garg, 2011) (g) defending roots against soil-borne pathogens (Dehne, 1982) and (h) improving host physiological processes like water absorption capacity of plants by increasing root hydraulic conductivity and favourably adjusting the osmotic balance and composition of carbohydrates (Ruiz-Lozano and Azcòn, 2000; Ruiz-Lozano, 2003, Evelin et al., 2009). This may lead to increased plant growth and subsequent dilution of toxic ion effect (Juniper and Abbott, 1993). As a result they are considered suitable candidates for bioamelioration of saline soils.

6.2 Increased Soil aggregation and Stability

Soil structure exerts important influences on the functioning of soil, its ability to support plant and animal life, controlling environmental quality with special emphasis on soil C sequestration, nutrient and gas fluxes and water quality. Soil structure is often expressed as the degree of stability of aggregates being a major factor which moderates physical, chemical and biological processes leading to soil dynamics (Bronick and Lal, 2005). Soil aggregates result from a combination of primary mineral particles with organic and inorganic materials. The process is complex and is influenced by the interaction of several biotic and abiotic factors. Among these, the most significant are soil moisture availability, mineral composition, soil texture, the quality and quantity of soil organic matter (SOM), microbial and enzyme activities, mineral nutrients, etc. Soil aggregation begins with the consolidation of soil particles into microaggregates (<250 μm) and further formation of macroaggregates (>250 μm) (Borie et al., 2008).

Soil aggregation is essential for maintaining soil porosity, allowing gaseous exchange and water infiltration, and facilitating biogeochemical cycling (Diaz-Zorita et al., 2002). Soil structure is also crucial to the success of sustainable agriculture and erosion resistance. Much of the focus for sustainable agriculture has shifted towards managing for well aggregated soil. Hyphae of arbuscular mycorrhizal fungi are considered to be primary soil aggregators for several reasons. During development of AM, the fungal hyphae grows out from the mycorrhizal root to develop a complex, ramifying network into the surrounding soil which can reach up to 30 m of fungal hyphae per gram of soil (Cavagnaro et al., 2005; Wilson et al., 2009). This net work can make up to 50 percent of fungal mycelium in soil (Rilling et al., 2002). Thereby representing a major part of soil microbial biomass (Rilling and Allen 1999; Leake et al., 2004), as obligate biotrophs, these fungi do not need to compete with saprobes for soil carbon and AM fungi hyphae are more resistant to fungivory than saprobic fungi (Klironomos and Kendrick, 1996). The mycelial network can have a binding action on the soil and improve soil structure.

The extramatrical hyphae of AM fungi support physical protection of soil organic matter (SOM) by promoting soil aggregation which is especially important in sandy and loamy soils where other binding agents are scarce and these hyphal contribution to physical protection gains higher relative importance (Miller and Jastrow, 2000). The soil aggregates are bound by glues that are produced by AM fungus especially on their hyphae and spores that are abundant in the rhizosphere of their host plants and are named glomalin, a sticky hydrophobic glycoprotein detected in large amount in diverse soils as Glomalin-Related Soil Protein (GRSP) which acts as the key factor to stabilize aggregates (Gadkar and Rilling, 2006) and water retention (Bedini et al., 2009). The combination of extrametrical hyphae and secretion of glomalin is considered to be an important element in helping to stabilise soil aggregates (Andrade et al.,1998; Rilling and Mummey, 2006), thereby leading to increased soil structural stability and quality (Bedini et al., 2009; Caravaca et al., 2006).

6.3 Phytoremediation of Pollutants

Ecosystems, due to various human and natural activities, have been contaminated with heavy metals. Sewage sludge, industrial effluents, fertilizers, pesticides, municipal wastes, burning of fossil fuels, military activities, pigments, batteries, inefficient mining activities, etc., have contributed towards contamination of large areas with heavy metals. Some of these metals, such as Zn, Cu, Mn Ni and Co are micronutrients necessary for plant growth, while others, such as Cd, Pb and Hg have no known biological significance. Heavy metal contaminated areas may jeopardise human health and ecology (Adriano et al., 1998) and, therefore should be managed in a way which reduces the possibilities of any risk concerning contamination of food chain, groundwater or air by wind erosion.

Numerous remediation technologies exist to treat heavy metal contaminated soils. Physio-chemical extraction in the core technique used, but have limitations. It is extremely expensive and usually destroys all soil microbes existing within the substratum. **Phytoremediation**, the use of plants to remediate or clean up contaminated soil, is an alternative. Three key approaches can be considered in the use of plants to clean the polluted soil. **Phytostabilisation** is a process in which pollutants are immobilised by the plant activity, resulting in reduction of the wind and soil erosion, and run-off processes into the groundwater or air (Losi et al., 1994). **Phytodegradation** covers the whole range of metabolic processes in which plants, usually assisted by microorganisms, degrade organic compounds, such as hydrocarbons, pesticides, explosives, etc. **Phytoextraction** involves metal hyperaccumulating plants, which can contain more than 1 per cent of metals in harvestable tissues. The technique consists of growing the plants on polluted soil, and then collecting, drying and finally burning the plant materials (Kumar et al., 1995; Brooks, 1997).

Irrespective of the earlier techniques successfully employed to remediate contaminated soil; it is evident that AM symbiosis must be taken into account in phytostabilisation. In places deficient of AMF, introduction of inoculum offers a fascinating perception for phytostabilisation techniques. The process may be stimulated by suitably selected fungal strains. The source of fungal inoculums production is very vital, as they should be adapted not only to toxicity, but also to climatic conditions. Fungal ecotypes from contaminated sites appear to be more successful that those from uncontaminated sites (Galli et al., 1994; Leyval et al., 1995; Jeffries et al., 2003). Extrametrical hyphae developed within the soil may play a role in heavy metal sequestration and thus in easing of their toxicity (Jeffries et al., 2003). The mycelium of the metal tolerant *Glomus mosseae* was shown to be several times more efficient in heavy metal adsorption that non-tolerant fungi and over ten times more effective than *Rhizopus arrhizus* used as biosorption organism (Joner et al., 2000). Although the *G. mosseae* strain was shown to transport Cd from the soil into the fungal structure within the roots of clover, Cd was immobilised there and its transfer to plant tissue was restricted (Joner and Leyval, 1997). Similar results were obtained on the Cd-and Zn-binding capacity of a fungal strain colonizing clover roots isolated from an absolute metallophyte plant, *Viola calaminaria* (Tonin et al., 2001). Inoculation of maize, lucerne (alfalfa), barley and others by fungal isolate obtained from *V. calaminaria* roots has been shown to attenuate heavy metal stress (Hildebrandt et al., 1999). This demonstrates the contribution of AMF to the accumulation of heavy metals in a non-toxic form within plant roots (Tonin et al., 2001).

Biological techniques used for removal/degradation of organic pollutants from soil mainly employ bacteria and saprophytic fungi (Wilson and Bradley, 1996; Bezalel et al., 1997, Schützendübel et al., 1999). The positive effects of plants on the degradation of polycyclic aromatic hydrocarbons (PAHs) have also been reported (Schwab and Banks, 1994; Reilley et al., 1996). In such cases plants that produce high root surface area are extensively used. Well developed associations with soil microorganism are an alternative for using chemicals to increase the availability of toxic substances (Schwab and Banks, 1994; Shann and Boyle, 1994). Mycorrhizal fungi are amongst the organisms that could effectively increase the area of activity. A range of mycorrhizal fungi are involved in the degradation of organic pollutants and thus they may also be potentially useful in phytodegradation (Jeffries et al., 2003). Recently, the promoted degradation of organic pollutants in soils in the presence of AM was observed, and AM bioremediation (AMB) is becoming a promising and perspective remediation technique for organic pollutants-contaminated soils. Although colonization by AM was negatively affected by PAH levels in soil (Leyval and Binet, 1998), it can still enhance plant survival and growth by decreasing phosphorus deficiency (Joner and Leyval, 2001) and water stress (Sanchez-Diaz and Honrubia, 1994), improving membrane integrity (Grahman et al., 1981) or by stimulation of oxidative enzyme production (Salzer et al., 1999). These phenomena are all responsible for reduction of stress due to pollution. The mechanisms of AMB mainly include the metabolism of organic pollutants by AM fungi, the degradation of these pollutants by the enzymes derived from AM exudation and by the enhanced root exudation and rhizospheric microbial activity in the presence of AM, and the removal of the pollutants by plant uptake and accumulation (Li et al., 2006).

The presence of an arbuscular mycorrhizal symbiosis was also demonstrated in hyperaccumulating plants, which are being used in phytoextraction. Recently, Turnau and Mesjasz-Przybylowicz (2003) have shown this for members of Composite family that hyperaccumulated nickel. Plants cultivated under greenhouse conditions and inoculated with native fungi had a higher shoot biomass and Ni content than noninoculated plants. The accumulation of P, K, Mn and Zn in the cortical layer of lateral roots of inoculated plants confirmed the important role of AMF in uptake and accumulation of these elements. Higher concentration of P, K, Fe, Ni, Cu and Zn in the vascular stele in roots of AMF-inoculated plants than in the non-inoculated ones indicates more efficient translocation of these elements to the aboveground parts of the plant (Kamal et al., 2010). These findings indicate the necessity of including the influence of AMF in studies on the uptake of elements by plants and in industrial use of *Berkheya coddii* for Ni extraction from polluted soils (Orlowska et al., 2013).

6.4 Nutrient Uptake

In countless environments, plant growth is restricted by insufficient nutrient supply (Elser et al 2007; Hell and Hillebrand, 2001). This condition is alleviated by a mutualistic association with soil fungi in the order of Glomeromycota which provide the plant host with diverse mineral nutrients in substitute for assimilates (Smith and Read, 2008).

The fungal hyphae emanating from the root system reaches far beyond the rhizosphere and therefore can acquire nutrients from soil volumes to which roots have no access (Friese and Allen, 1991). Furthermore, fungal hyphae is much thinner than roots (Bago et al., 1998), allowing them to explore small cracks on the micrometer range that are unreachable to roots.

The principal nutrient element acquired through AM is phosphorus (P), which is delivered to the plant in its inorganic oxidised form (Pi) (Karandashov and Bucher, 2005). Additionally, nitrogen (N) can be transferred to the AM plants through the fungal symbiont (Fellbaum et al., 2012; Hodge and Fitter 2010; Mader et al., 2000; Muller et al., 2013) involving a complex pathway that has been elucidated in substantial details in recent years (Govindarajuluv et al., 2005; Guether et al., 2009; Tian et al., 2010). Apart from these essential macronutrients, sulphur (Allen and Shachar-Hill, 2009; Sich et al., 2013) and some micronutrients can be delivered to the plants through AM fungi (Clark and Zeto, 2000; George, 2000).

6.4.1 *Phosphorus*

The most excellent known feature of AM functioning is the role played by AM fungi in enhancing soil P-scavenging skill of the plant. This skilfulness is set against transfer of sugars in the form of recent photosynthates from plants to AM fungus, so that balance of trade of organic C for P is a major influence on the outcome of the symbiosis.

Phosphorus, one of the essential macronutrients required for all organisms including plants and fungi, though present in relatively large amounts is difficult to obtain from soil (Schachtman et al., 1998; Smith and Smith, 2011, 2012; Smith et al., 2011). The preferred form of phosphorus taken up by plant is orthophosphates (Vance, 2000). The orthophosphate (Pi, $H_2PO_4^-$) absorbed by the root epidermal cells including root hairs, reacts strongly with iron and aluminium in soil of low pH and with calcium at high pH. Hence, available pH is much lower than the total P and Pi concentration in the soil (less than 10 μM). This Pi concentration is about 1000-fold lower than cellular concentration. Further, because plant and fungi scavenge P as negatively charged Pi ($H_2PO_4^-$) and the cell membranes have electrical potential differences that are inside negative, there is a steep electrochemical potential gradient against which uptake occurs (Smith and Smith, 2012). Therefore, in plant and fungi P uptake is strongly energy dependent (active). Furthermore, plants have and additional problem in that uptake of inorganic phosphate by root often occurs at a faster rate than it can be replaced by diffusion from soil. This imbalance leads to depletion of Pi in the rhizosphere in the order of millimetres around P-absorbing roots (Hinsinger et al., 2005) that further limit Pi uptake. As a result the rate at which roots grown into undepleted soil plays a major role in determining the effectiveness of P acquisition (Tinker and Nye, 2000). It is therefore not surprising that plants have evolved a number of strategies that can either increase the effectiveness with which they can scavenge P from soil or release P from inorganic sources. The most common and widespread of the strategies that can enhance the P scavenging capability of plants is AM symbiosis. AM colonization might boost the rate of phosphorus accumulation further than the restrictions which can be currently utilized, thus dipping the rate of phosphorus utilization effectiveness. Such momentary "luxury consumption" of phosphorus might, however, serve a storage function and be utilized subsequently, allowing mycorrhized plants ultimately to out-perform non-mycorrhized plants (Koide, 1991). Other strategies, scattered throughout the plant kingdom that enhances Pi availability or uptake are formation of dense cluster roots or ecto and ericoid mycorrhiza, rapid root growth or root hair formation (Lambers et al., 2008, 2011; Cheng et al., 2011).

The major advantage of AM symbiosis for plants in acquiring P is that AM fungi provides an effective pathway (AM pathway) by which P is scavenged from bulk soil and rapidly delivered to the cortical cells within the root, by passing direct pathway (by root epidermal cells and by root hairs if formed). The AM pathway (via AM fungi) can reduce the impact of Pi depletion in the rhizosphere and thus improve Plant P nutrition and growth.

Arbuscular mycorrhizas have three indispensable components: the root itself, plus AM fungal mycelia that develop inside and outside the root. These mycelia live in distinct environment. Inside the root, the intraradical fungal infection unit grows within the root apoplast (free diffusional space outside the plasma membrane within the plant, it is interrupted by Casparian strip in roots, by air spaces between plant cells and by the plant cuticle), both in intercellular spaces and in specialized compartment formed by invagination of cortical cell plasma membranes. The fungal structure forms symbiotic interfaces plant across which nutrient transfer occurs. These interfaces are composed of the plasma membrane of both symbionts and apoplastic compartment between them. Outside the root in the soil extensive, branched external mycelium grows from the infection units (Smith and Read, 2008). A single root axis might contain one or more infection units per cm on an average (Harley and Smith, 1983). Each infection unit produces external myceliums that extend several cm beyond the rhizosphere and potentially colonize different parts of the same roots system and of the other plants. External mycelium plays a critical role in scavenging Pi from the soil, based on several key features. The hyphal length of the mycelium in the soil (HLD mg^{-1}soil) normally considered greater than the root length density (RLD). Hyphae are much narrower (approx 2-20 μm) than roots (>300 μm) and of similar diameter to root hairs, but of much greater length, these have several vital consequences. Production of absorbing length (an important characteristic in uptake of immobile nutrients) by hyphae is much less pricey than root hairs. This narrow diameter also means that hyphae can also grow into soil pores of considerably smaller diameters than roots are able to do so and hence can exploit the P in soil solution in a given soil volume more efficiently and at lower soil moisture. In addition, narrower diameter reduces the problem, AM plant grows on soil with poorly available P, their responsiveness to colonization is higher than when grown in soil with equivalent amount of readily available P. The effect may be due to better access to available P by AM fungal hyphae campared with roots (Bolan et al., 1987). The possibility that AM fungal hyphae have strong potential to hydrolyze organic P is still puzzling (Joner et al., 2000; Smith and Smith, 2012). All these factors coupled with marked efficiency in translocation of P from sites of absorption to transfer sites within the root cortex render the external mycelium efficient P uptake and delivery system.

AM plants obtain nutrients (P and other) from soil through two different pathways. Direct uptake and AM pathway. Both pathways involve different cell types, different Pi transporters and P access from different regions and volumes of soils. Direct uptake from rhizosphere by root epidermal cells includes root hairs if formed. Expression of Pi transporter in these cells is maximal in the root apex and root hairs (Gordon-Weeks et al., 2003) and declines in more mature regions. Expression is often reduced with high supply and AM colonization (Javot et al., 2007). These reductions will lower direct uptake in the older region of root, but their relative importance is not clear.

Potential operation of AM pathway via AM fungi occurs behind the root apex. AM fungi grow extensively in soil to form a well developed hyphal network that absorbs Pi up to several centimetres from the root surface and can markedly extend the depletion zone. P is translocated rapidly to the roots overcoming the slow diffusion that occurs in soil solution. AM plants possess Pi transporters that are specifically or preferentially expressed in root cortical cells when these are colonized (Bucher, 2007; Javet et al., 2007), providing markers for AM pathway and good circumstantial evidence for the site of P delivery via fungal pathways and transfer to the plants.

The two pathwatys, direct and AM pathway in an AM root bare potentially independent because they involve uptake from different soil locations and expression of different Pi transporters in different cell types. However, this does not mean that the two pathways do not function autonomously.

6.4.2 Nitrogen

Arbuscular mycorrhizal fungi in addition to P acquirement from soil, also plays an important role in take up and transfer of significant amount of nitrogen to their host plants (He 2003). The availability of nitrogen frequently limits plant growth, and depending upon the soil conditions, nitrogen transfer by mycorrhizal fungi can represent a significant route of uptake by plants (Ames et al., 1983; He et al., 2003). Arbuscular mycorrhizal fungi have been strongly implicated in the transfer of nitrogen from one plant to another (He et al., 2003), can increase the utilization of different forms of nitrogen by plants (Hodge et al., 2001) and have been shown to take up nitrogen directly and transfer it to the host roots (Bago et al., 1996; Johansen et al., 1996; He et al., 2003). Numerous mechanisms have been suggested to explain AM effect on N uptake. The mechanisms involved include: i) direct uptake of combined nitrogen by AM (Frey and Schuepp, 1993), ii) the direct improvement of biological nitrogen fixation: an indirect activity based on the supply of PO_4 for N_2 fixation (Barea and Azcon-Aguilar, 1983), iii) facilitated 'N transfer' a process by which part of nitrogen, a biologically fixed benefits the nonfixing plants growing nearby (Hamel, 1996; Hamel et al., 1996) and iv) enzymatic activities involved in N metabolism (Harley and Smith, 1983). The AM hyphae have the capacity to take up and transport N from soil to roots (Bago et al., 1996).

Nitrate is the most important source of nitrogen for many higher plants growing in wide geographic areas (Schmidt, 1982). The process of nitrate absorption by plants has a fundamental biological importance because N is a vital element for life. Nitrate assimilation may be as important for plant life as the CO_2 assimilation in photosynthesis (Marschner, 1986). Investigational evidence has been reported that AM fungi can reduce nitrate by assimilatory reduction pathway, indicating that AM fungi have gene set for assimilatory nitrate reduction (Ho and Trappe, 1975; Kaldorf et al., 1994; Azcón and Tobar 1998). In many soils, nitrate is the main constituent of the mineral N pool. In recent years, several studies have demonstrated that nitrate can be mobilized from the soils and transferred to the root cells by AM external hyphae, improving the inflow of N to the AM plants (Johansen et al., 1993; Tobar et al., 1994; Azcón et al., 1996). Most of the evidence concerning the effect of AM fungi on nitrate acquisition has been attributed to an indirect P-mediated mechanism since the enzymatic system for nitrate reduction requires phosphate (Hoff et al., 1992). But the direct uptake and transport of nitrate by extraradical AM mycelium have been confirmed (Johansen et al., 1993; Tobar et al., 1994).

6.4.3 *Other Elements*

Arbuscular mycorrhizal fungus mycelia have been shown to increase uptake of many nutrients along with trace elements apart from P and N (Clark and Zeto, 2000; Alizadeh et al 2010). Information on acquisition of macronutrient cations (Ca, Mg, K) by AM plants has been relatively contradictory in that increases, decreases and no effects have been reported (Lambert et al 1979; Azcón and Barea, 1992). Enhanced acquisition of Ca, Mg and K was more in AM plants grown in acidic soils than in neutral to alkaline soils when colonized with selected AM isolates (Siqueira et al., 1990). In contrast to AM plants grown in acidic soil, those grown in neutral to alkaline soils, generally have limited enhancement of Ca, Mg and K acquisition (Bethlenfalvay et al., 1989; Hamel and Smith, 1991; Khalil et al., 1994). This also depends on AM isolates, and plant species or type of plant.

Micronutrients, copper and zinc have been shown to be consistently at higher concentration in AM plants. Acquisition of Cu and Zn was enhanced in AM plants (Cooper and Tinker, 1978; Lambert et al., 1979; Gildon and Tinker, 1983; Pacovsky, 1986), but in considerably smaller quantities than P (Cooper and Tinker, 1978). Enhanced acquisition of Zn and Cu in shoots was reduced when P was increased in soil (Lambert et al., 1979), although an enhanced acquisition of these nutrients occurred even at high levels of soil P (Raju et al., 1990). Contribution of hyphae to enhanced acquisition of Zn was calculated to be 48 per cent of total plant Zn (Kothari et al., 1990).

6.5 Reduction of Environmental Stress

Agricultural soils rarely provide ideal conditions for growth and survival of plants, and soil microbes. Since soil conditions are frequently changing, the soil environment may favor development of AMs at one point in time, and inhibit them at another time. AM fungi play an important role in the promotion of biological and chemical properties of plants under stressed conditions. AM facilitating plants to adapt and resist a wide range of biotic and abiotic stresses encountered in the environment. Adequate soil moisture and temperature may favor development of arbuscular mycorrhizas. However, when soil moisture and temperature are too high or too low, mycorrhizal formation may be adversely affected.

AM fungi may affect host plant function and productivity under both high and low moisture conditions (Schenck and Smith, 1982; Subramanian and Charest, 1999). AM symbiosis may alleviate plant responses to moderate moisture deficit by several mechanisms including water uptake from soil by hyphae, altered hormonal levels, causing changes in stomatal conductance, increased turgor by lowering leaf osmotic potential, improved nutrition of the host, and soil-root continuum (Entry et al., 2002).

AM fungi may alter the metal concentration in plants by metal immobilization in intra- or extra-radical hyphal cell wall components, metal chelation by fungally secreted compound, such as glomalin, or by metal compartmentalization inside fungal cells. Thus, AMF acts as metal sinks, reducing local concentrations in soils and creating more suitable environment for plants growing in soils with high metal contents (Mohammadi et al., 2011).

6.6 Plant-Water Relationship

Arbusular mycorrhizal fungi are essential for sustainable agriculture production system because they improve plant-water relations and thus increase the drought resistance of the host plants (Allen and Allen, 1986; Nelsen, 1987). Improved plant-water status and changes in water relations have been ascribed to a variety of mechanisms, including some mechanisms not directly related to phosphorus nutrition or water uptake (Davies et al., 1992). AM fungi infection has been reported to increase nutrient uptake in water stressed plants (Busse and Ellis, 1985), enabling plants to use water more efficiently. In fact, AM plants are more tolerant to drought and salt stresses than non-AM plants (Rosendahl and Rosendahl, 1991; Khalvati et al., 2005; Porcel et al., 2006; Aroca et al., 2007). AM plants are able to take up more water from soil under water deficient conditions (Marulanda et al., 2003; Khalvati et al., 2005). However, this ability depends on the fungal species (Marulanda et al., 2003).

The role played by AM fungi in alleviating water stress plants has been investigated and it appears that drought resistance is enhanced. The precise mechanism underpinning this is still in doubt, but may be attributed to a variety of mechanisms including increased water uptake from soil by extraradical mycelium hyphae, enhanced efficiency with which plant use water, increased development of water stable aggregates, increased hydraulic conductivity of host plants by reducing the frequency of nonconductive suberized (mature and hardened) root surface tissues, bridging the gap in water flow that can develop when roots extract water from adjacent soil at a rate much faster than for replacement by water from bulk soil farther away from root, altered hormonal levels, causing changes in stomatal conductance, increased turgor by lowering leaf osmotic potential, regulating root hydraulic properties and enhancing root hydraulic conductance, improved nutrition of the host, and soil-root continuum (Klepper, 1990; Ruiz-Lozano and Azcón, 2000; Hamblin, 1995: Al-Karaki, 1998; Espeleta et al., 1999; Auge, 2001; Entry et al., 2002; Kaya et al., 2003; Aroca et al., 2007).

6.7 Protection Against Root Pathogens

Plants are exposed to a variety of organisms in their environment: some of these interactions are beneficial like symbiotic fungi and insect pollinators, while some are detrimental, e.g., pathogenic microorganisms. These interactions between below-ground and above-ground organisms are receiving tremendous attention because they may influence plant defences against biotic and abiotic stresses (van Dam et al., 2003). Several investigations have demonstrated that AM symbiosis improved resistance to biotic stress caused by below ground root pathogens (Jaizme-Vega et al., 1997, 1998; Elsen et al., 2001; Wipps, 2004; St-Arnaud and Vujanovic, 2007; Akhtar and Siddiqui, 2008).

Most studies on interactions between AM fungi and phytopathogenic fungi show reduction in disease incidence (Catska, 1994; Matsubara et al., 2001; Thygesen et al., 2004); pathogen development (Cordier et al., 1996); and disease severity (Matsubara et al., 2001). The extent of AM fungal induced protection of host plants against phytoparasitic fungal attack ranges from complete protection (Torres-Baragan et al., 1996) to the more common partial protection (Matsubara et al., 2001). The degree of partial protection is AM fungal species and cultivar dependent (Yao et al., 2002). At the same time, there are evidences where AM fungi either did not provide protection to the host plant against fungal pathogens (Yao et al., 2002) or even enhanced pathogenicity (Ross, 1972). Thus, the effects of AM in this regard, while generally favourable, are variable and contingent on determinants not yet known.

A number of mechanisms have been proposed to explain AM fungal protection of host plants against soil-borne fungal pathogens (Linderman, 2000; Sylvia and Chellemi, 2001). These are (a) enhanced P nutrition; (b) competition for nutrients and infection sites; (c) changes in chemical composition of plant tissues; (d) changes in the microbial composition of the mycorizosphere; (e) biochemical changes in plant; and (f) alleviation of abiotic stress.

AM fungi evolved in regulating microbial populations in the mycorrhizosphere may in part explain phytoprotection against soil-borne fungal pathogens (Catska, 1994; Newsham et al., 1994). AM fungal regulation of plant pathogenic fungi in the mycorrhizosphere by antagonist microbes represent the best explanation for the protection of the associated hosts against soil-borne fungal pathogens when AM fungi are abundant and active.

AM fungi and root parasitic nematodes frequently interact because both are rhizosphere organisms and often inhabit the same plant tissues? It has been demonstrated that root colonization by AM fungi can reduce the penetration of nematodes into roots (Smith et al., 1986) and hamper their development subsequent to their penetration into roots (Habte et al., 1999). As a result AM fungi either eliminate or reduce damage caused by plant parasitic root nematodes (Pinochet et al., 1998; Habte et al., 1999; Hol and Cook, 2005; Elsen et al., 2008). Because of this fact and predictability with which AM fungi can be established on host roots, they have the potential for suppressing nematode population and disease severity.

Several mechanisms have been proposed to be involved in suppression of nematodes by AM fungi, proposed mechanisms include competition for host resources in root tissues (Hooker et al., 1994), biochemical changes in the plant and anatomical changes in the roots (Benhamou et al., 1994; Hooker et al., 1994), direct or indirect competition in the rhizosphere (St Arnaud et al., 1994), activation of the plant defence system and disease resistance (Benhamou et al., 1994; St Arnaud et al., 1994), and finally improved plant vigor through an enhanced nutrient uptake by nematode damaged roots (Pinochet et al., 1995). Further, AM fungi could affect nematode activity indirectly by enhancing the efficacy of other soil microorganisms antagonistic to nematodes (Rao and Gowen, 1998). Better P nutrition uptake alone does not account for alteration of nematode damage by AM Fungi, however (Strobel et al., 1982; Garg and Chandel, 2010).

AM fungi have been shown to increase resistance to root-invading nematodes and root infecting pathogenic fungi. The reasons for increased resistance to root invading nematodes and root pathogenic fungi are not well understood, but a well established AM infection is a prerequisite for protection. The current emphasis on low in put based agrotechnology for crop production systems has stimulated the used of AM fungi as bioprotectors, phytostimulators or biofertilizers against plant diseases caused by soil-borne pathogens.

6.8 Boost Plant Quality for Human Health

Recent evidences suggest that AM symbiosis can stimulate plant secondary metabolites. These metabolites are not only indispensable for increased plant tolerance to abiotic and biotic stresses, but also advantageous to human health through their antioxidant activity (Seeram, 2008). Secondary metaobites and mineral contents of crop used as food or medicinal remedies can be beneficial in preventing life threatening diseases, such as cancer, cardiovascular and neurodegenerative diseases or microbial infection (Seeram, 2008; Cummings and Kovacic, 2009). Although AM can encourage synthesis of phenolic compounds (Phenolic acids, flavonoids) and activate the carotenoid pathway in roots (Harrison and Dixon, 1993; Morandi, 1996; Strack and Fester, 2006; Schliemann et al., 2008), only a few analyses have targeted final crop products (leaves, roots or fruits) of mycorrhizal plants used as food medicinal remedies (Gianinazzi et al., 2010). These studies indicate that activation of plant secondary metabolites in response to AM fungi can result in increase in essential oil concentration of plant tissues.

7. CONCLUSION

Sustainable farming system makes use of natural process to achieve satisfactory levels of productivity and food quality, while minimizing the undesirable environmental impacts.

Several soil fertility factors contribute to sustainable agriculture through control of soil-borne plant diseases, including increased soil microbial activity leading to increased competition and parasitism within the rhizosphere. Since the first green revolution less attention has been give to sustainable agriculture. Increased environmental awareness and health hazard has led to shift from high intensity agriculture to low input sustainable agriculture production system. Intensively managed agriculture systems are inefficient and lead to reduced ecosystem functioning and environmental degradation. In low input agricultural production technology, natural activities of microbes contribute to the biocontrol, improve nutrient supply and soil health, thus improving the plant health and production. Plant growth promoting microbes promise to replace or supplement many destructive high intensity agriculture. Arbuscular mycorrhizal fungi that form symbiosis with more than 80 per cent plants are the main beneficial soil microbiota. This widespread and beneficial microorganisms of phylum Glomeromycota, despite its substantial influence on plant productivity is yet to be fully utilized in agricultural production system.

References

Abrol JP (1986). Fuel and Forage production from salt affected wasteland in India. Reclam Reveg Res 5: 65–74.

Abdel Latif AA, and Chaoxing H (2014). Does inoculation with Glomus mosseae improve salt tolerance in pepper plants? J Plant Growth Regul 33: 644–653.

Adriano DC, Chlopecka A, and Kaplan DI (1998). Role of soil chemistry in soil remediation and ecosystem conservation. (Soil chemistry and ecosystem health special publication no52) Soil Science Society of America, Madison, Wis. USA.

Akhtar MS, and Siddiqui ZA (2008). Arbuscular mycorrhizal fungi as potential bioprotectants against plant pathogens. In: Siddiqui ZA, Akhtar MS, Futai K (eds) Mycorrizae: Sustainable Agricuture and Forestry. Springer Science pp. 61–97.

Alizadeh O, Sharafzadeh S, and Medizadeh N (2010). Evaluation of mycorrhiza biological fertilizers on some macro ekements absorption in water stress conditions in sorghum. Adv. Environ. Biol. 4:428–432.

Al-Karaki GN (1998). Benefit/cost analysis and water use efficiency of arbuscular mycorrhizal association with wheat under drought stress. Mycorrhiza 8:41–45.

Al-Karaki GN (2000). Growth of mycorrhizal tomato and mineral acquisition under stress Mycorrhiza 10: 51–54.

Al-Karaki GN (2006). Nursery inoculation of tomato with arbuscular mycorrhizal fungi and subsequent performance under irrigation with saline water. Scientia Horticulturae 109: 1–7.

Allen EB, and Allen MF (1986). Water relation of xeric grasses in field: interaction of mycorrhizas and competition. New Phytol 104: 559–571.

Allen JW, and Shachar-Hill Y (2009). Sulphur transfer through arbuscular mycorrhiza. Plant physiol. 149: 549–560.

Ames RN, Reid CP, Porter LK, and Cambardella C (1983). Hyphal uptake and transfer of nitrogen from two ^{15}N-labelled sources by *Glomus mosseae,* vesicular arbuscular mycorrhizal fungus. New Phytol 95: 381–396.

Andrade G, Mihara KL, Linderman RG, and Bethlenfalvay GJ (1998). Soil aggregation status and rhizobacteria in the mycorrhizosphere. Plant Soil 202: 89-96.

Aroca R, Porcel R, and Ruíz-Lozano JM. (2007). How does arbuscular mycorrhizal symbiosis regulate root hydraulic properties and plasma membrane aquaporins in Phaseolus vulgaris under drought, cold or salinity stresses? New Phytol 173: 808-816.

Asghari HR, Marschner P, Smith SE, and Smith FA (2005). Growth response of Atriplex nummularia to inoculation with arbuscular mycorrhizal fungi at different salinity levels. Plant Soil 373: 245–256.

Auge RM (2001). Water relations, drought and vesicular-arbuscular mycorrhizal symbiosis. Mycorrhiza 11: 3–42.

Azcón R, and Barea JM (1992). Nodulation, N_2 fixation and N nutrition relationships in mycorrhizal and phosphate amended alfalfa plants. Symbiosis 12: 33–41.

Azcón R, Gómez M, and Tobar RM (1996). Physiological and nutritional responses by *Lactuca sativa* L. to nitrogen sources and mycorrhizal fungi under drought conditions. Biol Fertil Soils. 22: 156–161.

Azcón R, and Tobar RM (1998). Activity of nitrate reductase and glutamine synthetase in shoot and root of mycorrhizal *Allium cepa.* Effects of drought stress. Plant Sci 133: 1–8.

Bago B, Azcon-Aguilar C, Goulet A and Piche Y (1998). Branched absorbing structure (BAS): a feature of the extra radical mycelium of symbiotic arbuscular mycorrhizal fungi. New Phytol 139: 375-388.

Bago B, Pfeffer PE, Abubaker J, Jun J, Allan JW, Brouillette J, Douds DD, Lammers PJ, and Shachar-Hill Y (2003). Carbon export from arbuscular mycorrhizal roots involves the translocation of carbohydrates as well as lipid. Plant Physiol 131: 1496–1507.

Bago B, Pfeffer PE, and Shachar-Hill Y (2000). Carbon metabolism and transport in arbuscular mycorrhizas. Plant Physiol 124: 949–958.

Bago B, Vierheilig H, Piché Y, and Azcon-Aguilar C (1996). Nitrate depletion and pH changes induced by extraradical mycelium of arbuscular mycorrhizal fungus *Glomus intraradices* grown in monoculture New Phytol 133: 273–280.

Bao YY, and Yan W (2004). Arbuscular mycorrhizae and their structural types on common plants in grasslands of Midwestern inner Mongolia. Biodiversity Science 12: 501–508 (In Chinese).

Barea JM, and Azcon-Aguillar (1983). Mycorrhizas and their significance in nodulating nitrogen-fixing plants. Adv Agron 36: 1–54.

Benhamou N, Fortin JA, Hamel C, St-Arnaud M and Shatilla A (1994). Resistence responses of mycorrhizal Ri T-DNA transformed roots to infection by IFusarium oxysporum f.sp chrysanthemi. Phytopathol 84: 958-968.

Bedini S, Pellegrino E, Avio L, Pellegrino S, Bazzoffi P, Argese E and Giovannetti M (2009). Change in soil aggregration and glomalin-related content as affected by arbuscular mycorrhizal fungal species Glomus mosseae and Glomus intraradices. Soil Biol. Biochem 41: 1491-1496.

Bethlenfalvay GJ, Franson RL, Brown MS, and Mihara KL (1989). The *Glycine-Glomus-Bradyrhizobium* symbiosis IX Nutritional, Morphological and Physiological responses of nodulated soybean to geographic isolates of the mycorrhizal fungus *Glomus mosseae*. Physiologia Plantarum 76: 226–232.

Bezalel L, Hadar Y, and Cerniglia CE (1997). Enzymatic mechanisms involve in phenanthrene degradation by the white rot fungus *Pleurotus ostreatus*. Appl Environ Microbiol 63: 2495–2501.

Bezdicek DF, and Power JP (1984). Organic Farming: Current Technology and its Role in Sustainable Agriculture. Am. Soc Agron Special Publication 46, Madison.

Bolan NS, Robson A, and Barrow NJ (1987). Effects of vesicular arbuscular mycorrhiza on the availability of iron phosphates to plants. Plant Soil 99: 401–410.

Bonfante P, and Perotto S (1995). Strategies of arbuscular mycorrhizal fungi when infecting host plants. New Phytol 130: 3–21.

Bronick C J and Lal R (2005). Soil structure and management: A Review. Geoderma 124: 3-22.

Borie F, Rubio, R., and Morales, A (2008). Arbuscular mycorrhizal fungi and soil aggregation. J Soil Sci Plant Nutr 8: 9-18.

Brooks RR (1997). Plant accumulators of metals and their role in mineral exploration, archaeology, and land accumulation. In: Iskandar IK, Adriano DC (ed) Remediation of metal-contaminated-soils. Science Reviews, Northwood, UK, pp. 123–133.

Bucher M (2007). Functional biology of plant phosphorus uptake at root and mycorrhiza interfaces. New Phytol 173: 11–26.

Busse MD, and Ellis JR (1985). Vesicular arbuscular mycorrhizal (*Glomus fasiculatum*) influence on soybean drought tolerance in high phosphorus soil. Can J Bot 63: 2290–2294.

Cantrell IC, and Lindermann RG (2001). Preinoculation of lettuce and onion with VA Mycorrhizal fungi reduces deleterious effects of soil salinity. Plant Soil 233: 269–281.

Caravaca F, Alguacil MM, Azcon R and Roldan A (2006). Formation of stable aggregate in rhizosphere soil of *Juniperus oxycedrus*: Effect of AM fungi and organic amendments. Appl Soil Ecol 33: 30-38.

Catska V (1994). Interrelationship between vesicular arbuscular mycorrhiza and rhizosphere microflora in apple replant disease. Boil Plantarum 36: 99–104.

Cavagnaro TR, Smith FA. Smith SE and Jakabsen I (2005). Functional diversity in arbuscular mycorrhizas: exploitation of soil patches with different phosphate enrichment differ àmong fungal species. Plant cell Environ 28:642-650.

Chen X, Fang ZG, and Tang JJ (2001). Investigations on host plants of vesicular arbuscular mycorrhiza fungi (VAMF) within weed communities in Agricultural slope land in the red soil area of southern eastern China. Biodiversity Science 9: 122–128.

Cheng L, Bucciarelli B, Shen J, Allan DL, and Vance CP (2011). Update on white lupin cluster roots acclimation to phosphorus deficiency. Plant Physiol I56: 1025–1032.

Clark RB, and Zeto SK (2000). Mineral acquisition by arbuscular mycorrhizal plants. Journal of Plant Nutrition 23: 867–902.

Cooper KM, and Tinker PB (1978). Translocation and transfer of nutrients in vesicular-arbuscular mycorrhizas. Uptake and translocation of phosphorus, zinc and sulphur. New Phytol 81: 43–52.

Copeland RH, Martin CA, and Stutz JC (1996). Tomato growth in response to salinity and mycorrhizal fungi from saline or non saline soils. Horticultural Science 31: 341–344.

Cordier C, Gianinazzi S, and Gianinazzi-Pearson V (1996). Colonisation patterns of root tissues by *Phytophthora nicotianne* var *parasitica* related to reduced disease in mycorrhiza tomato. Plant soil 185: 223–232.

Cummings JA, and Kovacic JP (2009). The ubiquitous role of zinc in health and disease. J Vet Emerg Crit Care 19: 215–240.

da Silva Campos MA, da Silva FSB , Yano-Melo AM, de Melo N, F Pedrosa EMR, and Maia LC (2013). Responses of guava plants to inoculation with arbuscular mycorrhizak fungi in soil infested with *Meloidogyne enterolobii*. Plant pathol J 29: 242–248.

Davies FT, Potter JR, and Linderman RG (1992). Mycorrhiza and repeated drought exposure affect drought resistance and extraradical hyphae development of pepper plants independent of plant size and nutrient content. J Plant Physiol 139: 289–294.

Diaz-Zorita M, Perfect E, and Grove JH (2002). Disruptive methods for assessing soil structure. Soil and Tillage Research 64: 3–22.

Dehne HW (1982). Interaction between vesicular arbuscular mycorrhizal fungi and plant pathogens. Phytopathology 72: 1115–1119.

Demchenko K, Winzer T, Stougaard J, Parniske M, and Pawlowski K (2004). Distinct roles of *Lotus japonicas* SYMRK and SYM15 in root colonization and arbuscule formation. New Phytol 16: 381-392.

Elsen A, Declerck S, and De Waele D (2001). Effects of *Glomus intraradices* on the reproduction of the burrowing nematode (*Radopholus similis*) in dixenic culture. Mycorrhiza 11: 49–51.

Elser JJ, Bracken MES, Cleland EE, Gruner DS, Harpole WS, Hillebrand H, Ngai JT, Seabloom EW, Shurin, JB, and Smith JE (2007). Global analysis of nitrogen and phosphorus limitation of primary producers in freshwater, marine, and terrestrial ecosystems. Ecology Letters 10: 1135–1142.

Entry JA, Rygiewiez PT, Watrud LS, and Donnelly PK (2002). Influence of adverse soil conditions of the formation and function of Arbuscular mycorrhiza. Adv Environ Res 7: 123–138.

Espeleta JF, Eissenstat DM and Graham JH (1999). Citrus root responses to localized drying soil. A new approach to studying mycorrhizal rffects in roots of mature trees. Plant Soil 206: 1-10.

Evelin H, Giri B, and Kapoor R (2012). Contribution of *Glomus intraradices* inculation to nutrient acquisition and mitigation of ionic imbalance in NaCl-stressed *Trigonellafoenun-graecum*. Mycorrhiza 22: 203–217.

Evelin H, Kapoor R, and Giri B (2009). Arbuscular mycorrhizal fungi in alleviation of salt stress: A review. Ann Bot 104: 1263–1280.

Fellbaum CR. Gachomo EW, Beeseny Y, Choudhari S, Strahan GD, Pfeffer PE, Kiers ET, and Bucking H (2012). Carbon availability triggers fungal nitrogen uptake and transport in arbuscular mycorrhizal symbiosis. Pro Natl Acad Sci USA 109:2666–2671.

Finlay RD (2008). Ecological aspects of mycorrhizal symbiosis: with special emphasis on the functional diversity of interactions involving the extraradical mycelium. J Exp Bot 59: 1115–1126.

Fitter AH (2005). Darkness visible: reflections on underground ecology. J Ecol 93: 231–243.

Friese CF, and Allen MF (1991). The spread of VA mycorrhizal fungal hyphae in the soil-Inoculum types and external hyphal architecture. Mycologia 83: 409–418.

Frey B, and Schuepp H (1983). the relationship between photosynthesis and tuber growth in Solanum tuberosum L. J Expt Bot 28:999-1007.

Gadkar V and Rillig MC (2006). The arbuscular mycorrhizal fungal protein glomalin is a putative homolog of heat shock protein 60. FEMS Microbial Lett 263:93-101.

Gai JP, Tian H, Yang FY, Christie P, Li XL, and Klironomos JN (2012). Arbuscular mycorrhizal fungal diversity along a Tibetan elevation gradient. Pedobiologia 55: 145–151.

Galli U, Schuepp H, and Brunold C (1994). Heavy metal binding by mycorrhizal fungi. Physiol Plant 92: 364–368.

Gandolfi A, Sanders IR, Ross V, and Menozzi P (2003). Evidence of recombination in putative ancient asexual. Mol Biol Evol 20: 754–761.

Garg N, and Chandel S (2010). Arbuscular mycorrhizal networks: process and function. A review. Agron Sustain Dev 30: 581–599.

George F (2000). Nutrition uptake-contributions of arbuscular mycorrhizal fungi to plant mineral nutrition. In: Kapulnik Y and Douds DD (eds) Arbuscular mycorrhizas: Physiology and Function. Dordrecht: Kluwer Academic Publishers 307–343.

Gerdmann JW, and Trappe JM (1974). The Endogynaceae in the Pacific North West. Mycologia Memoir 5: 76.

Gianinazzi S, Gollettr A, Binet MN, van Tuien D, Redecker D, and Wipf D (2010). Agroecology: the key role of arbuscular mycorrhizas in ecosystrm services. Mycorriza 20: 519–530.

Gildon A, and Tinker PB (1983). Interactions of vesicular arbuscular mycorrhizal infection and heavy metals in plants. The effects of infection on uptake of copper. New Phytol. 95: 263–268.

Giovannetti M, Sbrana C, Avio L, and Strain P (2004). Patterns of below-ground plant interconnections established by means of arbusculr mycorrhizal networks. New Phytol 164: 175–181.

Giri B, Kapoor R, and Mukerji (2003). Influence of arbuscular mycorrhizal fungi and salinity on growth, biomass and mineral nutrition of *Acacia auriculiformis.* Biol and Fert Soils 38: 170–175.

Goltapeh EM, Danesh YR, Prasad R and Varma A (2008). Mycorrhizal Fungi: what we know and what should we know? In: Mycorrhiza 3rd edition (ed. Varma A), Springer-Verlag Berlin Heidelberg, 3–27.

Gordon-Weeks R, Tong YP, Davies TGE, and Leggewie G (2003). Restricted spatial expression of a high-affinity phosphate transporter in patato roots. J Cell Sci 116: 3134–3144.

Govindarajulu M, Pfeffer PE, Jin HR, Abubaker J, Douds DD, Allen JW, Bucking H, Lammers PJ, and Shachar-Hill Y (2005). Nitrogen transfer in arbuscular mycorrhizal symbiosis. Nature 435: 819–823.

Grahman JH, Leonard RT, and Menge JA (1981). Membrane-mediated decrease in root exudation responsible for phosphorus inhibition of vesicular-arbuscular mycorrhiza formation. Plant Physiol 68: 549–552.

Guether M, Neuhauser B, Balestrini R, Dynowski M, Landewig U, and Bonfante P (2009). A mycorrhizal specific nitrogen transporterfrom *Lotus japonicas* acquires nitrogen released by arbuscular mycorrhizal fungi. Plant Physiol 150: 73–83.

Habte M, Zhang YC, and Schmitt DP (1999). Effectiveness of *Glomus* species in protecting white clover against nematode damage. Can J Bot 77: 135–139.

Hajiboland R, Aliasgharzadeh N, Laiegh SF, and Poschenreider C (2010). Colonization with arbuscular mycorrhizal fungi improves salinity stress in tomato (*Solanum lycopersicum* L.) plants. Plant Soil 331: 313–327.

Hamblin AP (1995). Land quality indicators:who, how and for whom? Ecol Econ. 12: 341-348.

Hamel C (1996). Prospects and problems pertaining to the managementof arbuscular mycorrhizae in agriculture. Agr Ecos Environ 60: 197-210.

Hamel C, Dalpe Y, Lapierre C, Simard RR and Smith DL (1996). Endomycorrhizae in a newly cultivated acidic mesdow: Effects of three years of barley cropping, tillage, lime and phosphorus on root colonization and soil infectivity. Biol. Fertil Soils 21: 160-165.

Hamel C, and Smith DL (1991). Interspecific N-transfer and plant development in a mycorrhizal field-grown mixture. Soil Bio Biochem 23: 661–665.

Harley JL, and Smith SE (1983). Mycorrhizal symbiosis. London: Academic Press.

Harwani D, Choudhary P, Dhaker S, Prasad DK, and Mahna SK (2008). Tripartite Symbiotic Association: Legume-Rhizobia- Mycorrhiza. In: Maheswari DK and Dubey RC (eds) Potential Microorganisms for Sustainable Agriculture-A Techno-Commercial Perspective. IK International Publishing House Pvt Ltd., New Delhi, India p. 406–435.

Harwood RR (1990). A history of sustainable agriculture. In: Lal R and Pierce FJ (eds) Soil Management for Sustainability, Soil and Water Conservation. Society of America. Ankeny pp. 3–19.

Harisson MJ, and Dixon RA (1993). Isoflavonoid accumulation and expression of defense gene transcripts during the establishment of vesicular arbuscular mycorrhizal associations in roots of *Medicago truncatula*. Mol Plant-Microbe Interact 6: 643–654.

He H, Critchley C, and Bledsoe C (2003). Nitrogen transfer within and between plants through common mycorrhizal networks (CMNs) Crit Rev Plant Sci 22: 531–567.

Hell R, and Hillebrand H (2001). Plant concept for mineral acquisition and allocation. Current Opinion in Biotechnology 12: 161–168.

Hildebrandt U, Kaldorf M and Bothe H (1999). The zinc violet and its colonizationby arbuscular mycorrhizal fungi. J Plant Physiol 154: 709-717.

Hildebrandt U, Quziad F, Marner FJ, and Bothe H (2006). The bacterium *Paenibacillus validus* stimulates growth of the arbuscular mycorrhizal fungus *Glomus intraradices* up to the formation of the fertile spores. FEMS Microbiol Lett 254: 258–267.

Hinsinger P, Gobran GR, Gregory PJ and Wenzel WW. (2005). Rhizosphere geometry and heterogeneity arising from root mediated physical and chemical processes. New Phytol 168: 293-303.

Ho I, and Trappe JM (1975). Nitrate reducing capacity of two vesicular –arbuscular fungi. Mycologia 67: 886–888.

Hodge A, Campbell CD, and Fitter AH (2001). An arbuscular mycorrhizal fungus accelerates decomposition and acquires nitrogen directly from organic material. Nature 413:297–299.

Hodge A, and Fitter AH (2010). Substantial nitrogen acquisition by arbuscular mycorrhizal fungi from organic materials has implications for N cycling. Proc Natl Acad Sci USA 107: 13754–13759.

Hoff T, Stummann BM, and Henningsen KW (1992). Structure, function and regulation of nitrate reductase in higher plants. Physiol Plant 84: 616–624.

Hol GWH and Cook R (2005). An overview of arbuscular mycorrhizal fungi-nematode interactions. Basic Appl. Ecol. 6: 489-503.

Hooker JE, jaizme-Vegas M, and Atkinson D (1994). Impacts of arbuscular mycorrhizal on sustainable agriculture and natural ecosystems (Gianinazzi S, Schepp H (eds), Birkhauser Verlag Basel, Switzerland pp. 191–199.

Hooper L, and Cassidy A (2009). A review of the health care potential of bioactive compounda. J Sci Food Agric 86: 1805–1813.

Jackson W (1980). New Roots for Agriculture, Friends of the Earth. San Franciso.

Jacquelinet-Jeanmougin J, Gianinazzi-Pearson V, and Gianinzzi S (1997). Endomycorrhizas in the Gentianaceae. II. Ultrstructural aspects of symbiont relationships in *Gentiana lutea* L. Symbiosis 3: 269–286.

Jaizme-Vega MC, Renoury P, Pinochet J, and Jaumot M (1997). Interactions between the root-knot nematode *Meliodogyne incognita* and *Glomus mosseae* in banana. Plant Soil 196: 27–35.

Jaizme-Vega MC, Rodriguez-Romero AS, and Nunez LAB (2006). Effect of the combined inoculation of arbuscular mycorrhizal fungi and plant growth-promoting rhizobacteria on papaya (*Carica papaya* L) infected with root-knot nematode *Meloidogyne incognita.* Fruits 61: 151–162.

Javot H, Pumplin N, and Harrison MJ (2007). Phosphate in the arbuscular mycorrhizal symbiosis transport properties and regulatory roles. Plant Cell Environ 30: 310–322.

Jeffries P, Gianinazzi S, Perotto S, Turnau K, and Barea JM (2003). The contribution of arbuscular mycorrhizal fungi in sustainable maintenance of plant health and soil fertility. Biol Fert Soils 37: 1–16.

Johansen A, Finlay RD, and Olsson PA (1996). Nitrogen metabolism of external hyphae of the arbuscular mycorrhizal fungus *Glomus intraradices.* New Phytol 133: 705–712.

Johnsen A, Jakobsen I, and Jensen ES (1993). Hyphal transport by avesicular-arbuscular mycorrhizal fungus of N applied to the soil as ammonium or nitrate. Biol Fert Soils 16: 66–70.

Johnson NC, Rowland DL, Corkidi L, Egerton-Warburton LM, and Allen EB (2003). Nitrogen enrichment alters mycorrhizal allocation at five mesic to semiarid grasslands. Ecology 84: 1895–1908.

Johnson NC, Wilson GWT, Bowker MA, Wilson JA, and Miller RM (2010). Resource limitation is a driver of local adaptation in mycorrhizal symbioses. Proc Natl Acad Sci USA 107: 2093–2098.

Joner EJ, and Leyval C (1997). Uptake of ^{109}Cd by roots and hyphae of *Glomus mossese Trifolium subterraneum* mycorrhiza from soil amended with high and low concentration of cadmium. New Phytol 135: 353–360.

Joner EJ, and Leyval C (2001). Influence of arbuscular mycorrhiza on clover and ryegrass grown together in a soil spiked with polycyclic aromatic hydrocarbons. Mycorrhiza 10: 155–159.

Joner EJ, Briones R, and Leyval C (2000). Metal binding capacity arbuscular mycorrhizal mycelium. Plant Soil 226: 227–234.

Joner EJ, van Aarle IM, and Vosatka M (2000). Phosphatase activity of extra radical arbuscular mycorrhizal hyphae: a review. Plant Soil 226: 199–210.

Juniper S, and Abbott IK (1993). Vesicular arbuscular mycorrhizas and soil salinity. Mycorrhiza 4: 45–57.

Kaldorf M, Zimmer W and Bothe (1994). Genetic evidence for occurrence of assimilatory nitrate reductase in arbuscular mycorrhizal fungi. Mycorrhiza 5: 23-28.

Kamal S, Prasad R and Varma A (2010). Soil microbial diversity in relation to heavy metals. In: Sherameti I and Varma A (eds) Soil Heavy Metals. Springer-Verlag Berlin Heidelberg 31–64.

Kapoor R, Evelin H, Mathur P, and Giri B (2013). Arbuscular mycorrhuza: Approaches for abiotic stress tolerance in crop plants for sustainable agriculture. In: Tuteja N and Gill SS (eds) Plant accumulation to environmental stress. Springer, LLC, pp. 359–401.

Karandashov V and Bucher M (2005). Symbiotic phosphate transport in arbuscular mycorrhizas. Trends Plant Sci 10: 22-29.

Kaya C, Higgs D, Kirnak H and Tas I (2003). Mycorrhizal colonization improves fruit yield and water use efficieny in water melon (Citrullus lanatus Thunb) grown under well watered and water stressed conditions. Plant Soil 253: 287-292.

Khalil S, Loynachan TE, and Tabatabai MA (1994). Mycorrhizal dependency and nutrient uptake by improved and unimproved corn and soybean cultivars. Agron J 86: 949–958.

Khalvati MA, Hu Y, Mozafar A, and Schmidhalter U (2005). Quantification of water uptake by arbuscular mycorrhizal hyphae and its significance for leaf growth, water relations, and gas exchange of barley subjected to drought stress. Plant Biology 7: 706–712.

Klepper B (1990). Root growth and water uptake. In: Irrigation of agricultural crops Agron. Mono. No-30, BA Stewart and DR Nielsen Eds. Am Soc Agron Crop Sci Soc Am and Soil Sci Soc Am Maldison, Wisconsin, pp. 281-322.

Klironomos JN, Hart MM, Gurney J, and Moutoglis P (2001). Interspecific differences in the tolerance of arbuscular mycorrhizal fungi to freezing and drying. Can J Bot 79: 1161–116.

Klironomos JN, and Kendrick B (1996). Palatability of microfungi to soil arthropods in relation to the functioning of arbuscular mycorrhizae. Biol Fert Soils 21: 43–52.

Koide RT (1991). Nutrient supply, nutrient demand and plant response to mycorrhizal infection, New Phytol 117: 265–286.

Koide RT, and Mosse (2004). A history of research on arbuscular mycorrhiza. Mycorrhiza 14: 145–163.

Kosuta S, Chabaud M, Lougnon G Gough C, Denarie J, Barker D, and Becard G (2003). A diffusible factor from arbuscular mycorrhizal fungi induces symbiosis-specific MtENODI I expression in roots of *Medicago truncatuda*. Plant Physol 131: 952–962.

Kothari SK, Marschner H, and Romheld V (1990). Direct and indirect effects of VA mycorrhizae and rhizosphere microorganisms on mineral nutrient acquisition by maize (*Zea mays*) in a calcareous soil. New phytol 116: 637–645.

Kuhn G, Hijri M, and Sanders IR (2001). Evidence for the evolution of multiple gemones in arbuscular mycorrhizal. Nature 414: 745–748.

Kumar P, Duschenkov V, Motto H, and Raskin I (1995). Phytoextraction: the use of plants to remove heavy metals frim soils. Environ Sci Technol 29: 1232–1238.

Lambert DH, Baker DE, and Cole Jr H (1979). The role of mycorrhizae in the interactions of phosphorus with zinc copper and other elements. Soil Sc Soc Am Jour 43: 976–980.

Lambers H, Finnegan PM, Laliberté E, Pearse SJ, Ryan MH, Shane MW, and Veneklaas EJ (2011). Phosphorus nutrition of Proteaceae in severely phosphorus-impoverished soils: are there lesson to be learned for future crops? Plant Physiol 156: 1058–1066.

Lambers H, Raven JA, Shaver GR, and Smith SE (2008). Plant nutrient acquisition strategies changed with soil age. Trends Ecol Evol 23: 95–103.

Leake JR, Johnson D, Muckle GF and Read DJ. (2004). Network of power and influence: the role of mycorrhizal mycelium in controlling plant coomunities and agroecosystem functioning. Can J Bot 82: 1016-1045.

Leyval C, and Binet P (1998). Effect of polyaromatic hydrocarbons in soil on mycorrhizal plants. J. Environ Qual 27: 402–407.

Leyval C, Singh BR, and Joner EJ (1995). Occurrence and infectivity of arbuscular mycorrhizal fungi in some Norwegian soil influenced by heavy metals and soil properties. Water Air Soil Pollut 83: 203–216.

Li Q, Ling W, Gao Y, Li F, and Xiong W (2006). Arbuscular mycorrhizal bioremediation and its mechanisms of organic pollutants-contaminated soils]. Ying Yong Sheng Tai Xue Bao 17: 2217–21.[Article in Chinese].

Li T, and Zhao ZW (2005). Arbuscular mycorrhiza in hot and arid ecosystem in southwest China. Applied Soil Ecology 29: 135–141.

Linderman RG (2000). Effects of mycorrhizas on plant tolerance to diseases. In Arbuscular Mycorrhizas: Physiology and Function. Kapulnik Y and Douds DD (Eds), Kluwer Academic Publisher, Boston, 345-365.

Liu RJ, Wang FY, and Meng XX (2002). Arbuscular mycorrhizal fungi on Islands of Bohai Bay. Mycosystema 21: 525–532 (in Chinese).

Losi ME, Amrhein C, and Frankenberger WT (1994). Bioremediation of chromate contaminated groundwater by reduction and precipitation in surface soils. J Environ Qual. 23: 1141-1150.

Mader P, Vierheilig H, Streitwolf-Engel R, Boller T, Frey B, Christie P, and Wiemken A (2000). Transport of N-15 from a soil compartment separated by a polytetrafluoroethylene membrane to plant roots via the hyphae of arbuscular mycorrhizal fungi. New Phytol 146: 155–161.

Maherali H, and Klironomos JN (2007). Influence of phylogeny on fungal community assembly and ecosystem functioning. Science 316: 1746–1748.

Mathur N, Singh J, Bohra S, and Vyas A (2007). Arbuscular mycorrhizal status of medicinal halophytes in saline areas of Indian Thar Desert. International Journal of Soil Science 2: 119–127.

Manchanda G, and Garg N (2011). Alleviation of salt-induced ionic, osmotic and oxidative stresses in *Cajanus cajan* nodules by AM inoculation. Plant Biosyst 145: 88–97.

Marschner H (1986). Mineral nutrition of higher plants Academic Press, London.

Marulanda A, Azcón R, and Ruiz-Lozano JM (2003). Contribution of six arbuscular mycorrhizal fungal isolates to water uptake by *Lactuca sativa* plants under drought stress. Physiol Plant 119: 526–533.

Matson PA, Parton WJ, Power AG and Swift MJ (1997). Agricultural intensification and ecosystem properties. Science 277: 504-509.

Matsubara Y, Ohba N, and Fukui H (2001). Effect of arbuscular mycorrhizal fungus infection on the incidence of fusarium root rot in asparagus seedlings. J Jap Soc Hort Sci 70: 202–206.

Miller RM and Jastrow JD (2000). Mycorrhizal fungi influence soil structure. In Arbuscular Mycorrhiza: Physiology and Funtion. Dordrecht (Netherland) Kluwer, pp. 3-18.

Miller RM, Reinhardt DR, and Jastrow JD (1995). External hyphal production of vesicular arbuscular mycorrhizal fungi in pasture and tallgrass praire communities. Oecologia 103: 17–23.

Millner PD, Mulbry WW, and Reynolds SL (2001). Taxon-specific oligonucleotide primers for detection of *Glomus etunicatum.* Mycorrhizae 10: 259–265

Mohammadi K, Khalesro S, Sohrabi Y, and Heidari G (2011) A review: Beneficial effects of mycorrhizal fungi for plant growth. J Appl Environ Boil Sci 1: 310–319.

Morandi D (1996). Occurrence of phytoalexins and phenolic compounds in endomycorrhizal interactions, and their potential role in biological control. Plant Soil 185: 241–452.

Morton JB, and Benny GL (1990). Revised classification of arbuscular mycorrhizal fungi (Zygomycetes): a new order, Glomales, two new suborders, Glomineae and Gigosporineae, and two new families, Acaulosporaceae and Gigasporaceae, with an emendation of Glomaceae. Mycotaxon 37: 471–491.

Morton JB, and Benny GL (2001). Two new families of Glomales, Archaeosporaceae and Paraglomaceae, with two new genera *Archaeospora* and *Paraglomus*, based on concordant molecular and morphological characters. Mycologia 93: 181–195.

Mouk BO, and Ishii (2006). Effect of arbuscular mycorrhizal fungi on tree growth and nutrient uptake of Sclerocarya birrea under water stress, salt stress and flooding. J Jap Soc Hortic Sci 75: 26–31.

Muller A, George F, and Gabriel-Neumanu F (2013). The symbiotic recapture of nitrogen from dead mycorrhizal and nonmycprrhizal roots of tomato plants. Plant soil 364: 341–355.

Munkvold L, Kjoller R, Vestberg M, Rosendahl S, and Jakobsen I (2004). High functional diversity within species of arbuscular mycorrhizal fungi. New Phytol 164: 357–364.

Muthukumar T, Sha LQ, Yang XD, Cao M, Tang JW, and Zheng Z (2003). Mycorrhiza of plants in different vegetation types in tropical ecosystems of Xishuangbanna, Southwest China. Mycorrhiza 13: 289–297.

Nageli C (1842). Pilze in Innern von Zellen. Linnaea 16: 278–285.

Nelsen CE (1987). The water relations of vesicular arbuscular mycorrhizal systems. In: Safir GR (ed) Ecophysiology of VA mycorrhizal plants. CRC Press, pp. 71–79.

Newsham KK, Fitter AH, and Watkinson AK (1994). Root pathogenic and arbuscular mycorrhizal fungi determine fecundity of asymptomic plants in the field. J Ecol 82: 805–814.

Oehl F and Sieverding E (2004). *Pacispora,* a new vesicular-arbuscular mycorrhizal fungal genus in the Glomeromycetes. Journal of Applied Botany and Food Quality 78: 72–82.

Oehl F, Wiemken A, and Sieverding E (2003). *Glomus spinuliferum*: a new ornamented species in the Glomales. *Mycotaxon* **86**: 157–162.

Oehl F, Sýkorová Z, Redecker D, Wiemken A, and Sieverding E (2006). *Acaulospora alpina,* a new arbuscular mycorrhizal fungal species characteristic for high mountainous and alpine regions of the Swiss Alps. Mycologia 98: 286–294.

Oehl F, Sieverding E, Palenzuela J , Ineichen K and da Silva A (2011a). Advances in Glomeromycota taxonomy and Classification. IMA Fungus 2: 191-199.

Oehl F, Silva GA, Goto BT and Sieverding E (2011b). Glomeromycetes: three new genera and glomoid species reorganized. Mycotaxon 116: 75-120.

Oehl F, Silva DKA, Mala LC, Sousa NMF, Vieira HEE, Silva GA (2011c). *Orbispora* gen. nov., ancestral in the Scutellosporaceae (Glomeromycetes). Mycotaxon 116: 161-169.

Oehl F, Silva GA, Goto BT, Maia LC and Sieverding E (2011d). Glomeromycota: two new classes and a new order. Mycotaxon 116: 365-379.

Oehl F, Silva GA, Sanchez-Castro I, Goto BT, Maia LC, Vieera HEE, Barea JM, Sieverding E and Palenzuela J(2011e). Revision od Glomeromycetes with entrophosporoid and glomoid spore formation with three new genera. Mycotaxon 117: 297-316.

Oehl F, Sýkorová Z, Błaszkowski J, Sánchez-Castro I, Coyne D, Tchabi A, Lawouin L, Hountondji FCC, and Silva GA (2011f). *Acaulospora sieverdingii,* an ecologically diverse new fungus in the Glomeromycota, described from lowland temperate Europe and tropical West Africa. Journal of Applied Botany and Food Quality 84: 47–53.

Oehl F, Silva GA, Palenzuela J, Sanchez-Castro I, Castillo C and Sieverding E (2011g). *Acaulospora punctata,* a new fungal species in the Glomeromycetes from mountainous altitudes of the Swiss Alps and Chilean Andes. Nova Hedwigia 93: 353-362.

Oehl F, Palenzuela J, Sánchez-Castro I, Kuss P, Sieverding E, and Silva GA (2012). *Acaulospora nivalis*, a new fungus in the Glomeromycetes, characteristic for high alpine and nival altitudes of the Swiss Alps. Nova Hedwigia 95: 105–122.

Oehl F, Tohabi A, Siva GA, Sanchez-Castro l, Palenzuela J, do Monte Junior IP, Lawouin LE, Coyne D, and Hou Hountondji FCC (2014). *Acaulospora spinosissima, a* new arbuscular mycorrhizal fungus from the southern Guinea Savanna in Benin. Sydowia 66: 29–42.

Orłowska E, Przybyłowicz W, Orlowski D, Mongwaketsi NP, Turnau K, and Mesjasz-Przybyłowicz J (2013). Mycorrhizal colonization affects the elemental distribution in roots of Ni-hyperaccumulator Berkheya coddii Roessler. Environ Pollut 175: 100–109.

Pacovsky RS (1986). Micronutrient uptake and distribution in mycorrhizal or pH fertilized soybeans. Plant soil 95: 179–188.

Pandey RK, Prasad R, Mangunath VG and Goswami BK (2010). Biotechnology of biocontrol based biopesticides: Core component of biological deterrents. In: Gupta VK, Tuohy M and Gaur RK. Fungal Biochemistry and Biotechnology. LAP LAMBERT Academic Publishing AG & Co. KG, Germany 214–244.

Pandey R and Prasad R (2015). Adverse effect of indiscriminate use of chemical pesticides and its alternative eco-friendly means for management of diseases and pests infecting agricultural crops. In: Environmental Microbiology (ed. Prasad R), IK International, New Delhi, India 166–209.

Pierce FJ, and Lal R (1991). Soil management in the 21st century. In: Lal R and Pierce FJ (eds) Soil Management for Sustainability, Soil and Water Conservation. Society of America, Ankeny pp175–179.

Pinochet J, Jaizme MC, Fernandez C, Jaumot M, and De Waele D (1998). Screening bananas for root-knot nematode (*Meloidogyne* spp.) and lesion (*Pratylenchus goodeyi*) resistance for Canary Island. Fund Appl Nematol 21: 17–23.

Porcel R, Aroca R, Azcón R, and Ruiz-Lozano JM (2006). PIP aquaporin gene expression in arbuscular mycorrhizal *Glycine max* and *Lactuca sativa* plants in relation to drought stress tolerance. Plant Mol. Boil. 60: 389–404.

Porcel R, Aroca R, and Ruiz-Lozano JM (2012). Salinity stress alleviation using arbuscular mycorrhizal fungi: A review. Agron Sustain Dev 32: 181–200.

Raju PS, Clark RB, Ellis JR, and Maranville JW (1990). Effects of the species of VA-mycorrhizal fungi on growth and mineral uptake of sorghum at different temperatures. Plant Soil 12: 165–170.

Rao MS and Gowen SR (1998). Biomanagement of *Meloidogyne incognita* on tomato by integrating Glomus deserticola and Pasteuria penetrans. J. Plant Dis Prot 105: 49-52.

Read DJ (1991). Mycorrhizas in ecosystems. Experientia 47: 376–391.

Redecker D, and Raab P (2006). Phylogeny of Glomeromycota (Arbuscular mycorrhizal fungi): recent developments and new bene markers. Mycologia 98: 885–895.

Redhead JF (1977). Endotrophic mycorrhiza in Nigeria: species of the endogonaceae and their distribution. Trans Brit Mycol Soc 69: 275–280.

Reganold JP, Papendick RI, and Parr JF (1990). Sustainable agriculture. Sci Am 62: 112–120.

Reilley KA, Banks MK, and Schwab AP (1996). Dissipation of polycyclic aromatic hydrocarbons in the rhizosphere. J Environ Qual 25: 212–219.

Rilling MC, and Allen MF (1999). What is the role of arbuscular mycorrhizal fungi in plant to ecosystem responses to elevated atmospheric CO_2? Mycorrhiza 9: 1–9.

Rillig MC and Mummey DL (2006). Mycorrhizas ans soil structure. New Phytol. 171: 41-53.

Rilling, MC, Wright SF, and Eviner VT (2002). The role of arbuscular mycorrhizal fungi and glomalin in soil aggregation: comparing effects of five plant species. Plant and Soil 238: 325–333.

Rodale JI (1983). Breaking new ground: The search for Sustainable Agriculture. The Futurist 1: 15–20.

Rosendahl CN, and Rosendahl S (1991). Influence of vesicular arbuscular mycorrhizal fungi (*Glomus* spp) on response of cucumber (*Cucumis sativus* L) to salt stress. Environ Exp Bot 31: 313–318.

Rosendahl S, and Stukenbrock EH (2004). Community structure of arbuscular mycorrhizal fungi in undisturbed vegetation revealed by analysis of LSU rDNA sequence. Mol Ecol 13: 3179–3186.

Ross JP (1972). Influence of Endogone mycorrhiza and phytophthora rot of soybean. Phytopathology 62: 896-897.

Ruiz-Lozano JM (2003). Arbuscular mycorrhizal symbiosis and alleviation of osmotic stress: a new perspectives for molecular studies. Mycorrhiza 13: 309–317.

Ruiz-Lozano JM, and Azcón R (2000). Symbiotic efficiency and infectivity of an autochthonous arbuscular mycorrhizal *Glomus* sp from saline soils and *Glomus deserticola* under salnity. Mycorrhiza 10: 137–143.

Saito M (2000). Symbiotic exchange of Nutrients in Arbuscular Mycorrhizas: Transport and Transfer of Phosphorus. In: Kupulik Y and Douds D. (Eds) Arbuscular Mycorrhizas: Physiology and Function. Kluwer Academic Pub., Dodrdecht, Netherland, pp. 85–102.

Salzer P, Corbere H, and Boller T (1999). Hydrogen peroxide accumulation in *Medicago truncatula* roots colonized by the arbuscular mycorrhiza-forming fungus *Glomus intraradices*. Planta 208: 319–325.

Sanchez-Diaz M, and Honrubia M (1994). Water relation and elleviation of drought stress in mycorrhizal plants. In: Gianinazzi S and Schüepp H (eds) Impact of arbuscular mycorrhiza on sustainable agriculture and natural ecosystems. Birkhauser, Basel, Switzerland, pp. 167–178.

Sanders FE, Tinker BP, Black RLB, and Palmerly (1977). The development of endomycorrhizal root systems I. Speed of infection and growth-promoting effects with four species of vesicular- arbuscular endophyte. New Phytol 78: 257–268.

Santos-Gonzalez JC, Finlay RD, and Tehler A (2007). Seasonal dynamics of arbuscular mycorrhizal fungal communities in roots in a seminatural grassland. Appl Environ Microbial 75: 5613–5623.

Schachtman DP, Read RJ, and Ayling SM (1998). Phosphorus uptake by plants from soil to cell. Plant Physiol 116: 447–453.

Schenck NC, and Smith GS (1982). Responses of six species of vesicular arbuscular mycorrhizal fungi and their effect on soybean at four soil temperatures. New Phytol 93: 193–201.

Schliemann W, Ammer C, and Strack D (2008). Metabolites profiling of mycorrhizal roots of *Medicago truncarula*. Phytochem 69: 112–146.

Schmidt E (1982). Nitrification in soil. In: Stevenson FJ (ed.) Nitrogen in agricultural soils. American Society of Agronomy. Madison. Wisc. pp. 253–587.

Schützendübel A, Majcherczyk A, Johannes C, and Huttermann A (1999). Degradation of fluorine, anthracene, phenanthracene, fluoranthene and pyrene lacks connection to the production of extracellular rnzymes by *Pleurotus ostreatus* and *Bjerkandera adusta*. Int Biodeterior Biodegrad 43: 93–100.

Schwab AP, and Banks MK (1994). Biologically mediated dissipation of polyaromatic hydrocarbons in root zone. In: Anderson TA and Coats JR (eds) Bioremediation through rhizosphere technology. American Chemical Society, Washington, DC, pp. 132–141.

Schüβler A, Schwarzott D, and Walker C (2001). A new fungal phylum, the Glomeromycota: phylogeny and evolution. Mycol Res 105: 1413–1421.

Seeram NP (2008). Berry fruits compositional elements, biochemical activities, and the impact of their intake on human health, performance and disease. J Agric Food Chem 56: 627–629.

Sèguin S, Dèsaulniers N, Dalpè Y, and Lèvesque CA (2003). Development of AMF strain specific primers for detection in field grown colonized roots. 4th Int Conf. Mycorrhizae. Montreal Quebec Canada, August 10-15 2003, No 444.

Shann JR. and Boyle JJ (1994). Influence of plant species on in situ rhizosphere degradation. In: Anderson TA and Coats JR (eds) Bioremediation through rhizosphere technology. American Chemical Society, Washington, DC, pp. 70–81.

Sich D, Watanabe M, Devers FA, Brueckner F, Hoefgen R, and Krajinski F (2013). The arbuscular mycorrhizal symbiosis influences sulphur starvation responses of *Medicago truncatula*. New Phytol 197: 606–616.

Sieverding E and Oehl F (2006). Revision of Entrophospora and description of Kuklospora and Intraspora, two new genera in the arbuscular mycorrhizal Glomeromycetes. Journal of Applied Botany and Food Quality 80: 69–81.

Siqueira JO, Rocha IWF, Oleivra E, and Colozzi-Filho A (1990). The relative between vesicular-arbuscular mycorrhiza and lime: associated effects on the growth and nutrition of brachiara grass (*Brachiara decumbens).* Biol Fert Soil 10: 65–71.

Smith SE, and Smith FA (1990). Structure and function of the interfaces in biotrophic symbioses as they relate to nutrient transport. New Phytol 114: 1–38.

Smith SE, and Smith FA (2011). Roles of arbuscular mycorrhizas in plant nutrition and growth: new paradigms from cellular to ecosystems scales. Annu Rev Plant Biol 63: 227–250.

Smith SE, Jakobsen I, Gronlund M, and Smith FA (2011). Roles of arbuscular mycorrhizas in plant phosphorus nutrition: Interactions between pathways of phosphorus uptake in arbuscular mycorrhizal roots have important implications for understanding and manipulating plant phosphorus acquition. Plant Physiol 156: 1050–1057.

Smith SE, and Read DJ (2008). Mycorrhizal Symbiosis. Academic Press, London.

Smith SE, and Smith FA (2012). Fresh perspectives on the roles of arbuscular mycorrhizal fungi in plant nutrition and growth. Mycologia 104: 1–13.

Solaiman MDZ, and Saito M (1997). Use of sugars by intraradical hyphae of arbuscular mycorrhizal fungi revealed by radiorespirometry. New Phytol 136: 533–538.

St-Arnaud M and Vujanovic V (2007). Effect of the arbuscular mycorrhizal symbiosis on plant diseases and pests. In: Hamel C, Plenchette C (eds). Mycorrhizae in crop production: applying knowledge. Haworth, Binghampton, N.Y. (in press).

St Arnaud M, Hamel C, Caron M and Fortin JA (1994). Inhibition of *Pythium ultimum* in roots and growth substrate of mycorrhizal Tagetes patula colonized wuth *Glomus intraradices*. Can J Plant Pathol. 16: 187-194.

Stewart BA, Lal R, and El-Swaify SA (1991). Sustaining the resource base on an expanding world agriculture. In: Lal R and Pierce FJ (eds) Soil Management for Sustainability. Soil and Water Conservation, Society of America, Ankeny pp. 125–144.

Strack D, and Fester T (2006). Isoprenoid metabolism and plastid reorganization in arbuscualar mycorrhizal roots. New Phytol 172: 22–34.

Strobel NE, Hussey RS and Roncadori RW (1992). Interactions of vesicular arbuscular mycorrhizal fungi *Meloidogyne incognita* and soil fertility on peach. Phytopathology 72: 690-694.

Subramanian KS, and Charest C (1999). Acquisition of N by external hyphae of an arbuscular mycorrhizal fungus and its impact on physiological responses in maize under drought stress and well watered conditions. Mycorrhiza 9: 69–75.

Sumbali G, and Mehrotra RS (2009). Microbial Interactions. Principles of Microbiology. Tata Mc Graw Hill Education Pvt Ltd., New Delhi, pp. 469–494.

Sykorova Z, Ineichen K, Wiemken A, and Redecker D (2007). The cultivation bias:different communities of arbuscular mycorrhizal fungi detected in roots from the field, from bait plants transplanted to the field and from a greenhouse trap experiment. Mycorrhiza 18: 1–14.

Sylvia DM, and Chellemi DO (2001). Interaction amiong soil inhabiting fungi and their implication for biological control of root pathogens. Adv Agron 73: 1–33.

Tamasloukht MB, Sejalon-Delmas N, Kluever A, Jauneau A, Roux C, Becard G, and Franken P (2003). Root factors induce mitochondrial-related gene expression and fungal respiration during the decelopment switch from asymbiosis to presymbiosis in the arbuscular mycorrhizal fungus *Gigaspora rosea.* Plant Physiol 131: 1468–1478.

Tester M and Davenport R (2003). Na+ tolerance and Na+ transport in higher plants. Annals of Botany 91: 503–527.

Thygesen K, Larsen J, and Bodker L (2004). Arbuscular mycorrhizal fungi reduce development of pes root-rot caused by *Aphanomyces euteiches* using oospores as pathogen inoculums. Eur J Plant Pathol 10: 411–419.

Tian CJ, Kasiborski B, Koul R, Lammers PJ, Bucking H, and Shachar-Hill Y (2010). Regulations of the nitrogen transfer pathway in the arbuscular mycorrhizal symbiosis: Gene characterisation and the coordination of expression with nitrogen flux. Plant physiol 153: 1175–1187.

Tinker PBH, and Nye PH (2000). Solute movement in the rhizosphere. Oxford, UK: Oxford Univ Press, p. 464.

Tobar RM, Azcón R, and Barea (JM) (1994). Improved nitrogen uptake and transport from ^{15}N-labeled nitrate by external hyphae of arbuscular mycorrhiza under water stressed conditions. New Phytol 126: 119–122.

Tommerup IC, and Sivasithamparam K (1990). Zygospores and asexual spores of *Gigaspora decipiens*, an arbucular mycorrhizal fungus. Mycological Research 94: 897–800.

Tonin C, Vandenkoornhuyse P, Joner EJ, Straczek J, and Leyval C (2001). Assessment of arbuscular mycorrhizal fungi diversity in the rhizosphere of *Viola calaminaria* and effect of these fungi on heavy metal uptake by clover. Mycorrhiza 10: 161–168.

Torres-Barragan A, Zavaleta-Mejia E, Gonzalez-Chavez C and Ferrera-Cerrato R (1996). The use of arbuscular mycorrhiza to control onion white root (*Sclerotium cepivarum Berk*) under field condition. Mycorrhiza 6: 253-258.

Turnau K, and Mesjasz-Przybylowicz J (2003). Arbuscular mycorrhiza of Berkheya coddii and other Ni-hyperaccumulating members of Asteraceae from ultramafic soils in South Africa. Mycorrhiza 13(4): 185–190.

van Dam NM, Harvey JA, Wackers FL, Bezemer TM, van der Putten WH, and Vet LEM (2003). Interactions between aboveground and belowground induced responses against phytophages. Basic Appl Ecol 4: 63–77.

Vance CP (2003). Phosphorus acquisition and use: critical adaptations by plants for securing a nonrenewable resource. New Phytol 157: 423-447.

Vandenkoornhuyse P, Ridgway KP, Watson IJ, Fitter AH, and Young JPW (2003). Co-existing grass species have distinctive arbuscular mycorrhizal communities. Mol Ecol 12: 3085–3095.

Walker C, and Schüßler A (2004). Nomenclatural clarifications and new taxa in the Glomeromycota. Mycological Research 108: 981-982

Wang FY, and Liu RJ (2002). *Glomus zaozhuangianus*, a new species arbuscular mycorrhizal fungi. Mycosystema 21: 196–202 (in Chinese).

Whipps JM (2004). Prospects and limitations for mycoorhizas in biocontrol of root pathogens. Can J Bot 82: 1198-1227.

Wilson NG Bradley G (1996). Enhanced degradation of petrol (Slovene diesel) in an aqueous system by immobilized *Pseudomonas fluorencens.* J Appl Bacteriol 80: 99–104.

Yang L, Wang GH, Ren LC, and Zhao ZW (2002). Arbuscular mycorrhizae of the family Amaranthaceae. Acta Botanica Yunnanica 24: 37–40 (in Chinese).

Yao MK, Tweddell, RJ and Désilets H (2002). Effect of two vesicular-arbuscular mycorrhizal fungi on the growth of micropropagated potato plantlets and on the extent of disease caused by *Rhizoctonia solani*. Mycorrhiza 12: 235-242.

Yokoyama K, Tateishi T, Marumoto T, and Saito M (2002). A molecular marker diagnostic of a specific isolate of an arbuscular mycorrhizal fungus *Gigaspora margarita.* FEMS Microbiol Lett 212: 171–175.

11

Compatible Solute in Relation to Microbes and Their Biotechnological Significance

Keshwanand Tripathi[1*], G Abraham[1], V Rai[2], DW Dhar[1], V Rashi[1] and RK Yadav[1]

[1]Centre for Conservation and Utilization of Blue Green Algae
Division of Microbiology, Indian Agricultural Research Institute, New Delhi–110012
[2]National Research Centre on Plant Biotechnology
Pusa Campus, New Delhi–110012
*Corresponding author: *tripathikn009@gmail.com*

1. INTRODUCTION

Ions play an important role in cellular metabolic activity of halophilic and halotolerant microbes. Among them, sodium is one of the most abundant and widely occurring cation in agro-ecosystem. Salinity induces the reduction of water potential, which leads increased Na^+ concentrations of the cells. Inorganic ions are necessary for the activity of intracellular enzymatic machinery. Microorganisms growing at high salt concentrations have to maintain its intracellular environment at least iso-osmotic with the salt concentration in its environment (Vreeland, 1987). Different strategies are used by diverse groups of microorganisms to achieve a high osmotic pressure in the cytoplasm while keeping the Na^+ concentration low. Compatible solute strategy is commonly employed by a vast range of moderately halophilic and halotolerant bacteria, cyanobacteria, yeasts, algae and fungi. Cells grown at saline environment accumulate osmolytes in their cytoplasm by balancing osmotic potential through the synthesis or transport of organic compatible solutes from external medium allowing these microbes to grow in a wide range of salt concentrations. Two fundamentally different strategies generally are adopted by halophilic and halotolerant prokaryotes to cope with the osmotic stress. The first strategy, used by the aerobic archaea and bacteria, is based on the accumulation of high concentrations of inorganic ions in the cytoplasm. In most of the cases, K^+ rather than Na^+ is the prominent intracellular cation and Cl^- is the dominant anion. Halophilic archaea maintained water potential upon exposure of salinity by the accumulation to potassium ions (Oren, 1999). Accumulation of sodium ions (molar range) in response to high salinity, but the ratio of cytoplasmic potassium to sodium is greater than one (Das et al., 2015; Yadav et al., 2016).

The salt-in strategy permits little flexibility and adaptability to changing conditions, as many salt-adapted enzymes and structural proteins require continuous presence of high salt for activity and stability. The second strategy is accumulation of salt concentrations in the cytoplasm, and maintaining enzyme activity and functional integrity of other proteins. Low-cytoplasmic ionic concentrations are maintained by the actively pumping of ions in extracellular medium. Equilibrium is maintained by organic solutes that are either synthesized by the cells or transported from the medium (Kempf and Bremer, 1998). The compatible solutes are sugars, amino acid, alcohol and their derivatives, such as glycine betaine, glycerol and ectoine (Roberts, 2005). Such compounds are low-molecular-weight organic compounds, soluble in water, and not inhibitory to enzymatic activities even in the molar range. The intracellular concentrations of the organic solutes are regulated according to the salinity of the external medium.

2. ADAPTIVE STRATEGY IN RESPONSE TO SALINITY

Microorganisms are directly affected by changes in external environmental conditions. Salinity is one of the major abiotic stresses limiting agricultural productivity (Shrivastava and Kumar, 2015). Plants and microbes employ osmoregulation which is one of the protective mechanisms against salinity. Two types of microbes exist in saline environments, namely, halophilic and halotolerant microorganisms. Halophilic microorganisms are capable of growing in moderate and high salt concentrations up to saturation level of NaCl. Salt is essential for growth and development of such types of microbes. Halophilic bacteria grow over an extended range of salt concentrations (5 M to saturation level of salt) (Oren, 2008). On the other hand, for halotolerant microorganisms' salt may not be essential for growth and development. They are found in all three domains of life: archaea, bacteria and eukarya (Andrew et al., 2015). To adjust the external NaCl, cells of all three domains, viz., archaea, bacteria and eukarya, accumulate a variety of small uncharged molecules in the cell cytoplasm to counteract the external osmotic pressure.

Adaptation of microorganisms to salinity stress is a complex phenomenon and involves metabolic physiological and molecular network. Extremophilic microbes, growing in adverse condition of high temperature or salinity, maintain optimum metabolic process like enzymatic activities and membrane stability. Under stress conditions, rhizobacteria produce osmoprotectants (glutamate, trehalose, proline, glycine beatine, proline betaine and ectoine, etc.) to modulate their cytoplasmic osmolarity (Tripathi et al., 1998). The periplasmic space of bacteria contains large quantities of highly anionic polysaccharides known as membrane derived oligosaccharides, which are too large to diffuse through the porin proteins and probably function to maintain the turgor pressure of the periplasm of these organisms (Sleator and Hill, 2001).

Truly halophilic microbes, halophilic archaea and extremely halophilic bacteria and cyanobacteria adopt the 'salt-in' strategy for survival. In this strategy, accumulation of K^+ and Cl^- ions are involved in the maintenance of osmotic balance. However, this mechanism is employed only by a limited number of halophiles. The aerobic halophilic *Archaea* of the order Halobacteriales accumulate K^+ at 100 fold higher as compared to surrounding medium (Oren, 2006). Bioenergetic calculations show that the mode of osmoadaptation is relatively little energy expensive. However, the presence of high concentrations of KCl in the cytoplasm requires far reaching adaptations of all proteins to enable all intracellular enzymatic systems to be active at high salt (Oren, 1999). They generally do not synthesize compatible solutes for the maintenance of osmotic adjustments. Adaptation involves the selective transport of K^+ ions into the cytoplasm and Na^+ into the medium. On the other hand, halophilic bacteria accumulate or synthesize osmolytes to maintain the osmotic balance in response to the external saline environment. These osmolytes are compatible with the cell machinery even at molar concentrations appropriately named as 'compatible solutes'. They maintain turgor pressure, electrolyte concentrations, which lead to hydration level of the cytoplasm for growth under high saline environments.

Compatible solutes are neutral charged, low-molecular weight, highly water soluble osmoregulatory compounds (Empadinhas and Costa, 2008). They are chemically diverse and, include as derivatives of alcohols, amino acids, quaternary ammonium compound glycine betaines and ectoine. The chemical structure of some of the commonly found compatible solute is shown in Fig. 11.1. Kushner (1978) classified three category of halophilic organisms based on NaCl requirements. Slight halophilic need to up to 3 percent of NaCl, moderate halophiles optimally grow at 3–15% w/v NaCl; extreme halophilic at 25% w/v NaCl.

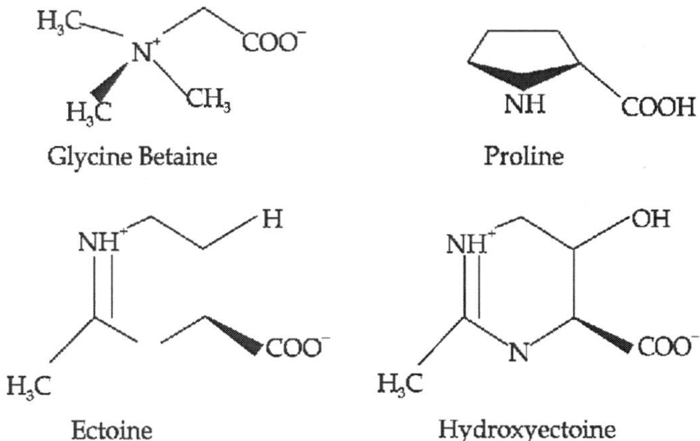

Fig. 11.1 Structures of some common compatible solutes

3. BIOSYNTHESIS OF COMPATIBLE SOLUTES

Biosynthesis of compatible solutes in salt-stressed cells of halophiles and halotolerant mainly due to maintenance of water potential and it helps the cells to take up water, establishing or maintaining turgor pressure (Galinski, 1993). The action of compatible solutes is best explained by the water hydration cell model (Bremer and Kramer, 2000). According to the model, compatible solutes do not directly interact with macromolecules, such as proteins and/or membranes, but it works as protective role for macromolecules optimum activity. According to the model, aqueous environments in or around the macromolecules prevent denaturation of cellular enzymes. Norton et al. (1982) revealed by the nuclear magnetic resonance (NMR) studies has alcoholic extract of osmolytes in cyanobacterial cells. Freshwater strains with low level of halotolerance generally accumulate sucrose and/or trehalose as their major compatible solute. However, moderately halophilic and halotolerant (marine) strains are accumulated glucosylglycerol (GG) or glucosylglycerate (GGA) as their main compatible solute. In most of the cases of halophilic/halotolerant strains that are able to grow in saturated salt concentrations usually synthesize glycine betaine or glutamate betaine (Hagemann, 2011). *Aphanothece halophytica* a halophilic unicellular cyanobacterium synthesized glycine betaine as main compatible solute at higher salinity (Waditee et al., 2003; Tripathi et al., 2013)

4. TREHALOSE AND SUCROSE

Trehalose (1–α-glucopyranosy (l→1)–α-d–glycopyranside) is an universal stress molecule, protecting cells and cellular biomolecules against abiotic stress, such as high osmolarity, heat, oxidation, desiccation and freezing. In addition to cellular protectants, trehalose is a source of carbon and energy, and a signaling molecule in specific metabolic pathways (Elbein et al., 2003; Paul, 2008). Sucrose (α-D-glucopyranosyl-(1→2)-β-D-fructofuranose) and trehalose are nonreducing disaccharides most commonly occurring in plants, fungi, halotolerant and halophilic bacteria/cyanobacteria. In contrast to other osmolytes, trehalose and sucrose can directly interact with cellular macromolecules. The steric orientation of the hydroxyl groups of these soluble carbohydrates plays an important role in the interaction with the head groups of glycerolipids. This in turn leads to trehalose and sucrose synthesis to replace at least part of the water molecule in the hydration shells of membranes (Malferrari et al., 2014). Synthesis of sucrose in cyanobacteria seems to be universal and at least one sucrose-synthesizing enzyme is encoded in almost all of the known genome sequences (Hageman, 2011).

5. BIOSYNTHESIS OF TREHALOSE

The mode of trehalose biosynthesis in cyanobacteria is not completely known. Five different enzymatic pathways have been reported for trehalose biosynthesis. The most common and well studied pathway in different species involves the enzyme trehalose-6-phosphate synthase, which catalyses the transfer of glucose molecules from UDP-glucose to glucose-6-phosphate, and formation of trehalose-6-phosphate. In the next step, trehalose-6-phosphate phosphatase catalyzes the breaking of the phosphate group from the intermediate disaccharide to generate trehalose. This pathway is reported in bacteria, archaea, yeasts, insects and plants (Paul et al., 2008). In the second pathway, trehalose is synthesized directly from maltose to trehalose synthase, which is encoded by treS. The TreS pathway has been reported in thermophilic bacterium *Thermus thermophilus* and the radiation-resistant bacterium *Deinococcus radiodurans* (Empadinhas and Costa, 2006). The third pathway, which is found in several bacteria and in the archaea converts the terminal unit of a glucose molecule to trehalose by the enzyme maltooligosyl trehalose synthase (Tsusaki et al., 1997) The fourth pathway, which is less studied, involves the conversion of ADP-glucose and glucose, instead of glucose-6-phosphate, into trehalose. Trehalose glycosyl transferring synthase (TreT), has been reported in thermophilic archaea *Thermococcus litoralis* and *Pyrococcus horikoshii* (Ryu et al., 2005). The alternative TreY/TreZ pathway for trehalose biosynthesis has been studied in model strain *Anabaena* 7120. Genome-wide microarray analysis shows, many stress induced genes in *Anabaena* strain. Some of the unregulated proteins were found to have similarities with TreY and TreZ. Studies conducted by Katoh et al. (2004) indicated that they were possibly involved in the synthesis of trehalose by transglucosidase reactions starting from glycogen.

6. BIOSYNTHESIS OF SUCROSE

Sucrose biosynthesis in microbes and plants has been less studied. Sucrose is one of the major products of photosynthesis and it is essential for growth, development, carbon storage, signal transduction and stress protection (Winter and Huber, 2000). Cardini et al. (1955) for the first time reported sucrose biosynthesis in plants. Sucrose-phosphate synthase (SPS) is a key enzyme, which catalyzes the conversion of UDP-glucose and fructose-6-phosphate into sucrose 6-phosphate, and sucrose-phosphatase (SPP), which converts sucrose via sucrose-6-phosphate intermediate. Under salinity, sucrose biosynthesis in cyanobacteria is catalysed by sucrose-phosphate synthase, which produces the intermediate sucrose-phosphate from UDP glucose and fructose-6-phosphate. Sucrose phosphate phosphatase catalyzes the hydrolysis of sucrose-6-phosphate to sucrose. The cyanobacterial SPS can also use ADP-, GTP-, and CTP-glucose as glucosyl donors in addition to UDP-glucose. Kaneko et al. (1996) identified the first cyanobacterial SPS gene, spsA in the genome of *Synechocystis* 6803 and observed a close similarity to previously characterized plant *sps* genes. Another pathway has been reported for sucrose biosynthesis, which is based on the enzyme sucrose synthase. Reversible conversion of UDP-glucose and D-fructose to sucrose and UDP has been reported. SuSy (Sucrose Synthase) activity was noticed in *Anabaena* 7119 (Porchia et al., 1999). Sucrose-phosphate synthase (SPS-I and SPS-II) and sucrose-phosphate phosphatase (SPP) were purified by ion exchange chromatography from *Anabaena* 7119 (Porchia and Salerno, 1996). Differences between SPSs of *Anabaena* and higher plants lie in substrate specificity, native molecular mass, oligomeric composition and effect of divalent cations on activity.

7. GLYCINE BETAINE

Glycine betaine (N,N,N-trimethyl glycine) (GB) is an amphoteric compound that is electrically neutral. It is one of the ubiquitous osmolytes synthesized in halophilic and halotolerant bacteria, cyanobacteria, archaea and transgenic plant chloroplast (Park et al., 2007). It is extremely soluble in water, but insoluble in non-polar hydrocarbon. GB stabilizes the structures and activities of enzymes and protein complexes, and maintains the integrity of membranes against the damaging effects of excessive salt, cold, heat and freezing (Sakamoto and Murata, 2002). In general, two enzymes are required for the production of GB in both, bacteria and higher plants through a two-step oxidation process of choline. However, in case of soil bacteria such as *Arthrobacter globiformis* and *Arthrobacter pascens*, GB synthesis occurs to direct the conversion of choline to GB through a single enzyme choline oxidase (Ikuta et al., 1977).

7.1 Biosynthesis of Glycine Betaine

Distribution of GB is seen in a wide variety of prokaryotes, eukaryotic microorganisms, higher plants and animals (Rhodes and Hanson, 1993). GB is synthesized in two different pathways by using two different substrates, such as choline and glycine. The conversion of choline to GB has been well studied in a series of organisms. The pathway involves one or two enzymes, depending on the oxidation of choline. The enzymatic pathway is widespread, occurring naturally in various microorganisms, plants and animals. In this pathway, GB is formed as the result of the two-step oxidation of choline via the toxic intermediate betaine aldehyde (Fig. 11.2). In higher plants, the reactions are catalysed by Choline monooxygenase (CMO) and betaine aldehyde dehydrogenase (BADH), enzymes localized in the stroma of chloroplasts. Rhodes and Hanson (1993) observed that the biosynthesis of GB is stress-inducible and the concentration of GB *in vivo* varies among plant species, ranging from 40 to 400 μmol per gram of dry weight.

Certain bacteria, such as *Escherichia coli* and animal cells are capable of synthesizing GB from choline in the presence of two dehydrogenases: a membrane-bound CDH and BADH (Lamark et al., 1991). In soil bacteria, in the genus *Arthrobacter* single-enzyme pathway has been reported (Ikuta et al., 1977).

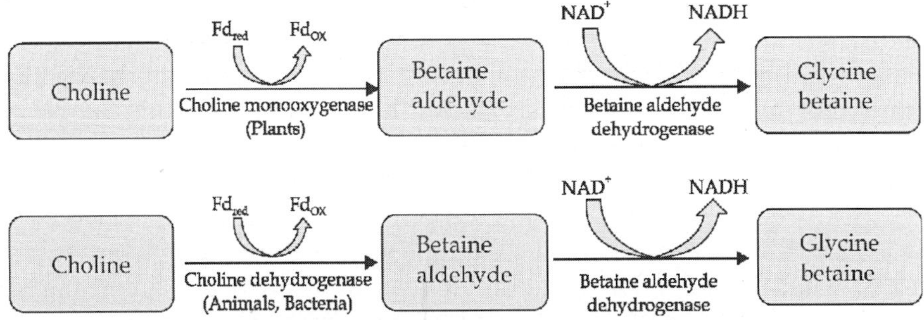

Fig. 11.2 Biosynthetic pathways of glycine betaine

8. ECTOINE

Ectoine is one of the most common osmolytes in the halophilic and halotolerant bacterium. It can be considered a heterocyclic amino acid or as a partially hydrogenated pyrimidine derivative (1,4,5,6-tetrahydro-2-methyl-4-pyrimidinecarboxylic acid (Galinski, 1985; Pastor et al., 2010). A host of genera *Halomonas, Pseudomonas, Actinobacterum* and *Bacillus* and others bacteria have been reported to accumulate ectoine (Kuhlmann and Bremer, 2002; Roberts, 2005).

8.1 Biosynthesis of Ectoine

Diverse spectrum of osmolytes are found in prokaryotes and eukaryotes. However, osmoprotectants belonging to different groups are specific for different halophilic and halotolerant microorganisms. Biosyntheses of glycerol, glycine betaine, and ectoine are the most energetically expensive process. Dominance as well as specificity of osmolytes is mainly determined by the energetic status of a cell and the availability of a carbon and nitrogen source (Mustakhimov et al., 2010). Aspartate semialdehyde is the initial molecule of ectoine biosynthesis, which is a linking molecule in amino acid metabolism. It converts to L-2,4-diaminobutyric acid, which is then acetylated to from N^{γ}-acetyldiaminobutyric acid (NADA). The final step is the cyclization of this solute to form ectoine. A set of three genes: *ectA, ectB, ectC* are involved in ectoine biosynthesis. The *ectA* gene encodes for diaminobutyric acid acetyltransferase; *ectB* codes for the diaminobutyric acid aminotransferase and *ectC* codes for ectoine synthase (Fig. 11.3).

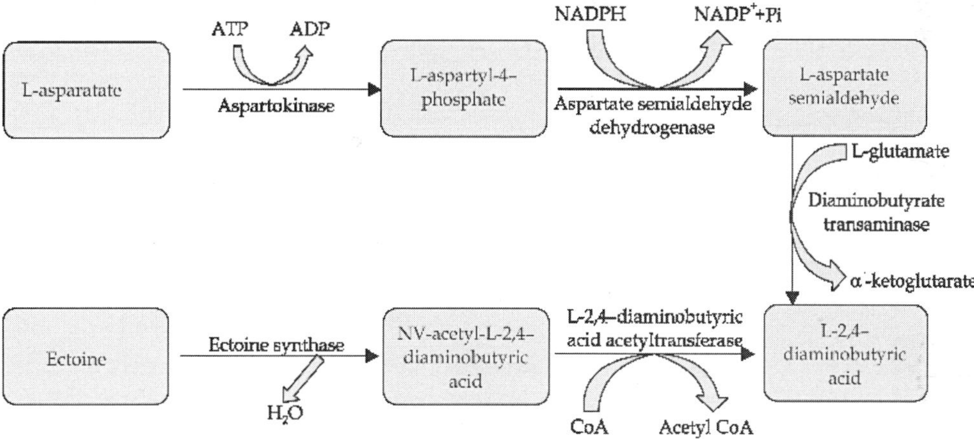

Fig. 11.3 Biosynthetic pathway of ectoine

9. PROLINE

The composition of amino acids, such as cysteine, arginine, and methionine, which constitute about 55 per cent of total free amino acids decreases under abiotic stress, such as salinity, whereas proline concentration rises (Shintinawy and Shourbagy, 2001). Proline is a proteinogenic amino acid, which is essential for primary metabolism. Kemble et al. (1954) first reported proline accumulation in the plant rye grass. Biosynthesis has been reported in group of bacteria, cyanobacteria and plants. Accumulation of proline in microbes and higher plants increases under different environmental stresses, such as drought, salinity, high light intensity and UV irradiation, heavy metals, oxidative stress (Choudhary et al., 2005) and biotic stresses (Haudecoeur et al., 2009). Precursor for proline synthesis is either glutamate or ornithine and under osmotic stress conditions, glutamate functions as the primary precursor (Pérez-Arellano et al., 2010). The biosynthetic pathway consists of two major enzymes, pyrroline carboxylic acid synthetase and pyrroline carboxylic acid reductase. Both are regulatory enzymes, used to overproduce proline in plants (Saran and Tyagi, 2004). Chris et al. (2006) reported protection against salinity by the accumulation of proline in the cyanobacterium *Cylindrospermum* sp. It is observed that addition of proline in nutritional medium enhanced salt tolerance in olive plants and also enhanced some of antioxidant enzyme, photosynthetic activity. Deivanai et al. (2011) has reported that proline improves salt tolerance in *Nicotiana tabacum* by enhancing the activity of antioxidant enzymes. Rice seedlings from seeds pretreated with 1 mM proline showed improvement in growth during salt stress (Hoque, 2008).

Proline can function as a molecular chaperone in stabilization of protein and in buffering the cytosolic pH and balance redox cell status (Maggio et al., 2002). Recently, a new procedure for markerless mutagenesis genes deletion in the *Halobacillus halophilus* for proHJA was developed. The mutant produced proline but the capability to act as osmoprotectant was lost. The loss in the production of proline was compensated by an increase in the production of the glutamate, glutamine and ectoine concentrations (Kocher et al., 2011).

9.1 Proline Biosynthesis

Several reports have been published in proline synthesis in halophilic members of the genus *Bacillus* as well other nonhalophilic and halotolerant bacteria. Compatible solute proline has also been reported in halophilic or halotolerant algae, mostly bacillariophyceae, and marine invertebrates. *Halobacillus halophilus,* a moderately halophilic bacterium synthesizes glutamate and glutamine at salinities at 1.0 M NaCl; and switches the osmolyte synthesis behavior mainly proline, when grown at 2.0 to 3.0 M NaCl (Saum and Müller, 2008). Accumulation of proline was maximum during the early exponential phases and is downregulated during the late exponential phase.

A cluster of three genes; *proH, proJ* and *proA* are involved in the biosynthesis of proline. which are organized in operon. The expression of this particular operon is correlated with the salinity of the medium, maximal being at 2.5 M NaCl. Sufficient synthesis of proline from glutamate is ensured by enzymes that are encoded genes, such as pyrroline-5-carboxylate reductase (*ProH*), a putative glutamate 5- kinase (*ProJ*) and a putative glutamate 5-semialdehyde dehydrogenase (*ProA*). The mRNA level of the *proJHA* genes was found to increase along with the proline content (highest after 6 h) suggesting the triggering of proline production with an increase in the salinity. In contrast to NaCl, the addition of Na-glutamate enhances the level of *proHJA* mRNA. Stimulation of *pro* gene transcription was observed at a minimal concentration of 0.2 M glutamate (Saum and Müller, 2007). These results suggest that internal glutamate concentration that increases with increasing salinity also function as signal that triggers the production of proline in response to NaCl. Glutamate is therefore considered a "second messenger" in *Halobacillus halophiles.* Proline is degraded by the enzymes proline dehydrogenase (ProDH) and D1-pyrroline-5-carboxylate dehydrogenase (P5CDH), which together form *put* operons, to glutamate via D1-pyrroline-5-carboxylate and was found to be upregulated by increasing salinity (Kocher et al., 2010).

10. POTENTIAL APPLICATION OF HALOPHILES AND HALOTOLERANT MICROBES

Microorganisms play a significant role in the management of abiotic stress, because of their unique properties of higher abiotic stress tolerance, huge genetic diversity and ubiquity (Choudhary, 2012). Hence, these organisms serve as important tools for understanding stress tolerance mechanisms that can be subsequently engineered into crop plants. Halophiles are important in biotechnologal research and applications. Halotolerance of many enzymes derived from halophilic microorganisms can be exploited wherever enzymatic transformations are required to function at low water activities, such as in the presence of high salt concentrations. Organic osmotic stabilizers produced by some of the halophiles and halotolerant microbes are found to be important (Oren, 2002). Some halophilic microorganisms may produce valuable compounds that can be used for development of biotechnological production processes. For example, several halophilic enzymes have been identified for potential biotechnological applications, including amylases, nucleases, and proteases. (Kulkurani et al., 1999; Moreno et al., 2013). Salt-dependent enzymes are found in mainly those groups of microorganisms that are able to maintain high intracellular ionic concentrations inside the cells. Such types of enzymes consist of acidic amino acid and are predominant in order *Halanaerobiales* (Oren, 2013). In addition, all exoenzymes excreted by halophiles have been found to be active under high salinity even when the organisms that produce them, may maintain low intracellular ionic concentrations. Dunaliella salina, a halophilic green algae; the commercial production of β-carotene and glycerol has been reported by Oren (2014). The halophilic/halotolerant microorganisms thus appear to be a reservoir of novel compounds/molecules which could be exploited further for enhancing abiotic stress tolerance in plants through biotechnological interventions.

11. CONCLUSION

Salinity is one of the major abiotic stresses, seriously affecting the crop yield and soil fertility. Certain microorganisms, such as halophilic and halotolerant microbes are important organisms to study the mechanism of salt tolerance. Generally, cells maintain high intracellular salt concentrations, osmotically at least equivalent to the external concentrations. Intracellular ionic concentrations similar to those of the surrounding medium were reported in aerobic extremely halophilic *Archaea*. In most cases which contain NaCl as the main salt the ionic composition of the cytoplasm generally differs greatly from that of the medium. Cells may maintain low salt concentrations within their cytoplasm. Compatible solutes are low-molecular-weight organic compounds, soluble in water at high concentrations, and even in the molar concentration range should not inhibit enzymatic activities. Glucosylglycerol (GG) is the main compatible solute in moderately halotolerant (marine) strains and sometimes glucosylglycerate (GGA) was also found as a secondary compatible solute. Halophilic strains are able to grow in saturated salt concentrations, usually synthesize glycine betaine or glutamate betaine.

Organic osmolytes are small solutes used by cells of numerous water-stressed organisms to maintain cell volume. Halophiles are best source for enzymes, enzyme stabilizer, cosmetics, medicine and fermented foods.

Acknowledgements

Financial assistance in the form of DST-Young Scientist project (SB/FT/LS-378/2012) to KT is gratefully acknowledged.

Conflict of interest

There is no conflict of interest amongst the authors.

References

Andrew S, Jonathan AC, Jim PW, Ricardo S, Richa S, Nils N, Colin DM, Irene RG, Jonathan DRH, John PQ, David JT, Satish VP, Rekha SS, Josefa A, Jan D, Ailsa DH, Bart L, Drauzio ENR, Mary AV, Nina G, Aharon O, Kenneth N, and Terry JM John E (2015). Is there a common water-activity limit for the three domains of life?. ISME Journal 9: 1333–1351.

Bremer E, and Kramer R (2000). Coping with osmotic challenges: osmoregulation through accumulation and release of compatible solutes in bacteria. In: Storz G, Hengge-Aronis R (eds) Bacterial Stress Responses, ASM Press, Washington DC, p. 501.

Cardini CE, Leloir LF, Chiriboga J (1955). The biosynthesis of sucrose. J Biol Chem 214: 149–155.

Cayley S, Lewis BA, and Record Jr MT (1992). Origins of the osmoprotective properties of betaine and proline in *Escherichia coli* K-12. J Bacteriol 174: 1586–1595.

Choudhary DK (2012). Microbial rescue to plant under habitat-imposed abiotic and biotic stresses. Appl Microbiol Biotechnol 96: 1137–1155.

Choudhary NL, Sairam RK, and Tyagi A (2005). Expression of delta1-pyrroline-5-carboxylate synthetase gene during drought in rice (*Oryza sativa* L.). Ind J Biochem Biophys 42: 366–370.

Chris A, Zeeshan M, Abraham G, and Prasad SM (2006). Proline accumulation in *Cylindrospermum* sp. Environ Expt Bot 57: 154–159.

Das P, Behera BK, Meena DK, Azami SA, Chatterjee S, Kanti M, and Sharma AP (2015). Salt stress tolerant genes in halophilic and halotolerant bacteria: Paradigm for salt stress adaptation and osmoprotection. Int J Curr Microbiol App Sci 4: 642–658.

Deivanai S, Xavier R, Vinod V, Timalata K, and Lim OF (2011). Role of exogenous proline in ameliorating salt stress at early stage in two rice cultivars. J Stress Physiol Biochem 7: 157–174.

Elbein AD, Pan YT, Pastuszak I, and Carrolll D (2003). New insights on trehalose: a multifunctional molecule. Glycobiol 13: 17R–27R.

Empadinhas N, and MS da Costa (2006). Diversity and biosynthesis of compatible solutes in hyperthermophiles. Int Microbiol 9: 199–206.

Empadinhas N, and MS da Costa (2008). Osmoadaptation mechanisms in prokaryotes: distribution of compatible solutes. Int Microbiol 11: 151–161.

Galinski EA (1993). Compatible solutes of halophilic eubacteria molecular principles, water-solute interaction, stress protection. Experientia 49: 487–96.

Galinski EA (1995). Osmoadaptation in bacteria. Adv Microb Physiol 37: 272–328.

Hagemann M (2011). Molecular biology of cyanobacterial salt acclimation. FEMS Microbiol Rev 35(1): 87–123.

Hagemann M, and Marin K (1999). Salt-induced sucrose accumulation is mediated by sucrose phosphate-synthase in cyanobacteria. J Plant Physiol 155: 424–430.

Haudecoeur E, Planamente S, Cirou A, Tannieres M, Shelp B J, Morera S, and Saure D (2009). Proline antagonizes GABA-induced quenching of quorum-sensing in Agrobacterium tumefaciens. Proc Natl Acad Sci U.S.A. 106: 14587–14592.

Hoque MA, Banu MNA, Nakamura Y, Shimoishi Y and Murata Y (2008). Proline and glycinebetaine enhance antioxidant defense and methylglyoxal detoxification systems and reduce NaCl-induced damage in cultured tobacco cells. J Plant Physiol. 165: 813–824.

Ikuta S, Matuura K, Imamura S, Misaki H, and Horiuti Y (1977). Oxidative pathway of choline to betaine in the soluble fraction prepared from *Arthrobacter globiformis*. J Biochem 82: 157–163.

Kaneko T, Sato S, Kotani H, Tabata S, et al. (1996). Sequence analysis of the genome of the unicellular cyanobacterium Synechocystis sp. strain PCC6803. II. Sequence determination of the entire genome and assignment of potential protein-coding regions. DNA Res 30: 3(3): 109-36.

Katoh H, Asthana RK, and Ohmori M (2004). Gene expression in the cyanobacterium *Anabaena* sp. PCC7120 under desiccation. Microb Ecol 47: 164–174.

Kemble AR, and MacPherson HT (1954). Liberation of amino acid sin perennial ray grass during wilting. Biochem J 58: 46–59.

Kempf B and Bremer E (1998). Uptake and synthesis of compatible solutes as microbial stress responses to high osmolarity environments. Arch Microbiol 170: 319-330.

Köcher S, Averhoff B, and Müller V (2011). Development of a genetic system for the moderately halophilic bacterium *Halobacillus halophilus*: generation and characterization of mutants defect in the production of the compatible solute proline 13: 2122–2131.

Kolman MA, Nishi NC, Perez-Cenci M, and Salerno GL (2015). Sucrose in cyanobacteria: from a salt-response molecule to play a key role in nitrogen fixation. Life 5: 102–126.

Kuhlman AU, and Bremer (2002). Osmotically regulated synthesis of the compatible solute ectoine in *Bacillus pasteurii* and related *Bacillus* spp. Appl Environ Microbiol (68) 2: 772–783.

Kulkurani N, Shendye A, and Rao M (1999). Molecular and biotechnological aspect of xylenase. FEMS Microbiology Letters 23 (4): 411–456.

Kushner DJ (1978). Life in high salt and solute concentrations: halophilic bacteria. In: Kushner DJ (ed) Microbial life in extreme environments. Academic Press Ltd., London, United Kingdom, pp. 317–368.

Lamark T, Kaasen I, Eshoo MW, Falkenberg P, McDougall J and Strøm AR (1991). DNA sequence and analysis of the bet genes encoding the osmoregulatory choline-glycine betaine pathway of *Escherichia coli*. Mol Microbiol 5: 1049-1064.

Malferrari M, Nalepa A, Malferrari M, Nalepa A, Venturoli G, Francia F, Lubitz W, Möbius K, Savitsky A, Venturoli G, Francia F, Lubitz W, Köbius M, and Savitsky A (2014). Structural and dynamical characteristics of trehalose and sucrose matrices at different hydration levels as probed by FTIR and high-field EPR. Phys Chem Chem Phys 16: 9831–9848.

Moreno ML,García MT, and Mellado E (2013). Halophilic Bacteria as a Source of Novel Hydrolytic Enzyme. Life 3: 38–51.

Mustakhimov II, Reshetnikov AS, Glukhov AS, KhmeleninaVN, and Kalyuzhnaya MG (2010). Identification and characterization of *EctR1*, a new transcriptional regulator of the ectoine biosynthesis genes in the halotolerant methanotroph *Methylomicrobium alcaliphilum*. J Bacteriol 192: 410–417.

Norton RS, Mackay MA, and Borowitzka LJ (1982). The physical state of osmoregulatory solutes in unicellular algae. A natural abundance carbon-13 nuclear-magnetic-resonance relaxation study. Biochem J 202: 699–706.

Oren A (1999). Bioenergetic Aspects of Halophilism. Mol Mar Biol Biotech 63(2): 334–348.

Oren A (2002). Diversity of halophilic microorganisms: Environments, phylogeny, physiology, and applications. J Ind Microbiol Biot 28: 56–63.

Oren A (2006). Life at high salt concentration. Prokaryotes 263–282.

Oren A (2008). Microbial life at high salt concentrations: phylogenetic and metabolic diversity. Saline Syst 2008: 4(2) 1–8.

Oren A (2013). Life at high salt concentration, intracellular KCl concentration, and acidic proteome. Front Microbiol 4: 315.

Oren A (2014). The ecology of *Dunaliella* in high-salt environments. JBR 21–23.

Park E-J, Jeknic Z, Pino1 M-T, Murata N, and Chen T H-H (2007). Glycine betaine accumulation is more effective in chloroplasts than in the cytosol for protecting transgenic tomato plants against abiotic stress. Plant Cell Environ 30: 994–1005.

Pastor JM, Salvador M, Argandoña M, Bernal V, Reina-Buenob M, Csonka LN, Iborra J, Vargas C, Joaquín J. Nieto, and Cánovas M (2010). Ectoines in cell stress protection: Uses and biotechnological production. Biotechnology Advances 28: 782–801.

Paul MJ, Primavesi LF, Jhurreea D, and Zhang Y (2008). Trehalose metabolism and signaling. Annu Rev Plant Biol 59: 417–441.

Pérez-Arellano I, Carmona-Álvarez F, Martínez,AI, Rodríguez-Díaz J, and Cervera J (2010). Pyrroline-5-carboxylate synthase and proline biosynthesis: From osmotolerance to rare metabolic disease. Protein Sci 19: 372–382.

Porchia A C, and Salerno GL (1996). Sucrose biosynthesis in a prokaryotic organism: Presence of two sucrose phosphate synthases in *Anabaena* with remarkable differences compared with the plant enzymes. Proc Natl Acad Sci USA 93(24): 13600–13604.

Porchia AC, Curatti L, and Salerno GL (1999). Sucrose metabolism in cyanobacteria: sucrose synthase from *Anabaena* sp. Strain PCC 7119 is remarkably different from the plant enzymes with respect to substrate affinity and amino-terminal sequence. Planta 210: 34–40.

Rhodes D, and Hanson AD (1993). Quaternary ammonium and tertiary sulfonium compounds in higher plants. Annu Rev Plant Physiol Plant Mol Biol 44: 357–384.

Roberts MF (2005). Organic compatible solutes of halotolerant and halophilic microorganisms. Saline Syst 1: 1–30.

Ryu SI, Park CS, Cha J, Woo EJ, and Lee SB (2005). A novel trehalose-synthesizing glycosyltransferase from Pyrococcus horikoshii molecular cloning and characterization. Biochem Biophys Res Commun 329: 429–36.

Sairam RK, and Tyagi A (2004). Physiology and molecular biology of salinity stress tolerance in plants. Curr Science 86 (3): 407–421.

Saum H and Muller V (2007). Salinity-dependent switching of osmolytes strategies in a moderately halophilic bacterium: glutamate induces proline biosynthesis in *Halobacillus halophillus*. J Bacterial 189: 6968-6975.

Saum H and Muller (2008). Regulation of osmoadaptation in the moderate halophilic *Halobacillus halophilus*: chloride, glutamate and switching osmolytes strategies 4-4.

Shintinawy FEl, and Shourbagy MNE (2001). Alleviation of changes in protein metabolism in NaCl-stressed wheat seedlings by thiamine. Biologia Plantarum 44(4): 541–545.

Shrivastava P, and Kumar R (2015). Soil salinity: A serious environmental issue and plant growth promoting bacteria as one of the tools for its alleviation. SJBS 22: 123–131.

Sleator RD, and Hil C (2001). Bacterial osmoadaptation: the role of osmolytes in bacterial stress and virulence, FEMS Microbial letters; 49-71.

Tripathi AK, Mishra BM, and Tripathi P (1998). Salinity stress responses in plant growth promoting rhizobacteria. J Biosci Bioeng 23: 463–471.

Tsusaki K, Nishimoto T, Nakada T, Kubota M, and Chaen H (1997). Cloning and sequencing of trehalose synthase gene from *Thermus aquaticus* ATCC 33923. Biochim Biophys Acta 1334: 28–32.

Vreeland RH (1987). Mechanisms of halotolerance in micro-organisms. CRC Crit. Rev Microbiol.14: 311 356.

Waditee R, Tanaka Y, Aoki K, Hibino T, Jikuya H, Takano J, Takabe T, and Takabe T (2003). Isolation and functional characterization of N-methyltransferases that catalyze betaine synthesis from glycine in a halotolerant photosynthetic organism *Aphanothece halophytica*. J Biol Chem 278: 4932–4942.

Winter H, and Huber S (2000). Regulation of sucrose metabolism in higher plants: localization and regulation of activity of key enzymes. Crit Rev Plant Sci 19: 31–67.

Yadav RK, Thagela P, Tripathi K, and Abraham G (2016). Physiological and proteomic analysis of salinity tolerance of the halotolerant cyanobacterium *Anabaena* sp. Wold. J of Microbiol and Biotechnol 32(9): 1-11.

12

Plant Disease Management through Microbial Agents for Sustainable Agriculture

SUSANTA BANIK
Department of Plant Pathology, SASRD
Nagaland University, Medziphema Campus
Nagaland-797106, India
Email: *susanta.iari@gmail.com*

1. INTRODUCTION

Agricultural production is marred by various biotic and abiotic stresses causing huge crop losses all over the world. Of the estimated average crop loss of 36.5 percent worldwide, 14.1 percent is caused by diseases, while 12.2 percent and 10.2 percent crop losses are caused by weeds and insects, respectively (Agrios, 2005). Among various biotic stresses, plant diseases caused by various groups of pathogens like fungi, bacteria, nematode, viruses, viroids, mollicutes, fastidious bacteria, phanerogams and algae apparently behave as silent killers. Several management strategies have been evolved and refined subsequently, but none prove to be panacea for plant disease. A combination of different management approaches are always recommended for minimizing crop losses. Though in present-day agriculture, synthetic chemicals dominate the plant protection arsenals, still with respect to plant pathogens, only fungal pathogens could be tamed to some extent with the latest discoveries of new chemical molecules. These chemicals after a few years of use are rendered ineffective due to evolvement of new population of fungal pathogens resistant to such chemicals. Moreover, indiscriminate use of the chemicals endangers the sanctity and abundance of natural resources; awareness of this prevents use of chemicals by many crop growers.

For managing prokaryotic diseases in plants (bacteria, fastidious bacteria, mollicutes), antibiotics are recommended. However, owing to the fear of development of antibiotic resistance and subsequent transfer of that antibiotic resistance to human pathogenic bacteria its use in agriculture is severely restricted (Janse, 2006).

Nematode diseases can be managed by application of appropriate chemicals in soil; however, large-scale application of nematicide in soil is not feasible. Lastly, at present there is no chemical measure for managing the submicroscopic entities like virus and viroid which are now getting a free run in the crop fields.

In this backdrop, natural control mechanisms like biological control or biocontrol is spreading its influence though in limited scale. Man has exploited natural biocontrol agents comprising mainly of antagonistic fungal and bacterial species for managing plant diseases that are caused chiefly by fungal and bacterial pathogens and to some extent nematodes. Search for biocontrol agents has resulted in an endless list of microorganisms which are antagonistic to one or many plant pathogens and the list is ever-growing with new reports coming in with the passage of time. There are some landmark developments in the arena of research for exploitation of natural microbial agents in plant disease management (Table 12.1). These biocontrol or microbial control agents have been successfully commercialized in many cases for managing plant diseases. Like other plant disease control methods, microbial control method has several merits over chemical method (Banik and Jambhulkar, 2014); however, it also has inherent drawbacks some of which seem to be insurmountable in near future. In this chapter, role of various groups of microbes in plant disease management along with their merits and demerits and mechanisms of action are discussed.

Table 12.1: Landmarks in biocontrol of plant pathogens

Year	Contributor	Contribution
1794	Persoon	Erected the genus *Trichoderma*
1874	William Roberts	Demonstration of antagonistic action of microorganisms (*Penicillium glaucum* and bacteria); introduction of the term 'antagonism'
1898	Guerin	Existence of fungal endophytes
1914	C.F.Von	Feasibility of biological control of plant diseases
1926	Sanford	Suppression of potato scab by green manuring
1932	Weindling R	Reporting parasitic nature of *Trichoderma* on plant pathogens (*Rhizoctonia solani*)
1934	Weindling R	Discovery of gliotoxin, the first antimicrobial from *Trichoderma*
1948-52	Grossbard	Antibiotic production by microbes in soil (*Aspergillus, Penicillium, Streptomyces, Trichoderma*)
1952-57	Wright	Antibiotic production by microbes in soil (*Aspergillus, Penicillium, Streptomyces, Trichoderma*)
1963	-	First example of biological control of a plant disease – *Heterobasidion annosum* controlled by *Phlebia gigantea*
1973	Allan Kerr	Use of *Agrobacterium radiobacter* K84 for control of crown gall disease. A. *radiobacter* K84 was discovered and developed in Australia. It is the first commercial biocontrol agent for any plant disease. *Agrobacterium radiobacter* K1026 is the first genetically engineered microorganism released for widespread use in the environment (for control of crown gall disease).
1974	Kenneth F. Baker and R. James Cook	Publication of the book 'Biological Control of Plant Pathogens'

1980	Kloepper	Importance of siderophore production by *Erwinia carotovora*
1983	R. James Cook and Kenneth F. Baker	Publication of the book 'The Nature and Practice of Biological Control of Plant Pathogens'
1989	-	Registration of *Trichoderma harzianum* ATCC 20476 for control of plant diseases

2. MICROBIAL CONTROL VERSUS CHEMICAL CONTROL

From ecological point of view, microbial control has been graded as superior to chemical control for the obvious reason that microbes are constituents of natural environment. With their introduction in the field, microbial dynamics change in favor of the introduced microbes. However, there are certain technological bottlenecks because of which microbial method could not be made at par with chemicals (Banik and Jambhulkar, 2014). All the pros and cons of microbial and chemical control methods are compared below (Table 12.2).

Table 12.2: Microbial control vs. chemical control of plant diseases

S. No.	Parameters	Microbial control	Chemical control
1	Action	Slow	Quick
2	Aim	Not to eradicate, but to keep the pathogen below damaging level	Aims at eradication of pathogens
3	Bioaccumulation	No	May occur
4	Cost	Cheaper, more cost effective as organisms multiply and maintain their population once introduced	Multiple applications at intervals, expensive
5	Development cost	Usually lower	Very high
6	Duration of control	Long term control	Short term control
7	Easiness	Easy to use	Laborious
8	Ecosystem	Preserved	Disturbed
9	Environment friendliness	Friendly	No
10	Human health	Apparently no side effect	Side effects
11	Impact	Often not visible	Dramatic
12	Induction of plant defense	Yes	Mostly no
13	Interaction	May degrade chemical pesticide	Kill beneficial microbes
14	Mode of action	Multiple and complex	Usually single or multiple sites of action, sometimes complex
15	Occurrence	Naturally occurring	Often synthesized

Contd...

16	Performance	Unpredictable in different ecological situation	Performance guaranteed
17	Plant growth promotion	Yes	No
18	Producer	Local	MNCs
19	Resistance development in pathogen	No	Yes
20	Shelf life of control agents	Less (usually six months)	High (may be up to 2 years)
21	Systemic	Cannot go systemic	Can be systemic
22	Target pathogen	Mainly soil-borne	All types of pathogens
23	Target specificity	Specific	Non-specific

3. MECHANISMS OF ACTION OF MICROBIAL ANTAGONISTS

The mechanisms employed by microbial antagonists to suppress or inhibit plant pathogens are hyperparasitism/mycoparasitism, competition, antibiosis, cross protection and induced resistance, lysis, plant growth promotion and hypovirulence. However, the antagonists can exhibit more than one mechanism to control a plant pathogen. For example, *Bacillus subtilis* suppresses club root disease on canola via antibiosis and induced host resistance (Lahlali et al., 2013). A glimpse of the microbes employed in plant disease management is presented in Table 12.3 and the mechanisms are described thereafter.

Table 12.3: Microbes employed for plant disease management

Bacteria	*Agrobacterium radiobacter* K84, K1026, *Bacillus subtilis, B. licheniformis, B. megaterium, B. pumilus, B. cereus, B. amyloliquifaciens* *Burkholderia cepacia* *Erwinia amylovora* harpin protein *Lysobacter* *Pantoea* *Pseudomonas aureofaciens, P. fluorescens, P. syringae, P. chlororaphis* *Streptomyces lydicus, S. griseoviridis*
Fungi	*Ampelomyces quisqualis* *Candida oleophila* *Coniothyrium minitans* *Dactylella* *Gliocladium catenulatum, G. virens* *Myrothecium verrucaria* *Paecilomyces lilacinus* *Trichoderma harzianum, T. viride, T. polysporum* *Pasteuria penetrans* *Fusarium oxysporum (non-pathogenic)* *Pythium oligandrum* *Phlebia gigantea*

Mycorrhizal fungi	*Pisolithus tinctorius, Paxillus involutus, Glomus fasciculatum, G. macrocarpum, G. intraradices*
Endophytic fungi and bacteria	*Piriformospora indica, Epichloë festucae, Neotyphodium* spp., *Chaetomium* spp., *Klebsiella* spp., *Azospirillum* spp., *Nocardia* spp., *Curtobacterium flaccumfaciens, Bacillus megaterium, Bacillus* spp.

3.1 Hyperparasitism

Hyperparasitism is considered the most direct form of antagonism employed by fungal biocontrol agents only. In this, hyphae of the fungal biocontrol agents (hyperparasite) grows towards the target fungus, comes in contact and coil around and finally dissolve the cell wall or membrane of the target fungus by enzymatic actions in order to get access to and absorb the hyphal content of the target fungus. *Trichoderma harzianum* shows hyperparasitism against *Rhizoctonia solani* (Altomare et al., 1999). Chitinase, β-1,3 glucanase are involved in enzymatic dissolution of hyphal wall by *Trichoderma* (Harman 2000). Some species of *Trichoderma* also produce cellulase. Role of L-amino acid oxidase identified from *T. harzianum* ETS 323 has been emphasized in parasitism against *B. cinerea* (Cheng et al., 2012). This phenomenon is also referred to as mycoparasitism. Biocontrol activity of *T. atroviride* P1 was enhanced against plant pathogens by overexpressing the *Saccharomyces cerevisiae DPM1* gene coding for dolichyl phosphate mannose (DPM) synthase. There was improved antifungal activity in *T. atroviride* P1 against *Pythium ultimum* due to elevated DPM synthase and cellulolytic activity owing to this transformation (Zembek et al., 2011). Other mycoparasitic fungi are *Ampelomyces quisqualis* (target: powdery mildew fungi – *Sphaerotheca fuliginea*), *Coniothyrium minitans* and *Sporidesmium sclerotiorum* (target: sclerotia forming fungi like *Sclerotinia sclerotiorum*), *Gliocladium* spp. (target- many soil-borne fungi), *Pythium* spp. (target: plant pathogenic *Pythium* spp.) and *Verticillium lecanii, Sphaerellopsis filum* and *Cladosporium* sp. (target: rust fungi). Parasitism is known to occur in bacteria as well. *Bdellovibrio bacteriovorus* is a Gram-negative soil bacterium parasitic on other Gram-negative bacteria. Isolates of Bacillus cause lysis of germ tubes arising from uredospores of cereal rust pathogens.

3.2 Competition

In the microbial niche in soil or plant surface nutrients and space constitute the factors for which microbes compete with each other and form various kinds of interactions. Biocontrol agents by virtue of their rapid and faster growth compared to plant pathogenic ones occupy and establish a niche thereby resulting in physical exclusion from a site. A classic example in this case is formed by the application of spore suspension of *Phlebia gigantea* on cut surface of pine stumps where rapid growth of the fungus *P. gigantea* prevent colonization by butt and stem rot fungus *Heterobasidion annosum*. Competition for micro-nutrients (iron and manganese) has been cited often as a mechanism for suppression of plant pathogens. It is reported that biocontrol agents have more efficient uptake systems than the pathogenic microbes (Nelson, 1990). Iron binding ligands are known to be produced by biocontrol agents to chelate iron (Kloepper et al., 1980) to form siderophore-

Fe-complex that is specifically recognized by membrane bound protein receptors and taken up by the biocontrol agents. This makes iron non-available to other microbes including pathogens thereby suppressing their growth. Some bacterial biocontrol agents can even utilize the siderophores produced by other bacteria (Loper and Henkels 1999). A few cases of competition by fungal biocontrol agents are cited in the table below (Table 12.4).

Table 12.4: Competition between antagonist and target pathogen

Biocontrol agent	Target pathogen	Competition for	Reference
Burkholderia Cepacia	*Pythium aphanidermatum* (pea)	Stimulants released from plant	Heungens and Parke, 2000
Candida guillermondii	*Penicillium expansum* (apple)	Nitrates	Scherm et al., 2003
Cryptococcus laurentii BSR-Y22 and *Sporobolomyces roseus* FS43–238	*Botrytis cinerea* (apple)	Fructose, glucose and sucrose	Filonow, 1998
Enterobacter cloacae	*Pythium ultimum* (many plants)	Stimulants released from plant	Kageyama and Nelson, 2003
Fusarium lateritium	*Eutypa armeniacae* (gummosis or die back of apricot)	Space (cut or pruning site or wounds); other mechanism antibiotic production	-
Non-pathogenic *Fusarium oxysporum*	*Fusarium oxysporum*	Carbon source (glucose)	Larkin and Fravel, 1999
Non-pathogenic strain of *Rhizoctonia*	*Rhizoctonia solani*	Infection sites and/or nutrients, in or on roots	Herr, 1995
Phlebia gigantea	*Heterobasidion annosum* (butt and stem rot of pine trees)	Space (cut surface on stumps); other mechanism antibiotic production	-
Pseudomonas fluorescens A506	*Erwinia amylovora* (fire blight of apple)	Space and nutrients	Lindow and Leveau, 2002
Pseudomonas spp.	*Fusarium oxysporum* f. sp. *radicis-lycopersici* (tomato)	Space and nutrients	Bolwerk et al., 2003
Pseudomonas spp.	*Pythium* and *Fusarium* species	Iron	Loper and Buyer, 1991; Duijff et al., 1993
Trichoderma spp.	*P. ultimum* and/or *Rhizopus oryzae* (cotton)	Stimulants released from plant	Howell, 2002
Agrobacterium radiobacter K84	*Agrobacterium tumefaciens* (crown gall of pome fruits)	Colonization sites and nutrients, and antibiotic production (agrocin 84)	-
Erwinia herbicola	*Erwinia amylovora* (fire blight of apple)	Nutrient and space, antibiotic production	-

Biocontrol agents are also reported to compete for certain stimulants like fatty acids or peroxidation product of fatty acid (Harman and Nelson, 1994), volatile compounds like ethanol and acetaldehyde which are required by some plant pathogens to overcome their dormancy and cause plant infection.

3.3 Antibiosis

Antibiosis refers to production of antibiotics or low molecular weight compounds by microorganisms to directly affect the growth of other microorganisms including plant pathogens. This mechanism of suppression of plant pathogens is commonly observed in dual culture technique *in vitro* using antagonists when inhibition zone is observed. Antibiosis play important role in biocontrol of plant pathogens by fungal and bacterial antagonists. Among three major types of antibiotics known till date (nonpolar/volatile, polar/non-volatile and water soluble), volatile compounds are more effective. An antagonist may produce more than one type of antibiotics (Tables 12.5 and 12.6). Besides production of antibiotics, some antagonists also produce factors (enzymes or proteins) that detoxify or degrade the virulence factor of the pathogen (Table 12.7).

Table 12.5: Various antibiotics/antimicrobial metabolites reported to be produced by biocontrol agents

Antibiotics/inhibitory compounds	Producer microbe	Reference
Amphisin, 2,4-diacetylphloroglucinol, hydrogen cyanide, oomycin A, phenazine, pyoluteorin, pyrrolnitrin, tensin, tropolone and cyclic lipopolysaccharides	*Pseudomonas* spp.	Défago, 1993; Nielsen et al., 2002; Raaijmakers et al., 2002; de Souza et al., 2003; Nielsen and Sørensen, 2003; Arseneault et al., 2013
Gramicidin S, oligomycin A, kanosamine, iturin, zwittermycin A and xanthobaccin	*Bacillus*, *Streptomyces* and *Stenotrophomonas* spp.	Milner et al., 1995, 1996; Hashidoko et al., 1999; Kim et al., 1999; Nakayama et al., 1999; Edwards and Seddon, 2001; Romero et al., 2007
Flocculosin	*Pseudozyma flocculosa*	Cheng et al., 2003
Gliotoxin and peptaibols	*Trichoderma* spp.	Wilhite et al., 1994; Wiest et al., 2002

Table 12.6: Antibiotic production by bacterial biocontrol agents

S. No.	Biocontrol agent	Antibiotic produced	Plant disease controlled	Reference
1	*Agrobacterium radiobacter*	Agrocin 84	Crown gall disease (*Agrobacterium tumefaciens*)	Kerr, 1980
2	*Bacillus amyloliquefaciens* FZB42	Bacillomycin, Fengycin	Fungal wilt (*Fusarium oxysporum*)	Koumoutsi et al., 2004
3	*Bacillus amyloliquefaciens* TrigoCor strain	Iturin	Fusarium head blight of wheat	Crane et al., 2013
4	*Bacillus cereus* UW85	Zwittermicin A	Damping off (*Phytophthora medicaginis* and *Pythium aphanidermatum*)	Smith et al., 1993
5	*Bacillus subtilis* AU195	Bacillomycin D	Aflatoxin contamination (*Aspergillus flavus*)	Moyne et al., 2001
6	*B. subtilis* BBG100	Mycosubtilin	Damping off (P. *aphanidermatum*)	Leclere et al., 2005
7	*B. subtilis* QST713	Iturin A	Damping off (*Botrytis cinerea* and *Rhizoctonia solani*)	Paulitz and Belanger, 2001; Kloepper et al., 2004
8	*Burkholderia cepacia*	Pyrrolnitrin Pseudane	Damping off and rice blast (*R. solani* and *Pyricularia oryzae*)	Homma et al., 1989
9	*Lysobacter* sp. strain SB-K88	Xanthobaccin A	Damping off (*Aphanomyces cochlioides*)	Islam et al., 2005
10	*Pantoea agglomerans* C9-1	Herbicolin	Fire blight (*Erwinia amylovora*)	Sandra et al., 2001
11	*Pseudomonas fluorescens* 2-79 and 30-84	Phenazines	Take-all (*Gaeumannomyces graminis* var. *tritici*)	Thomashow et al., 1990
12	*P. fluorescens* Pf-5	Pyoluteorin, Pyrrolnitrin	Damping off (*Pythium ultimum* and *R. solani*)	Howell and Stipanovic, 1980
13	*P. fluorescens* F113	2, 4- diacetyl-phloroglucinol (DAPG)	Damping off (*Pythium* spp.)	Shanahan et al., 1992
14	*P. putida* WCS358r	Phenazine, DAPG	Wheat diseases	Glandorf et al., 2001
15	*Pseudomonas* sp. LBUM223	Phenazine-1-carboxylic acid	Common scab of potato (*Streptomyces scabies*)	Arseneault et al., 2013

Table 12.7: Degradation/detoxification of virulence factors of pathogens by antagonists

Antagonistic microbe	Virulence degrading/ detoxifying factor produced	Action	Target pathogen	Reference
Pantoea dispersa	Esterase	Detoxification of the toxin albicidin	*Xanthomonas albilineans*	Zhang and Birch, 1996, 1997
Trichoderma harzianum	Protease	Degradation of hydrolytic enzymes	*Botrytis cinerea*	Kapat et al., 1998; Elad and Kapat, 1999
Klebsiella oxytoca	Protein	Detoxification of albicidin	*X. albilineans*	Walker et al., 1988
Alcaligenes denitrificans	Protein	Detoxification of albicidin	*X. albilineans*	Basnayake and Birch, 1995
Ralstonia solanacearum	Protein	Detoxification of fusaric acid	*Fusarium*	Toyoda et al., 1988
Cladosporium tenuissimum	Cladosporol	Inhibition of β-1,3-glucan biosynthesis	*Cronartium flaccidum* and *Periderimum pini*	Moricca et al., 2001

3.4 Cross Protection and Induced Resistance

The term cross protection has been used widely to denote first use of mild strain of plant pathogenic virus to provide protection to plants from future infection of severe strain of the same virus. Mild strain of tobacco mosaic virus and citrus tristeza virus are best known to give cross protection to plants against severe strains of the same viruses in tomato and citrus, respectively. In citrus, the cross protection effect of mild strain remains active for many years.

Resistance in plants can be induced or induced resistance can be achieved by restricted inoculation of pathogens, attenuated pathogens, non-pathogenic organisms or chemical substances. It provides broad spectrum protection in cereals, cucurbits, legumes, solanaceous crops and fruit plants. Prior inoculation of cucurbits with fungal pathogen like *Colletotrichum lagenarium* or viral pathogen like tobacco necrosis virus protects plants against several pathogens. Injection infiltration of heat killed cells of wilt causing bacterium *Ralstonia solanacearum* gives protection to the plants against the virulent pathogen. Microbes that exhibit induced resistance include *Bacillus* spp., *Lysobacter enzymogenes*, *Pseudomonas* spp. among bacteria and *Fusarium* spp., *Trichoderma* spp., *Rhizoctonia* spp., mycorrhizal fungi and *Pythium oligandrum*. *Trichoderma harzianum* T39 induces resistance in grapevine against downy mildew pathogen, but abiotic stresses affect induced resistance negatively (Roatti et al., 2013). *T. harzianum* NBRI-1055 was reported to trigger defense responses in an enhanced level in sunflower and provided protection against seedling blight caused by *Rhizoctonia solani* (Singh et al., 2013).

Some of the physiological changes induced by the microbes in plant are:

- Strengthening of epidermal and cortical wall.
- Deposition of callose, lignin and phenolics.
- Increased levels of enzymes: chitinases, β-1,3-glucanases, β-1,4-glucosidase, PR-1 protein, glucosidases, peroxidases, phenylalanine ammonia lyase, hydroperoxide lyase and an associated oxidative burst.
- enhanced phytoalexin production
- enhanced expression of stress-related genes and resistance to the oxidative burst

Elicitors of induced resistance are known mostly from bacteria. These are lipopolysaccharides, siderophores, flagella or the protein subunit flagellin, bacterial volatiles, salicylate, the cyclic peptide syringolin, antibiotics, such as 2,4-diacetylphloroglucinol and a *N*-trialkylated benzylamine derivative. A few of the elicitors like xylanase, serine proteinase, are known to be produced from *Trichoderma* spp.

3.5 Lysis

Antagonistic microorganisms, both fungi and bacteria can cause lysis of different structures or components of target pathogen through production of various lytic enzymes or other metabolites (Table 12.8) in the vicinity.

Table 12.8: Lytic agents secreted by antagonists/biocontrol agents

Antagonist	Lytic agent	Target pathogen	Reference
Serratia marcescens	Chitinase	*Sclerotium rolfsii*	Ordentlich et al., 1988
Lysobacter enzymogenes strain C3	Beta-1,3-glucanase	-	Palumbo et al., 2005
Pseudomonas fluorescens CHA0	Hydrogen cyanide	Black rot of tobacco (*Thielaviopsis basicola*)	Voisard et al., 1989
Enterobacter cloacae	Ammonia	*Pythium ultimum* (damping-off of cotton)	Howell et al., 1988
Serratia plymuthica C48	Chitinase	*Botrytis cinerea*	Frankowski et al., 2001
Paenibacillus sp. strain 300 and *Streptomyces* sp. strain 385	Chitinase, β-1,3-glucanase	*Fusarium oxysporum* f. sp. *cucumerinum*	Singh et al., 1999
Pseudomonas stutzeri	chitinase and laminarinase	*F. solani*	Lim et al., 1991
Serratia plymuthica IC14	Protease	*Sclerotinia sclerotiorum* and *B. cinerea*	Kamensky et al., 2003
Pseudomonas cepacia	β-1,3-glucanase	*R. solani, S. rolfsii* and *Pythium ultimum*	Fridlender et al., 1993

3.6 Plant Growth Promotion

Plant growth promotion helps the plants indirectly withstanding the onslaught of various plant pathogens. Better plant vigor and health makes the plant escape disease at certain stages of growth. Biocontrol agents when applied in soil bring about increased solubilization of nutrients, increased nutrient uptake and sequestration of nutrients. Promotion of mycorrhizal and rhizobial functions, regulation of ethylene production, release of plant growth hormone and decreased toxicity towards heavy metals are some other mechanisms to promote plant growth. *Trichoderma* spp. are probably the most well known among the fungal plant growth promoters (Harman, et al., 2004; Harman, 2006) and *Bacillus* and *Pseudomonas* sp. are typical bacterial agents as plant growth promoters.

3.7 Hypovirulence

Hypovirulence or reduced virulence has been observed with a few fungal pathogens till date. Such fungal isolates possess viruses (mycoviruses) with double stranded RNA genome. Such a virus infection in fungal pathogen makes the pathogen less fit to cause a plant disease. Inability to infect host properly, reduced sexual sporulation, female infertility and reduced pigmentation due to the presence of mycoviruses are the symptoms of hypovirulence. However, such viruses do not kill the host fungus. These hypovirulent isolates when introduced to a plant diseased with the virulent isolate of the same fungus, the mycoviruses get transmitted to the virulent isolate through hyphal fusion. Viruses spread to the virulent isolate making it hypovirulent and thus reducing the disease. Such hypovirulence has been used in Europe to control Chestnut blight disease caused by the fungus *Cryphonectria parasitica* (Heiniger and Rigling, 1994).

4. PLANT DISEASES CONTROLLED BY MICROBES

4.1 Bacterial Antagonists

Among bacterial antagonists *Bacillus* and *Pseudomonas* are most commonly exploited genera. Bacterial antagonists have been widely used for fungal as well as bacterial disease control in plants (Table 12.9). The genus *Bacillus* was established by Ferdinand Cohn in 1872. There are over 20 species under the genus. The genus is ubiquitously present in both rhizospheric and non-rhizospheric soil. It improves plant growth by several mechanisms like nitrogen fixation, phosphate solubilization, secretion of plant growth promoting substances, inhibition of soil-borne plant pathogenic fungi, and induction of disease resistance.

Table 12.9: Bacterial agents for control of plant diseases

S. No.	Name of microbe	Target pathogen(s)	Crop diseases managed
1	*Agrobacterium radiobacter* strain K84	*Agrobacterium tumefaciens*	Crown gall disease in ornamental crops and stone fruit crops
2	*Agrobacterium radiobacter* strain 1026	*A. tumefaciens*	Ornamentals, fruit and nuts
3	*Bacillus subtilis*	*Pythium* spp.	Damping off disease
4	*B. subtilis*	*Fusarium* spp., *Rhizoctonia* spp., *Alternaria* spp., *Aspergillus* spp., *Phytophthora* spp., *Verticillium* spp.	Various vegetables and ornamental plants
5	*B. subtilis*	*Nectria galligena*	Canker on apple trees
6	*B. licheniformis*	*Sclerotinia homeocarpa*	Ornamentals and turf grasses
7	*B. pumillus*	*Fusarium* spp., *Rhizoctonia solani*	Soybean
8	*B. cereus*	*Phytophthora medicaginis*	Damping off of alfalfa
9	*Burkholderia cepacia*	*Fusarium* spp., *nematode*, *Pythium* spp., *Rhizoctonia solani*	Wheat, maize, vegetables, leguminous crops and cotton
10	*Erwinia amylovora* harpin protein	Many	Field, vegetable and ornamental crops
11	*Erwinia carotovora* non-pathogenic	Bacterial soft rot	Vegetables, crucifers
12	*Erwinia herbicola*	*Erwinia amylovora*	Fire blight of pome fruits
13	*Pseudomonas aureofaciens*	Turf grass pathogens	Turf grass
14	*P. aureofaciens*	*Pythium* spp., *Rhizoctonia solani*	Vegetables, ornamentals
15	*P. fluorescens*	*Erwinia amylovora*	Potato, tomato and fruit crops
16	*P. syringae*	*Botrytis* spp., *Mucor* spp., and *Penicillium* spp.	Potato, fruit crops
17	*P. chlororaphis*	*Drechslera* spp., *Fusarium* spp., *Septoria* spp.	Seed-borne disease in cereals
18	*P. chlororaphis*	*Fusarium oxysporum, Pythium* spp., *Rhizoctonia solani*	Wilt disease, stem and root rot
19	*Streptomyces lydicus*	Soil-borne pathogens	Ornamental and other crops
20	*S. griseoviridis*	Soil-borne pathogens	Ornamental and vegetables
21	*Pseudomonas solanacearum* non-pathogenic	*P. solanacearum*	Rots in vegetables
22	*Paenibacillus polymyxa*	Damping off	Cucumber
23	*P. polymyxa*	*Meloidogyne javanica, Fusarium oxysporum, Aspergillus niger, A. oryzae, Leptosphaeria maculans*	Multiple crops

24	*Erwinia carotovora* non-pathogenic	*Erwinia carotovora*	Soft rot in vegetables
25	*Streptomyces colombiensis*	*Rhizoctonia solani*	Turf grass
26	*Streptomyces goshikiensis*	*R. solani*	Turf grass, rice
27	*Pseudomonas syringae* ESC-10, *Pseudomonas syringae* ESC-11	*B. cinerea, Helminthosporium* spp., *Fusarium* spp., *Penicillium* spp., *Mucor* spp., *Geotrichum candidum*	Storage rot of pome fruits, potato

Bacterial species like *Pseudomonas fluorescens* and *P. putida* are well known to produce siderophores which chelate iron and facilitate its transport from environment to microbial cell. Siderophores like pyoverdine and pseudobactin have high affinity for soluble iron (Fe^{+3}) and therefore deprive other microorganisms including pathogens of iron thereby inhibiting the growth and development of pathogenic microbes. It is one of the classic cases of nutrient competition among microbes implicated as one of the mechanisms of biological control of plant pathogens. *Pseudomonas* species are also known to compete for amino acids with *B. cinerea* thereby inhibiting conidial germination.

Agrobacterium radiobacter strain K84 has been commercialized for biocontrol of crown gall pathogen *Agrobacterium tumefaciens*. The strain K84 produces bacteriocin named as agrocin 84 which is inhibitory to the crown gall bacterium. It also competes with the pathogen for colonizing the sites for infection and for nutrients.

4.2 Fungal Antagonists

Among fungal antagonists *Trichoderma* is the dominant organism and it has proved itself as a gift of nature to the mankind and attained the status of superhero fungus (Mohiddin et al., 2010). Several other fungi have been evaluated, formulated and released commercially for a range of plant pathogens be it air-borne or soil-borne (Table 12.10). *Trichoderma*, an ascomycete is one of the most studied biocontrol agents acting against numerous plant pathogens (Table 12.11). It acts through a complex mechanism where different mechanisms like hyperparasitism, antibiotic production, lysis (Table 12.12) act synergistically. Many of the naturally existing microbes have been found promising and awaiting further research, refinement and release as commercial formulations (Table 12.13).

Table 12.10 Fungal agents for control of plant diseases

S. No.	Name of microbe	Target pathogen(s)	Crop diseases managed
1	*Ampelomyces quisqualis*	Powdery mildew pathogens (*Sphaerotheca fuliginea*)	Cucurbits, fruit crops, ornamentals
2	*Aspergillus flavus* (non-aflatoxigenic)	*Aspergillus flavus*	Cotton, groundnut, maize

Contd...

3	*Aspergillus niger* AN27	*Fusarium oxysporum* f.sp. *melonis*	Wilt of muskmelon and watermelon (Sen, 2000)
4	*Aureobasidium pullulans*	*Botrytis cinerea, Monilinia* sp., *Erwinia amylovora*	Blossom infection of strawberry, cherry and damson
5	*Aureobasidium pullulans*	*Pezicula* sp., *Nectria galligena, B. cinerea, Penicillium expansum, Monilinia fructigena*	Storage rot of apple
6	*Candida oleophila*	*Botrytis cinerea, Penicillium* spp.	Storage rot of citrus, pome fruits
7	*Chaetomium globossum*	*Venturia inaequalis*	Apple scab
8	*Cladosporium cladosporioides*	*Nectria galligena*	European canker of apple trees
9	*Coniothyrium minitans*	*Sclerotinia sclerotiorum, S. minor*	Canola, cucumber, groundnut, lettuce, capsicum, soybean, sunflower, tomato, ornamental and vegetable crops
10	*Coriolus versicolor*	*Armillaria mellea*	Protection of birch stump (Rishbeth, 1976)
11	*Cryptococcus albidicus*	*B. cinerea, Penicillium expansum*	Storage rot of apple and pear
12	*Fusarium lateritium*	*Eutypa armeniaceae*	Gummosis or dieback of apricot
13	*Fusarium oxysporum* (non-pathogenic)	*Fusarium oxysporum, F. moniliforme*	Basil, carnation, cyclamen, tomato, vegetables, potato
14	*Fusarium oxysporum* (non-pathogenic)	*Fusarium oxysporum*	Asparagus, basil, carnation, cyclamen, gerbera, tomato
15	*Gliocladium catenulatum*	*Pythium* spp., *Rhizoctonia solani, Botrytis* spp., *Didymella* spp.	Green house crops
16	*G. roseum*	*B. cinerea*	Gray mold of strawberry (Sutton et al., 1997)
17	*G. virens*	*Rhizoctonia solani* and *Pythium* spp.	Ornamental and food crop plants grown in greenhouses, nurseries, homes, etc.
18	*Metschnikowia fruticola*	*Aspergillus niger, Botrytis cinerea, Penicillium* spp., *Rhizopus stolonifera*	Storage rots of pome and stone fruit, grapes and strawberry
19	*Myrothecium verrucaria*	Parasitic nematode	Cole crops, grapes, ornamentals
20	*Nematopthora gynophila, Verticillium chlamydosporium*	Cyst nematode	Cereals
21	*Paecilomyces lilacinus*	Various nematodes	Banana, citrus, pineapple, potato, sugarcane, tomatoes, wheat and others
22	*Pasteuria penetrans*	Root knot nematode	Many vegetables

23	*Phlebia gigantea*	*Heterobasidion annosum*	Rot on pine and spruce trees
24	*Pichia guilliermondii*	*Botrytis* sp. and *Penicillium* sp.	Rot of apple and orange
25	*Pythium oligandrum*	*Pythium* spp., *Fusarium* spp., *Botrytis* spp., *Phytophthora* spp., *Aphanomyces* spp., *Alternaria* spp., *Tilletia caries, Pseudocercosporella herpotrichoides, Gaeumannomyces graminis, Rhizoctonia solani, Sclerotium cepivorum*	Vegetables, fruits, legumes, cereals, canola, forest nurseries and ornamental plants
26	*Sporothrix flocculosa*	*Sphaerotheca pannosa* var. *rosae*	Rose powdery mildew (Bélanger et al., 1994)
27	*Trichoderma atroviride*	*B. cinerea*	Tomato, vine
28	*Trichoderma harzianum*	Fungi causing wilt, take-all and root rot diseases	Flowers, fruit, ornamentals, turf and vegetables
29	*T. harzianum*	*Rhizoctonia solani, Pythium* spp., *Fusarium* spp., *Sclerotinia homeocarpa* and *Sclerotium rolfsii*	Trees, shrubs, transplants, all ornamentals, tomato, cucumber, bean, cabbage, maize, cotton, groundnut, potato, sorghum, soybean, sugarbeet, turf and greenhouse ornamentals
30	*T. harzianum*	*Primarily Botrytis cinerea,* also *Collectotrichum* spp., *Fulvia fulva, Monilinia laxa, Plasmopara viticola, Pseudoperonospora cubensis, Rhizopus stolonifer, Sclerotinia sclerotiorum*	Cucumber, grape, nectarine, soybean, strawberry, sunflower, tomato
31	*T. viride*	*Rhizoctonia* spp., *Pythium* spp., *Fusarium* spp., root rot, seedling rot, collar rot, red rot, damping-off, Fusarium wilt	Mustard, lentil, oilseeds, soybean, cotton, chillies, chickpeas, green pigeonpea, tobacco, cardamom, turmeric, green gram, black gram, tea, coffee, rubber, ginger, tomato, sugarcane, citrus, grapes, sunflower, cereals, vegetables and others
32	*Ulocladium oudemansii*	*B. cinerea, S. sclerotiorum*	Black currants, grape, kiwi fruit, ornamentals

Table 12.11: *Trichoderma* as biocontrol agent against fungal pathogens

Pathogen	Disease	Crops
Chondrostereum purpureum	Silver leaf disease	Apple, cherry, pear and plum
Fusarium oxysporum and *Fusarium solani*	Wilt	Chilli
Fusarium oxysporum f.sp. *carthami*	Wilt	Safflower
F. o. f.sp. *gladioli*	Corm rot	Gladiolus
F. o. f.sp. *lycopersici*	Wilt	Tomato
F. o. f.sp. *sesami*	Wilt	Sesame
F. o. f.sp. *cumini*	Wilt	Cumin
Fusarium spp.	Wilt	Tomato, potato, chilli
F. udum	Wilt	Pigeon pea
Macrophomina phaseolina	Dry root rot	Black gram
M. phaseolina	Charcoal rot	Sunflower
M. phaseolina	Root rot	Egg plant
Phytophthora cryptogea	Foot rot	Gerbera
Phytophthora spp.	Root rot	Black pepper, citrus, pineapple, rubber, tobacco
Pythium aphanidermatum	Damping off	Tomato, brinjal
Pythium spp., *Rhizoctonia*	Damping off	Cabbage, chilli, Chinese cabbage, groundnut, maize, potato, soybean, tomato
Rhizoctonia solani	Sheath blight	Maize, rice
R. solani	Root rot	Chickpea
R. solani	Dry root rot	Chilli
R. solani	Black scurf	Potato
R. solani	Seedling blight	Green gram
R. solani	Root rot	Soybean
Sclerotinia sclerotiorum	Cottony rot	Cabbage, Chinese cabbage
S. sclerotiorum	Head rot	Sunflower
Sclerotium rolfsii	Stem rot, root rot	Chilli, groundnut, potato, soybean, tomato

Table 12.12: Antibiotics and enzymes produced by Trichoderma

S. No.	Enzymes/Antibiotics	Reference
1	Beta-1,3-glucanase	Perez et al., 2001
2	Chitinase	Elad et al., 1982
3	Chitobiase	Ulhao and Peberdy, 1993
4	Endochitinase	Claydon et al., 1987
5	Gliotoxin, Trichodermin, Viridin	Haggag and Mohamed, 2002; Wilhite et al., 2001
6	Protease	Elad et al., 2000
7	Trichodermin	Godfredsen and Vangedal, 1965
8	Trichosetin	Marfori et al., 2002
9	Trichoviridin	Yamano et al., 1970
10	L-amino acid oxidase	Cheng et al., 2012

Table 12.13: Other potential microbes under study for plant disease management

S. No.	Name of the microbe	Target pathogens	Crops	Reference
1	*Bacillus cereus*	*Phytophthora Medicaginis* and *P. aphanidermatum*	Damping off	Smith et al., 1993
2	*Enterobacter cloacae*	*Pythium ultimum*	Damping off of cotton	Howell et al., 1988
3	*Glomus Fasciculatum* and *G. macrocarpum* (arbuscular mycorrhiza)	*Myxomycetes*	Apple replant disease	Catska, 1994
4	*G. intraradices* (arbuscular mycorrhiza)	*Radopholus similis* and *Pratylenchus coffeae*	Nematodes in banana	Elsen et al., 2008
5	*Lysobacter* sp.	*Aphanomyces cochlioides*	Damping off	Islam et al., 2005
6	*Paecilomyces lilacinus*	*R. similis*	Burrowing nematode in banana	Mendoza et al., 2004
7	*P. lilacinus*	*Meloidogyne incognita*	Root knot nematode in tomato	Kiewnick and Sikora, 2003
8	*Pantoea agglomerans*	*Erwinia amylovora*	Fire blight of pome fruits	Sandra et al., 2001; Johnson and Temple, 2013
9	*Paxillus Involutus* (ectomycorrhiza)	*Fusarium oxysporum* and *F. moniliforme*	Root rot in red pine	Ross and Marx, 1972
10	*Pisolithus tinctorius* (ectomycorrhiza)	*Phytophthora cinnamomi*	Root rot in trees	Ross and Marx, 1972

Contd...

11	*Pseudomonas fluorescens*	*Thielaviopsis basicola*	Black rot of tobacco	Voisard et al., 1989
12	*Scytalidium uredinicola*	*Endocronartium harknessii*	Western gall rust on *Pinus banksiana* and *P. contorta*	
13	*Serratia marcescens*	*Sclerotium rolfsii*	Various	Ordentlich et al., 1988
14	*Tuberculina maxima*	*Cronartium ribicola*	Blister rust of white pine	
15	*Verticillium lecanii*	Powdery mildew and rust fungi	Various	Heintz and Blaich, 1990
16	*V. lecanii*	*Uromyces dianthi*	Rust in carnation	
17	*Bacillus firmus* GB-126 and *P. lilacinus* 251	*Rotylenchulus reniformis*	Reniform nematode in cotton	Castillo et al., 2013
18	*Agrobacterium vitis* strain F2/5	*Agrobacterium vitis*	Grape crown gall	Kaewnum et al., 2013
19	*Streptomyces* sp. (n=17 isolates)	*Streptomyces scabies* (n=15 isolates)	Scab of potato	Otto-Hanson et al., 2013

5. OTHER GROUPS OF MICROBES IN PLANT DISEASE MANAGEMENT

5.1 Mycorrhiza

Mycorrhiza provides both physical and chemical protection to host plants. Mycorrhizal fungi

- produce antibiotics inhibitory to plant pathogens;
- increase absorption of nutrients and water as fungal hyphae behave in a similar way as root hairs. Fungal hyphae can cover more area than root hairs, are more attracted to nutrients than root hairs and can interiors of soil space where root hairs cannot reach. Hyphae increase the amount of soil the plant can access by upto 100,000 times;
- produce plant growth hormone cytokinins and gibberellins and promote plant growth;
- increase uptake of phosphorus, potassium, copper, iron, nickel, sulfur and zinc
- enhanced colonization of nitrogen fixing bacterium *Rhizobium* in legumes due to increased phosphorus availability;
- protect plant from heavy metal toxicity;
- promote growth of beneficial microbe and change microbial flora in rhizosphere;

- improve physical characteristics of soil by producing compounds like glomalin, a glycoprotein that promote formation of soil aggregates; and
- induce resistance in host plant; in the mycorrhiza colonized plants there is accumulation of phenolic compounds, and local and systemic defense responses occur (Huang et al., 2003).

Plant diseases are suppressed due to improvement in host nutrition and increase in plant vigor by the mycorrhizal fungi. Healthy hosts can be better in position to resist to tolerant the soil-borne pathogens like *Fusarium, Phytophthora, Pythium, Rhizoctonia, Verticillium*, etc. Moreover, protective sheath or the mantle formed on root surface by the ectomycorrhizal fungi physically block the infection sites to protect the root from fungal, bacterial as well as nematode infection.

5.2 PGPR

Plant growth promoting rhizobacteria (PGPR) is a sub-type under plant growth promoting bacteria (PGPB) that also include bacteria which establish endophytic population inside the plant. Thus, the PGPB whether free-living (PGPR), or endophytic though differ in terms of their ecological niches share almost same mechanisms for plant growth promotion and biocontrol of plant pathogens. These are competition for space and/or nutrition, production of inhibitory compounds and induction of systemic resistance to a wide variety of biotic and abiotic stresses. PGPB has been shown to quench quorum-sensing capacity of bacterial pathogens by degrading autoinducer signals and thereby blocking expression of virulence genes (Dong et al., 2000; Molina et al., 2003; Uroz et al., 2003; Dong et al., 2004; Morello et al., 2004; Newton and Fray, 2004; Shrivastava et al., 2014; Prasad et al., 2015).

- Competition for nutrients in the nutrient rich niches along the root surface
- Competition for iron and production of siderophore under iron-limiting environment
- Antibiotic production
- Lytic enzyme production
- Detoxification and degradation of virulence factors
- Induced systemic resistance

5.3 Endophytes

Microorganisms living within plant tissues, seeds, root, stems, leaves for all or part of their life cycle without causing any visible symptoms of their presence are defined as endophytes. Both bacteria (Table 12.14) and fungi (Table 12.15) are extensively present as endophytes in various plants. These endophytic microbes benefit plants in various ways:

- promote plant growth through secretion of plant growth regulators, phosphate solubilizing activity, enhancing mycorrhizal colonization and hyphal growth, and supplying biologically fixed nitrogen and essential vitamins to plants;

- biocontrol of various pathogens through production of antifungal and antibacterial compounds, siderophore production, nutrient competition, induction of systemic acquired resistance;
- activate stress response system of plant more rapidly and strongly; and
- osmotic adjustment, stomatal regulation, modification of root morphology, enhanced uptake of minerals.

Table 12.14: Examples of endophytic bacteria from some crops

Crop	Endophytic bacteria
Rice	*Azorhizobium caulinodans, Bradyrhizobium japonicum, Rhizobium leguminosarum, Sphingomonas paucimobilis, Azoarcus* sp., *Chromobacterium violaceum, Herbaspirillum seropedicae, Klebsiella* sp., *Klebsiella variicola, Pantoea* sp., *Serratia marcescens, Sphingobacterium* sp.
Maize	*Burkholderia pickettii, Enterobacter* spp., *Bacillus megaterium, Arthrobacter globiformis, Microbacterium testaceum*
Wheat	*Streptomyces, Mycobacterium* sp., *Klebsiella* sp.
Banana	*Azospirillum brasilense, Azospirillum amazonense, Burkholderia* sp., *Herbaspirillum seropedicae, Citrobacter* sp., *Klebsiella variicola*
Sugarcane	*Klebsiella variicola, Herbaspirillum rubrisulbalbicans, Herbaspirillum seropedicae, Gluconacetobacter diazotrophicus*
Soybean	*Enterobacter sakazakii, Enterobacter agglomerans, Erwinia* sp., *Klebsiella pneumoniae, Klebsiella oxytoca, Pantoea* sp., *Pseudomonas citronellolis*
Citrus	*Nocardia* sp., *Curtobacterium flaccumfaciens, Bacillus megaterium, Bacillus* sp., *Pantoea agglomerans, Enterobacter cloacae, Burkholderia cepacia, Methylobacterium mesophilicum*

Table 12.15: Endophytic fungi in plant disease management

Fungal endophytes	Plant pathogens inhibited	Reference
Chaetomium and *Phoma* of wheat	*Puccinia* and *Pyrenophora* spp. of wheat	Dingle and McGee, 2003; Istifadah and McGee, 2006
Piriformospora indica	Wilt pathogen *Fusarium culmorum* and a leaf pathogen *Blumeria graminis* of barley	Waller et al., 2005
Acremonium strictum from *Dactylis glomerata*	Potato pathogen *Helminthosporium solani*	Rivera Varas et al., 2007
Epichloë festucae	*Sclerotinia homoeocarpa* in *Festuca rubra*	Clarke et al., 2006
Non-pathogenic strains of *Fusarium oxysporum*	*Meloidogyne incognita*	Hallmann and Sikora, 1996
Neotyphodium spp. in *Lolium pretense*	*Barley yellow dwarf virus* (BYDV); incidence of aphid vector of the virus is low in endophyte-infected plant	Lehtonen et al., 2006

Cryptosporiopsis quercina	*Pyricularia oryzae*	Strobel et al., 1999
Colletotrichum sp. from *Artemisia annua*	*Phytophthora capsici, Rhizoctonia cerealis, Gaeumannomyces graminis*	Lu et al., 2000
Non-pathogenic *Fusarium oxysporum* from healthy banana roots	*Fusarium oxysporum* f.sp. *cubense* (Fusarium wilt of banana in greenhouse)	Belgrove et al., 2011
Endophytic strain *Fusarium oxysporum* Fo47	*F. oxysporum* f.sp. *lycopersici* (Fusarium wilt in tomato)	Aime et al., 2013
Piriformospora indica	Root rot in barley (*Fusarium culmorum*)	Harrach et al., 2013

Endophytic *Fusarium* spp. were found to decrease the activity of burrowing nematode *R. similis in vitro* as well as in soil in absence of plants (Vu et al., 2004). There are many reports of inhibition of plant pathogens by endophytic bacteria such as against oak wilt pathogen *Ceratocystis fagacearum* (Brooks et al., 1994), take-all of wheat pathogen *Gaeumannomyces graminis* (Coombs et al., 2004), *Rhizoctonia solani* (Parmeela and Johri, 2004), *Erwinia* and *Xanthomonas* (Sessitsch et al., 2004).

5.4 Nematophagous Fungi

Nematophagous (nematode destroying) fungi are the natural enemies of nematodes. There are more than 200 species of fungi which can attack living nematodes in juvenile/adult/egg stages and derive nutrient from them. They form various infection structures (Table 12.16) to trap vermiform nematodes or use hyphal tips to attack eggs/cysts. The nematophagous fungi can be categorized into three groups: nematode trapping fungi (e.g., *Arthrobotrys* spp.), endoparasitic fungi (*Drechmeria coniospora, Hirsutella rhossiliensis, Catenaria anguillulae* etc.) and egg- and cyst-parasitic fungi (*Pochonia chlamydosporia*).

Table 12.16: Nematophagous fungi and their infection structures

Nematophagous fungi	Infection structure
Arthrobotrys oligospora	Adhesive nets
A. conoides	
A. musiformis	
A. superb	
Duddingtonia flagrans	
Monacrosporium gephyropagum	Adhesive branches
M. ellipsosporum	Adhesive knobs
M. haptotylum	
A. dactyloides	Constricting rings
A. brochopaga	

Nematoctonus concurrens	Adhesive knobs and adhesive spores
N. leiosporus	Adhesive spores
Hirsutella rhossoliensis	
Drechmeria coniospora	
Stylopage hadra	Adhesive hyphae
Cystopage cladospora	
Pochonia chlamydosporia	Appressoria
Harposporium anguillulae	Ingested spore
Catenaria anguillulae	Zoospores
Haptoglossa dickii	
Pleurotus ostreatus	Toxic droplets

5.5 Bacteriophages

There are limited options (antibiotics and copper compounds) to effectively control bacterial diseases of plants. Biological control strategies like use of non-pathogenic strains or attenuated strains of pathogen, saprophytic bacteria, PGPR have been used to manage plant pathogenic bacteria with variable results. The idea of using bacteriophage for plant disease control came soon after its discovery by Twort (1915) and d'Herelle (1917) (Moore, 1926). But its practical use has been unsuccessful (Goto, 1992). However, time to time there have been some encouraging results claimed by researchers that are presented below (Table 12.17).

Table 12.17 Bacteriophages in management of plant bacterial diseases

Plant disease	Causal bacteria	Reference	Remarks
Black rot of cabbage	*Xanthomonas Campestris* pv. *campestris*	Bergamin Filho and Kimati 1981	When applied three days before to one day after inoculation of pathogen with maximum reduction on application of phage on the same day as pathogen
Bacterial spot of pepper	*Xanthomonas campestris* pv. *vesicatoria*	Bergamin Filho and Kimati 1981	When applied three days before to the day of application of the pathogen with maximum reduction on application of phage on the same day as pathogen

Bacterial wilt of tobacco	*Ralstonia solanacearum*	Tanaka et al., 1990	Mixture of avirulent *R. solanacearum* and phage had synergistic effect
Fire blight of apple blossom	*Erwinia amylovora*	Schnabel et al., 1999	Effective when applied at the same time as the pathogen
Tomato bacterial spot	*Xanthomonas* sp.	Obradovic et al., 2004, 2005	Phage in combination with SAR inducer gave reliable disease control
Fire blight of pear	*E. amylovora*	Svircev et al., 2006	Mixture of *Pantoea agglomerans* and phages had synergistic effect
Xanthomonas leaf blight of onion	*Xanthomonas* sp.	Lang et al., 2007	Phage in combination with acibenzolar-S-methyl (SAR inducer) gave enhanced control

6. CONCLUSION

Damaging effects of inorganic agricultural practices in the overall environment have forced to ponder and bring out the idea of sustainable agriculture which encompasses the agricultural technologies and practices that would have no adverse effect on the environment, be effective and accessible by the farmers and lead to increased productivity. Such agricultural technologies in the plant protection sector must depend on biological and ecological processes. Several of the natural organisms have been exploited for their biocontrol potential and many have been in use already for quite some time. However, biocontrol agents are not true substitute for chemical pesticides and there are some inherent bottlenecks associated with biocontrol agents. There is need to address the concerns in the use of such biocontrol agents and improving their efficacy. With biotechnological approach, the natural mechanisms of biocontrol are being investigated in molecular detail which would one day further lessen the use of chemical pesticides to meet the goals of sustainable agriculture.

References

Agrios GN (2005). Plant Pathology. Elsevier Academic Press, USA.

Aime S, Alabouvette C, Steinberg C, and Olivain C (2013). The endophytic strain *Fusarium oxysporum* Fo47: a good candidate for priming the defense responses in tomato roots. Mol Plant Microbe Interact 26(8): 918–926.

Altomare C, Norvell WA, Bjorkman T, and Harman GE (1999). Solubilization of phosphate and micro nutrients by the plant growth promoting fungus *Trichoderma harzianum* Riafi. Appl Environ Microbiol 65: 2926–2933.

Arseneault T, Goyer C, and Filion M (2013). Phenazine production by *Pseudomonas* sp. LBUM223 contributes to the biological control of potato common scab. Phytopathology 103(10): 995–1000.

Banik S, and Jambhulkar PP (2014). Biological control of diseases in field crops: status and concerns. In: Rodriguez HN, Sarkar NC, Ramanjaneyulu AV, Ratikanta Maiti (eds) Advances in Agro-technology-A Text Book. Vol. II. Puspa Publishing House, Kolkata, pp. 335–351.

Basnayake WVS, and Birch RG (1995). A gene from *Alcaligenes denitrificans* that confers albicidin resistance by reversible antibiotic binding. Microbiology-UK141: 551–560.

Bélanger RR, Labbé C, and Jarvis WR (1994). Commercial-scale control of rose powdery mildew with fungal antagonist. Plant Disease 78: 420–424.

Belgrove A, Steinberg C, and Viljoen A (2011). Evaluation of nonpathogenic *Fusarium oxysporum* and *Pseudomonas fluorescens* for panama disease control. Plant Disease 95(8): 951–959.

Bergamin Filho A, and Kimati H (1981). Estudos sobre um bacteriofago isolado de *Xanthomonas campestris*. II. Seu emprego no controle de *X. campestris* e *X. vesicatoria*. Summa Phytopathologica 7: 35–43.

Bolwerk A, Lagopodi AL, Wijfjes AHM, Lamers GEM, Chin-A-Woeng TFC, Lugtenberg BJJ, and Bloemberg GV (2003). Interactions in the tomato rhizosphere of two *Pseudomonas* biocontrol strains with the phytopathogenic fungus *Fusarium oxysporum* f.sp *radicis-lycopersici*. Mole Plant Microbe Interact 16: 983–993.

Brooks DS, González CF, Appel DN, and Filer TH (1994). Evaluation of endophytic bacteria as potential biological control agents for oak wilt. Biol Control 4: 373–381.

Castillo JD, Lawrence KS, and Kloepper JW (2013). Biocontrol of the reniform nematode by *Bacillus firmus* GB-126 and *Paecilomyces lilacinus* 251 on cotton. Plant Disease 97(7): 967–976.

Catska V (1994). Interrelationship between vesicular-arbuscular mycorrhiza and rhizosphere microflora in apple replant disease. Biologia Plant 36: 99–104.

Cheng CH, Yang CA, and Peng KC (2012). Antagonism of *Trichoderma harzianum* ETS 323 on *Botrytis cinerea* mycelium in culture conditions. Phytopathol 102: 1054–1063.

Cheng YL, McNally DJ, Labbe C, Voyer N, Belzile F, and Bélanger RR (2003). Insertional mutagenesis of a fungal biocontrol agent led to discovery of a rare cellobiose lipid with antifungal activity. Appl Environ Microbiol 69: 2595–2602.

Clarke BB, White JF, Hurley H, Torres MS, Sun S, and Huff DR (2006). Endophyte-mediated suppression of dollar spot disease in fine fescues. Plant Dis 90: 994–998.

Claydon N, Allen M, Hanson JR, and Avent AG (1987). Antifungal alklyl pyrones of *Trichoderma harzianum*. Trans Br Mycol Soc 88: 505–513.

Cook RJ, and Baker KF (1983). The Nature and Practice of Biological Control of Plant Pathogens. American Phytopathological Society, St Paul, Minnesota, USA, p. 589.

Coombs JT, Michelsen PP, and Franco CMM (2004). Evaluation of endophytic actinobacteria as antagonists of *Gaeumannomyces graminis* var. *tritici* in wheat. Biol Control 29: 359–366.

Crane JM, Gibson DM, Vaughan RH, and Bergstrom GC (2013). Iturin levels on wheat spikes linked to biological control of Fusarium head blight by *Bacillus amyloliquefaciens*. Phytopathol 103(2):145–155.

de Souza JT, de Boer M, de Waard P, van Beek TA, and Raaijmakers JM (2003). Biochemical, genetic, and zoosporicidal properties of cyclic lipopeptide surfactants produced by *Pseudomonas fluorescens*. Appl Environ Microbiol 69: 7161–7172.

Défago G (1993). 2,4-Diacetylphloroglucinol, a promising compound in biocontrol. Plant Pathology 42: 311–312.

Dingle J, and McGee PA (2003). Some endophytic fungi reduce the density of pustules of *Puccinia recondita* f.sp. *tritici*in wheat. Mycol Res 107: 310–316.

Dong YH, Xu JL, Li XZ, and Zhang LH (2000). AiiA, an enzyme that inactivates the acylhomoserine lactone quorum-sensing signal and attenuates the virulence of *Erwinia carotovora*. Proc Natl Acad Sci USA 97: 3526–3531.

Dong YH, Zhang XF, Xu JL, and Zhang LH (2004). Insecticidal *Bacillus thuringiensis* silences *Erwinia carotovora* virulence by a new form of microbial antagonism, signal interference. Appl Environ Microbiol 70: 954–960.

Duijff BJ, Meijer JW, Bakker PAHM, and Schippers B (1993). Siderophore-mediated competition for iron and induced resistance in the suppression of Fusarium wilt of carnation by fluorescent *Pseudomonas* spp. Netherlands Journal of Plant Pathology 99: 277–289.

Edwards SG, and Seddon B (2001). Mode of antagonism of *Brevibacillus brevis* against *Botrytis cinerea in vitro*.J Appl Microbiol 91: 652–659.

Elad Y, Chet I, Doyle P, and Henis Y (1982). Parasitism of Trichoderma spp. on *Rhizoctonia solani* and *Sclerotium rolfsii*-scanning microscopy and fluorescent microscopy. Phytopathology 72: 85–88.

Elad Y, Freeman S, and Monte E (2000). Biocontrol agents: Mode of action and interaction with other means of control. Vol 24 IOBC, Sevilla, Espana.

Elad Y, and Kapat A (1999). The role of *Trichoderma harzianum* protease in the biocontrol of *Botrytis cinerea*. Eur J Plant Pathol 105: 177–189.

Elsen A, Gervacio D, Swennen R, and De Waele D (2008). AMF-induced biocontrol against plant parasitic nematodes in *Musa* sp.: a systemic effect. Mycorrhiza 18(5): 251–256.

Filonow AB (1998). Role of competition for sugars by yeasts in the biocontrol of gray mold of apple. Biocontrol Science and Technology 8: 243–256.

Frankowski J, Lorito M, Scala F, Schmidt R, Berg G, and Bahl H (2001). Purification and properties of two chitinolytic enzymes of *Serratia plymuthica* HRO-C48. Arch Microbiol 176: 421–426.

Fridlender M, Inbar J, and Chet I (1993). Biological control of soil-borne plant pathogens by a β-1,3-glucanase-producing *Pseudomonas cepacia*. Soil Biol Biochem 25: 1211-1221.

Glandorf DC, Verheggen P, Jansen T, Jorritsma JW, Smit E, Leefang P, Wernars K, Thomashow LS, Laureijs E, Thomas-Oates JE, Bakker PA, and Van Loon LC (2001). Effect of genetically modified *Pseudomonas putida* WCS358r on the fungal rhizosphere microflora of field-grown wheat. Appl Environ Microbiol 67: 3371–3378.

Godfredsen WO, and Vangedal S (1965). Trichodermin, a new sequiterpene antibiotic. Acta Chem Scand 19: 1088–1102.

Goto M (1992). Fundamentals of Bacterial Plant Pathology.San Diego, CA: Academic Press.

Haggag, WM, and Mohamed HAA (2002). Enhancement of antifungal metabolites production from gamma ray induced mutants of some *Trichoderma* species for control of onion white rot disease. Plant Pathol Bull 11: 45–56.

Hallmann J, and Sikora RA (1996). Toxicity of fungal endophyte secondary metabolites to plant parasitic nematodes and soil-borne plant pathogenic fungi. Eur J Plant Pathol 102: 155–162.

Harman GE (2000). Myths and dogmas of Biocontrol: Changes in the perceptions derived from research on *Trichoderma harzianum* T-22. Plant Dis84:377–393.

Harman GE (2006). Overview of mechanisms and uses of *Trichoderma* spp. Phytopathology 96: 190–194.

Harman GE, Howell CR, Viterbo A, Chet I, and Lorito M (2004). *Trichoderma* species-opportunistic, avirulent plant symbionts. Nat Rev Microbiol 2: 43–56.

Harman GE and Nelson EB (1994). Mechanisms of protection of seed and seedlings by biological control treatments: Implications for practical disease control. In: Martin T., (ed.), Seed Treatment: Progress and Prospects. BCPC, Farnham, UK, pp. 283-292.

Harrach BD, Baltruschat H, Barna B, Fodor J, and Kogel KH (2013). The mutualistic fungus *Piriformospora indica* protects barley roots from a loss of antioxidant capacity caused by the necrotrophic pathogen *Fusarium culmorum*. Mol Plant Microbe Interact 26(5): 599–605.

Hashidoko Y, Nakayama T, Homma Y, and Tahara S (1999). Structure elucidation of xanthobaccin A, a new antibiotic produced from *Stenotrophomonas* sp. strain SB-K88. Tetrahedron Letters 40: 2957–2960.

Heiniger U, and Rigling D (1994). Biological control of chestnut blight in Europe. Annu Rev Phytopatho l32: 581–599.

Heintz C, and Blaich R (1990). *Verticillium lecanii* als Hyperparasit des Rehmchltaus (*Uncinula necator*). Vitis 29: 229–232.

Herr LJ (1995). Biological control of *Rhizoctonia solani* by binucleate *Rhizoctonia* spp. and hypovirulent *R. solani* agents.Crop Protection 14: 179–186.

Heungens K, and Parke JL (2000). Zoospore homing and infection events: Effects of the biocontrol bacterium *Burkholderia cepacia* AMMDR1 on two oomycete pathogens of pea (*Pisum sativum* L.). Appl Environ Microbio l66: 5192–5200.

Homma Y, Kato Z, Hirayama F, Konno K, Shirahama H, and Suzui T (1989). Production of antibiotics by *Pseudomonas cepacia* as an agent for biological control of soil-borne plant pathogens. Soil Biol Biochem 21: 723–728.

Howell CR (2002). Cotton seedling pre-emergence damping-off incited by *Rhizopus oryzae* and *Pythium* spp. and its biological control with *Trichoderma* spp. Phytopathology 92: 177–180.

Howell CR, Beier RC, and Stipanovic RD (1988). Production of ammonia by *Enterobacter cloacae* and its possible role in the biological control of *Pythium* pre-emergence damping-offby the bacterium. Phytopathology 78: 1075–1078.

Howell CR, and Stipanovic RD (1980). Suppression of *Pythium ultimum* induced damping -off of cotton seedlings by *Pseudomonas fluorescens* and its antibiotic, pyoluterin. Phytopathology 70: 712–715.

Huang J, Luo S, and Zeng R (2003). Mechanisms of plant disease resistance induced by arbuscular mycorrhizal fungi. Ying Yong Sheng Tai Xue Bao 14(5): 819–822.

Islam TM, Hashidoko Y, Deora A, Ito T, and Tahara S (2005). Suppression of damping-off disease in host plants by the rhizoplane bacterium *Lysobacter* sp. strain SB-K88 is linked to plant colonization and antibiosis against soil-borne peronosporomycetes. Appl Environ Microbiol 71: 3786–3796.

Istifadah N, and McGee PA (2006). Endophytic *Chaetomium globosum* reduces development of tan spot in wheat caused by *Pyrenophora tritici-repentis*. Australas Plant Path 35: 411–418.

Johnson KB, and Temple TN (2013). Evaluation of strategies for fire blight control in organic pome fruit without antibiotics. Plant Dis 97(3): 402–409.

Janse JD (2006). Phytobacteriology: Principles and Practice. 1st edn., CABI Publishing. Oxfordshire.

Kaewnum S, Zheng D, Reid CL, Johnson KL, Gee JC, and Burr TJ (2013). A host-specific biological control of grape crown gall by *Agrobacterium vitis* Strain F2/5: its regulation and population dynamics. Phytopathol 103(5): 427–435.

Kageyama K, and Nelson EB (2003). Differential inactivation of seed exudate stimulation of *Pythium ultimum* sporangium germination by *Enterobacter cloacae* influences biological control efficacy on different plant species. Appl Environ Microbio l69: 1114–1120.

Kamensky M, Ovadis M, Chet I, and Chernin L (2003). Soil-borne strain IC14 of *Serratia plymuthica* with multiple mechanisms of antifungal activity provides biocontrol of *Botrytis cinerea* and *Sclerotinia sclerotiorum* diseases. Soil Biol Biochem 35: 323–331.

Kapat A, Zimand G, and Elad Y (1998). Effect of two isolates of *Trichoderma harzianum* on the activity of hydrolytic enzymes produced by *Botrytis cinerea*. Physiol Mol Plant Path 52: 127–137.

Kerr A (1980). Biological control of crown gall through production of agrocin 84. Plant Dis 64: 25–30.

Kiewnick S, and Sikora RA (2003). Efficacy of *Paecilomyces lilacinus* (strain 251) for the control of root-knot nematode. Commun Agric Appl Biol Sci 68(4 Pt A):123–128.

Kim B, Moon SS, and Hwang BK (1999). Isolation, identifi cation, and antifungal activity of a macrolide antibiotic, oligomycin A, produced by *Streptomyces libani*. Can J Bot 77: 850–858.

Kloepper JW, Leong J, Teintze M, and Schroth MN (1980). *Pseudomonas* siderophores: A mechanism explaining disease suppression in soils. Curr Microbiol 4: 317–320.

Kloepper JW, Ryu CM, and Zhang S (2004). Induce systemic resistance and promotion of plant growth by *Bacillus* spp. Phytopathology 94: 1259–1266.

Koumoutsi A, Chen XH, Henne A, Liesegang H, Gabriele H, Franke P, Vater J, and Borris R (2004). Structural and functional characterization of gene clusters directing nonribosomal synthesis of bioactive lipopeptides in *Bacillus amyloliquefaciens* strain FZB42. J Bact 186: 1084–1096.

Lahlali R, Peng G, Gossen BD, McGregor L, Yu FQ, Hynes RK, Hwang SF, McDonald MR, and Boyetchko SM (2013). Evidence that the biofungicide Serenade (*Bacillus subtilis*) suppresses club root on canola via antibiosis and induced host resistance. Phytopathology 103(3): 245–254.

Lang JM, Gent DH, and Schwartz HF(2007). Management of Xanthomonas leaf blight of onion with bacteriophages and a plant activator. Plant Dis 91: 871–878.

Larkin RP, and Fravel DR (1999). Mechanisms of action and dose-response relationships governing biological control of fusarium wilt of tomato by nonpathogenic *Fusarium* spp. Phytopathology 89: 1152–1161.

Leclere V, Bechet M, Adam A, Guez JS, Wathelet B, Ongena M, Thonart P, Gancel F, Chollet-Imbert M, and Jacques P (2005). Mycosubtilin overproduction by *Bacillus subtilis* BBG100 enhances the organism's antagonistic and biocontrol activities. Appl Environ Microbiol 71: 4577–4584.

Lehtonen PT, Helander M, Siddiqui SA, Lehto K, and Saikkonen K (2006). Endophytic fungus decreases plant virus infections in meadow ryegrass (*Loliumpratense*). Biol Lett 2: 620–623.

Lim HS, Kim YS, and Kim SD (1991). *Pseudomonas stutzeri* YPL-1 genetic transformation and antifungal mechanism against *Fusarium solani*, an agent of plant root rot. Appl Environ Microbiol 57: 510–516.

Lindow SE, and Leveau JH (2002). Phyllosphere microbiology. Curr Opinin Biotechnol 13: 238–243.

Loper JE, and Buyer JS (1991). Siderophores in microbial interactions on plant surfaces. Mol Plant Microbe Interact 4: 5–13

Loper JE, and Henkels MD (1999). Utilization of heterologous siderophores enhances levels of iron available to *Pseudomonas putida* in the rhizosphere. Appl Environ Microbiol 65: 5357–5363.

Lu H, Xou WX, Meng JC, Hu J, and Tan RX (2000). New bioactive metabolites produced by *Colletotrichum* sp., an endophytic fungus in *Artemisia annua*. Plant Science 151: 67–73.

Marfori EC, Kajiyama S, Fukusaki E, and Kobayashi A (2002). Trichosetin, a novel tetramic acid antibiotic produced in dual culture of *Trichoderma harzianum* and *Catharanthus roseus* callus. Z Naturforschung 57: 465–470.

Mendoza A, Sikora RA, and Kiewnick S (2004). Efficacy of *Paecilomyces lilacinus* (strain 251) for the control of *Radopholus similis* in banana. Commun Agric Appl Biol Sci 69(3): 365–372.

Milner JL, Raffel SJ, Lethbridge BJ, and Handelsman J (1995). Culture conditions that influence accumulation of zwittermicin A by *Bacillus cereus* UW85. App Microbiol Biotechnol 43: 685–691.

Milner JL, Silo-Suh L, Lee JC, He HY, Clardy J, and Handelsman J (1996). Production of kanosamine by *Bacillus cereus* UW85. Appl Environ Microbiol 62: 3061–3065.

Mohiddin FA, Khan MR, Khan SM, and Bhat BH (2010). Why *Trichoderma* is considered super hero (super fungus) against the evil parasites? Plant Pathology Journal 9: 92–102.

Molina L, Constantinescu F, Reimmann C, Duffy B, and De´fago G (2003). Degradation of pathogen quorum-sensing molecules by soil bacteria: a preventive and curative biological control mechanism. FEMS Microbiol Ecol 45: 71–81.

Moore ES (1926). D'Herelle's bacteriophage in relation to plant parasites.South African Journal of Science 23: 306.

Morello JE, Pierson EA, and Pierson LS (2004). Negative cross-communication among wheat rhizosphere bacteria: effect on antibiotic production by the biological control bacterium *Pseudomonas aureofaciens* 30-84. Appl Environ Microbiol 70: 3103–3109.

Moricca S, Ragazzi A, Mitchelson KR, and Assante G (2001). Antagonism of the two-needle pine stem rust fungi *Cronartium flaccidum* and *Peridermium pini* by *Cladosporium tenuissimum in vitro* and *in planta*. Phytopathology 91: 457–468.

Moyne AL, Shelby R, Cleveland TE, and Tuzun S (2001). Bacillomycin D: an iturin with antifungal activity against *Aspergillus flavus*. J Appl Microbiol 90: 622–629.

Nakayama T, Homma Y, Hashidoko Y, Mizutani J, and Tahara S (1999). Possible role of xanthobaccins produced by *Stenotrophomonas* sp. strain SB-K88 in suppression of sugar beet damping-off disease. Appl Environ Microbiol 65: 4334–4339.

Nelson EB (1990). Exudate molecules initiating fungal responses to seed seeds and roots. Plant and Soil 129: 61–73.

Newton JA, and Fray RG (2004). Integration of environmental and host-derived signals with quorum sensing during plant-microbe interactions. Cell Microbiol 6: 213–224.

Nielsen TH, Sørensen D, Tobiasen C, Andersen JB, Christophersen C, Givskov M, and Sorensen J (2002). Antibiotic and biosurfactant properties of cyclic lipopeptides produced by fluorescent *Pseudomonas* spp. from the sugar beet rhizosphere. Appl Environ Microbiol 68: 3416–3423.

Nielsen TH, and Sørensen J (2003). Production of cyclic lipopeptides by *Pseudomonas fluorescens* strains in bulk soil and in the sugar beet rhizosphere. Appl Environ Microbiol 69: 861–868.

Obradovic A, Jones JB, Momol MT, Balogh B, and Olson SM (2004). Management of tomato bacterial spot in the field by foliar applications of bacteriophages and SAR inducers. Plant Dis 88: 736–740.

Obradovic A, Jones JB, Momol MT, Olson SM, Jackson LE, Balogh B, Guven K, and Iriarte FB (2005). Integration of biological control agents and systemic acquired resistance inducers against bacterial spot on tomato. Plant Dis 89: 712–716.

Ordentlich A, Elad Y, and Chet I (1988). The role of chitinase of *Serratia marcescens* in the biocontrol of *Sclerotium rolfsii.* Phytopathol 78: 84–88.

Otto-Hanson LK, Grabau Z, Rosen C, Salomon CE, and Kinkel LL (2013). Pathogen variation and urea influence selection and success of *Streptomyces* mixtures in biological control. Phytopathol 103(1): 34–42.

Palumbo JD, Yuen GY, Jochum CC, Tatum K, and Kobayashi DY (2005). Mutagenesis of beta-1,3-glucanase genes in *Lysobacter enzymogenes* strain C3 results in reduced biological control activity toward Bipolaris leaf spot of tall fescue and Pythium damping-off of sugar beet. Phytopathology 95: 701–707.

Parmeela and Johri BN (2004). Phylogenetic analysis of bacterial endophytes showing antagonism against *Rhizoctonia solani.* Curr Sci 87: 687–692.

Paulitz TC, and Belanger RR (2001). Biological control in greenhouse systems. Annu Rev Phytopathol 39: 103–133.

Perez C, de-La Fuente L, Arias A, Altier N, and de-La Fuente L (2001). Use of native fluorescent *Pseudomonas* for controlling seedling diseases of *Lotus cornicalatus* L. Agrociencia-Montevidea 5: 41–47.

Prasad R, Kumar M, and Varma A (2015). Role of PGPR in soil fertility and plant health. In:Egamberdieva D, Shrivastava S, Varma A (eds) Plant Growth-Promoting Rhizobacteria and Medicinal Plants. Springer International Publishing Switzerland 247–260.

Raaijmakers JM, Vlami M, and de Souza JT (2002). Antibiotic production by bacterial biocontrol agents. Anton Leeuw Int J Gen Mol Microbiol 81: 537–547.

Rishbeth J (1976). Chemical treatment and inoculation of hardwood stumps for control of *Armillaria mellea*. Ann Appl Biol 82: 57–70.

Rivera Varas VV, Freeman TA, Gusmestad NC, and Secor GA (2007). Mycoparasitism of *Helminthosporium solani* by *Acremonium strictum*. Phytopathology 97: 1331–1337.

Roatii B, Perazzolli M, Gessler C, and Pertot I (2013). Abiotic stresses affect *Trichoderma harzianum* T39-induced resistance to downy mildew in grapevine. Phytopathology 103(12): 1227–1234.

Romero D, de Vicente A, Rakotoaly RH, Dufour SE, Veening JW, Arrebola E, Cazorla FM, Kuipers OP, Paquot M, and Perez-Garcia A (2007). The iturin and fengycin families of lipopeptides are key factors in antagonism of *Bacillus subtilis* toward *Podosphaera fusca*. Mol Plant Microbe Interact 20: 430–440.

Ross EW, and Marx DM (1972). Susceptibility sand pine to *Phytophthora cinnamomi*. Phytopathology 62: 1197–1200.

Sandra AI, Wright CH, Zumoff LS, and Steven VB (2001). *Pantoca agglomerans* strain EH318 produces two antibiotics that inhibit *Erwinia amylovora in vitro*. Appl Environ Microbiol 67: 282–292.

Scherm B, Ortu G, Muzzu A, Budroni M, Arras G, and Migheli Q (2003). Biocontrol activity of antagonisticyeasts against *Penicillium expansum* on apple. J Plant Pathol 85: 205–213.

Schnabel EL, Fernando WGD, Meyer MP, Jones AL, and Jackson LE (1999). Bacteriophage of *Erwinia amylovora* and their potential for biocontrol. Acta Horticulturae 489: 649–654.

Sen B (2000). Biological control: A success story. Ind Phytopathol 53: 243–249.

Sessitsch A, Reiter B, and Berg G (2004). Endophytic bacterial communities of field-grown potato plants and their plant-growth-promoting and antagonistic abilities. Can J Microbiol 50: 239–249.

Shanahan P, O'Sullivan DJ, Simpson P, Glennon JD, and O'Gara F (1992). Isolation of 2,4-Diacetylphloroglucinol from a fluorescent pseudomonad and investigation of physiological parameters influencing its production. Appl Environ Microbiol 58: 353–358.

Shrivastava S, Prasad R, and Varma A (2014). Anatomy of root from eyes of a microbiologist. In: Root Engineering. Springer-Verlag Berlin Heidelberg 40: 3–22.

Singh BN, Singh A, Singh BR, and Singh HB (2013). *Trichoderma harzianum* elicits induced resistance in sunflower challenged by *Rhizoctonia solani*. J Appl Microbiol DOI:10.1111/jam.12387.

Singh PP, Shin YC, Park CS, and Chung YR (1999). Biological control of *Fusarium* wilt of cucumber by chitinolytic bacteria. Phytopathology 89: 92–99.

Smith KP, Havey MJ, and Handelsman J (1993). Suppression of cottony leak of cucumber with *Bacillus cereus* strain UW85. Plant Dis 77: 139–142.

Strobel GA, Miller RV, Martinez-Miller C, Condron MM, Teplow DB, and Hess WM (1999). Cryptocandin, a potent antimycotic from the endophytic fungus *Cryptosporiopsis* cf. *quercina*. Microbiology 145: 1919–1926.

Sutton JC, Gang Peng DWL, Yu H, Zhang P, and Valdebenito-Sanhueza RM (1997). *Gliocladium roseum* a versatile adversary of *Botrytis cinerea* in crops. Plant Dis 81: 316–328.

Svircev AM, Lehman SM, Kim WS, Barszcz E, Schneider KE, and Castle AJ (2006). Control of the fire blight pathogen with bacteriophages. In: Zeller W, Ullrich C (eds) Proceedings of the 1st International Symposium on Biological Control of Bacterial Plant Diseases, October 23-26, 2005. Berlin, Germany: Die Deutsche Bibliothek - CIP-Einheitsaufnahme, 259–261.

Tanaka H, Negishi H, and Maeda H (1990). Control of tobacco bacterial wilt by an avirulent strain of *Pseudomonas solanacearum* M4S and its bacteriophage. Ann Phytopathol Soc Japan 56: 243–246.

Thomashow LS, Weller DM, Bonsall RF, and Pierson LS III (1990). Production of the antibiotic phenazine-1-carboxylic acid by fluorescent pseudomonas in the rhizosphere of wheat. Appl Environ Microbiol 56: 908–912.

Toyoda H, Hashimoto H, Utsumi R, Kobayash H, and Ouchi S (1988). Detoxification of fusaric acid by a fusaric acid resistant mutant of *Pseudomonas solanacearum* and its application to biological control of Fusarium wilt of tomato. Phytopathology 78: 1307–1311.

Ulhao CJ, and Peberdy JF (1993). Effect of carbon sources on chitobiose production by *Trichoderma harzianum*. Mycol Res 97: 45–48.

Uroz S, D'Angelo-Picard C, Carlier A, Elasri M, Sicot C, Petit A, Oger P, Faure D, and Dessaux Y (2003). Novel bacteria degrading *N*-acylhomoserine lactones and their use as quenchers of quorum-sensing-regulated functions of plant-pathogenic bacteria. Microbiology 149: 1981–1989.

Voisard C, Keel C, Haas D, and Defago G (1989). Cyanide production by *Pseudomonas fluorescens* helps suppress black root of tobacco under gnotobiotic conditions. EMBO J 8: 351–358.

Vu TT, Sikora RA, and Hauschild R (2004). Effects of endophytic *Fusarium oxysporum* towards *Radopholus similis* activity in absence of banana. Commun Agric Appl Biol Sci 69(3):381–385.

Walker MJ, Birch RG, and Pemberton JM (1988). Cloning and characterization of an albicidin resistance Gene from *Klebsiella oxytoca*. Mol Microbiol 2: 443–454.

Waller F, Achatz B, Baltruschat H, Fodor J, Becker K, Fischer M, Heier T, Hückelhoven R, Neumann C, Von Wettstein D, Franken P, and Kogel KH (2005). The endophytic fungus *Piriformospora indica* reprograms barley to salt stress tolerance, disease resistance, and higher yield. PNAS USA 102: 13386–13391.

Wiest A, Grzegorski D, Xu BW, Goulard C, Rebuffat S, Ebbole DJ, Bodo B, and Kenerley C (2002). Identification of peptaibols from *Trichoderma virens* and cloning of a peptaibol synthetase. J Biol Chem 277: 20862–20868.

Wilhite SE, Lumsden RD, and Straney DC (1994). Mutational analysis of gliotoxin production by the biocontrol fungus *Gliocladium virens* in relation to suppression of Pythium damping-off. Phytopathology 84: 816–821.

Wilhite SE, Lunsden RD, and Strancy DC (2001). Peptide synthetase gene *in Trichoderma virens*. Appl Environ Microbiol 67: 5055–5062.

Yamano T, Hemmi S, Yamatnoto I, and Tsubaki K (1970). Trichoviridin a new antibiotic. Japanese Kokai 70: 5435.

Zembek P, Perlinska-Lenart U, Brunner K, Reithner B, Palamarczyk G, Mach RL, and Kruszewska JS (2011). Elevated activity of dolichyl phosphate mannose synthase enhances biocontrol abilities of *Trichoderma atroviride*. Mol Plant Microbe Interact 24(2): 1522–1529.

Zhang L, and Birch RG (1996). Biocontrol of sugar cane leaf scald disease by an isolate of *Pantoea dispersa* which detoxifies albicidin phytotoxins. Lett Appl Microbiol 22: 132–136.

Zhang LH, and Birch RG (1997). The gene for albicidin detoxification from *Pantoea dispersa* encodes an esterase and attenuates pathogenicity of *Xanthomonas albilineans* to sugarcane. Proc Nat Acad Sci USA 94: 9984–9989.

13

Micro-algal and Physicochemical Studies of Wetland Soil in Keoladeo National Park, Bharatpur, Rajasthan (India)

NIVEDITA SHARMA [1,4*], NARENDRA KUMAR[1,2] AND SURENDRA SINGH[3]

[1]School of Studies in Microbiology, Jiwaji University
Gwalior (M.P.) India

[2]Department of Biotechnology, IMS Engineering College
Ghaziabad (U.P.) India

[3]Centre of Advanced Study in Botany, BHU
Varanasi, India

[4]Department of Biodiversity, Atal Bihari Vajpayee Hindi Vishwavidyalay
Bhopal (M.P.) India

*Email: *niveditachaturvedi241023@gmail.com*

1. INTRODUCTION

Wetlands are regarded as important ecosystem which plays a transitional role between open water and terrestrial ecosystems. They are endowed with specific structural and functional attributes performing major ecological roles in the biosphere (Edwards, 1990). They have been recognised as peatbogs, grass and sedge marshes, floodplains consisting of recent alluvial deposits bordering rivers, inland valleys, shallow ponds, mudflats and littoral areas of larger bodies of water which can be grouped in to naturally occurring and man-made or anthropogenic wetlands (Gopal et al., 1982; Eshiett, 1992). Most of the studies have reported microbial communities of forest soils and water (He et al., 2006; Hackl et al., 2005; Laverman et al., 2005). The structural diversity of microbial communities in soils is tightly related to the above-ground plant species composition, thus providing an important link between the above and the below-ground processes in the terrestrial ecosystem (Pankhurst, 1997).

Microbes are continuously depleting from the biosphere due to changes in the ecosystem caused by natural and human activities. Increase in anthropogenic activities is, however, resulting into degradation of wetlands. Keoladeo National Park (KNP), the site of the present study, is a designated Ramsar site and also a world heritage site. Since, no study has so far been conducted on the microbial aspect of the Keoladeo National Park;

the present study was therefore conducted to have an insight of the soil cyanobacterial biodiversity of KNP.

Regarding microbial diversity, it is suggested that less than 5 percent of the total number of microbes have so far been discovered and major portion of the diversity in these organisms are yet to be explored and identified in future. There are ~1.4 million organisms that have been described (Wilson, 1988; Minelli, 1993) and <10,000 of these are popular enough to be known (Pimm et al., 1995). The grand total has been estimated (by adding the organisms yet to be described) to be 10^7 to 10^8 species (Pimm et al., 1995; May, 1992). The various forms of microorganisms are bacteria, algae, fungi, protozoa and viruses which are found everywhere on the planet, in the soil, water, air and in the sea. Among the natural habitats, soil system support a good number of microorganisms. Recent developments using modern technology in microbial diversity research indicate that more than 95 percent of the naturally occurring microorganisms worldwide are undiscovered and their ecological role is therefore, currently unknown. The vastness of their number is only now appreciated; less than 5 percent are named at present. However, our knowledge on diversity of microorganisms as well as their role in sustaining global life support systems is rather meagre.

Cyanobacteria, also referred to as blue-green algae, are the pioneer oxygenic phototrophs on earth whose distribution around the world is surpassed only by bacteria (Adams, 2000). Fossil evidence points to their presence in geographically diverse regions during the Precambrian (2 to more than 3.5 billion years ago). They are a large and morphologically diverse group of phototrophic prokaryotes, which occur in almost every habitat on earth. This versatility may explain the remarkable lack of morphological (and presumably physiological) change seen in 3.5-billion-year-old fossilized cyanobacteria and their modern-day counterparts (Adams, 2000). Their long evolutionary history has been marked by key geochemical and biotic transitions, including the creation of oxygenic photosynthesis (Holland, 1977), a prerequisite for the development and proliferation of metabolically complex microbial and higher eukaryotic life forms. Indeed, the plastids of higher plants and other photosynthetic eukaryotes are thought to have possibly arisen from a single common ancestor, which is a result of an endosymbiosis between a phagotrophic host and a cyanobacterium (Loffelhandt and Bohnert, 1977). Cyanobacteria were originally considered algae because of their microscopic morphology, pigmentation and oxygen evolving photosynthesis in which photosystems, PS II and PS I are connected in series. The genome size of cyanobacteria, representative of all major taxonomic groups, lies in the range, 1.6 × 109 to 8.6 × 109 Da which is comparable to that of other bacteria (1.0 to 3.6 × 109 Da) (Herdman et al., 1979) . Cyanobacteria are known to occur in oxic and anoxic environments. Several species can switch to the typical bacterial anoxygenic photosynthesis using sulphide as electron donor; other species assimilate sugars and organic compounds in presence of light (photo-heterotrophy) (Rippka, 1972). In dark, cyanobacteria gain energy by respiring endogenous carbohydrates, which are accumulated in the light. However, under anoxic conditions some species maintain this requirement by fermentation whereas in a few cases chemo-organotrophy is found. Cyanobacteria can grow under very low water potential; such species can resist desiccation and grow in arid environments (deserts) and or can tolerate high salinity to grow in hypersaline ponds (Thajuddin and Subramanian, 1992). In many environments, cyanobacteria are the primary producers at the base of the food web of the ecosystem, viz., marine waters, hypersaline, brackish waters, soda lakes, freshwater, paddy fields, soils, deserts, cave walls, hot springs, polar regions and other extreme environments. Cyanobacteria are symbionts of a variety of other organisms, viz., the marine diatom Rhizosolenia, leaves of Azolla and the roots of Cycas (Rai, 1990).

Cyanobacteria are a group of Gram-negative, morphologically diverse, aerobic phototrophs whose distribution is ubiquitous in nature and are found everywhere including places with extreme climatic conditions, such as in Antarctica, in hot springs (Whitton and Potts, 2000) and even in oxic and anoxic environments (Thajuddin and Subramanian, 2005). They flourish in places with nitrogen deficient environment. They are important as they contribute towards carbon and nitrogen economy of different ecological habitats and can grow in purely inorganic medium using light as energy and CO_2 and N_2 as sole carbon and nitrogen sources, respectively (Wyatt and Silvey, 1969). The role of nitrogen fixing cyanobacteria in enhancing soil fertility has been long known and is well documented (Singh, 1961; Venkataraman, 1981). Cyanobacteria contribute to overall soil health not only by its ability to perform biological nitrogen fixation, but also because of its ability to produce polysaccharides and other bioactive compounds which have growth stimulating effect on plants, as well as important in maintaining soil quality and preventing erosion. However, in nature cyanobacteria are constantly challenged by different factors and stresses including fluctuations in temperature, pH, moisture content (droughts and floods), salinity and so forth. These factors influence distribution and diversity of cyanobacteria in a region. Among these, pH is an important determinant as cyanobacteria are known to prefer neutral to alkaline pH for their optimum growth (Singh, 1961; Kaushik, 1994). In recent times, studies on cyanobacterial diversity has received due attention (Thajuddin and Subramanian, 2005; Prasanna et al., 2009).

Soil is the most important non-aqueous habitat for algae and cyanobacteria (blue-green algae) (Zenova et al., 1995). Due to their high capacity for morphological and physiological adaptations to different environments, both algae and cyanobacteria often act as pioneer microorganisms in terrestrial ecosystems (Metting, 1981; Hoffmann, 1989). In farmlands, soil algae and cyanobacteria play an important role because they both have tremendous potential to serve as sources of nitrogen and carbon for other organisms. Soil fertility is generally improved by the organic matter produced by these organisms (Mishra and Pabbi, 2004). Soil algae excrete growth-promoting substances, such as hormones, vitamins, amino acids, and organic acids that affect other organisms in many ways (Roger and Reynaud, 1982; Wilson, 2006). The presence of soil algae also stabilizes the soil surface and thus reduces erosion (Evans and Johansen, 1999; Hu et al., 2004). The polysaccharides produced by some soil algae increase soil porosity, aggregation, and water-holding capacity (Roger and Reynaud, 1982; Choudhary et al., 2007). Soil algae are also ameliorators in the reclamation of saline and metals, thereby improving soil quality (Rai et al., 1998; Whitton and Potts, 2000).

2. MATERIALS AND METHODS

2.1 Study Area

The study was carried out in KNP (Fig. 13.1), Bharatpur , Rajasthan, declared in 1956 as a protected area and bird sanctuary and later upgrade to a National Park, in 1981 (Sharma and Praveen, 2002). The 29 km² park (27° 7.6′ – 27° 12.2′ N and 77° 29.5′ - 77° 33.5′ E, almost equidistant, about 180 km, between Delhi and Jaipur) is one of the important waterfowl habitats in the country (Azeez et al., 2007), and one of the early Ramsar sites (Mathur et al., 2005). The park is segmented into 15 blocks, named alphabetically from A to O, separated by earthen dykes or mud trails, for ease of management and tourism. Of these 15 blocks in the park, 6 blocks were selected for the periodic collection of soil samples. These sites represent the entire habitat of the park. The sites selected for the periodic collection of soil samples are named as follows: A block, B block, D block, E block, K block and L block.

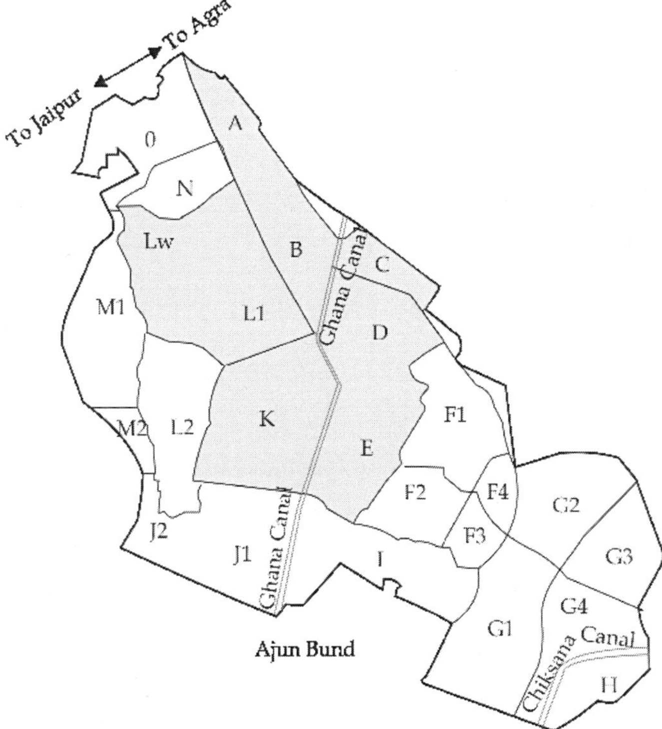

Fig. 13.1 Different blocks (A, B, D, E, K and L) of Keoladeo National Park

2.2 Sample Collection

Soil samples were collected up to 0–6 cm depth from the surface with the help of sterilized iron borer, in sterilized polyethene bags previously marked with adhesive stickers following the method of Johnson and Curl (1972), and taken in ice-packed coolers to the laboratory for bacteriological and physicochemical analysis.

2.3 Soil pH

Calomel electrode method was used for determining the soil pH (Piper, 1966). Soil was dispersed in the distilled water in a ratio of 1:5 (w/v) and was shaken vigorously for one hour on an electric shaker. The pH of the soil suspension was recorded by using pH meter.

2.4 Physicochemical Studies of Wetland Soils

Organic carbon in soil samples was analyzed by following the method of Datta et al., (1962), inorganic phosphorus was analyzed by following the method of Fiske and Subbarow (1925). Sodium and potassium contents were estimated by using Flame Photometer (Doll and Lucas, 1973), Calcium was estimated by EDTA method as described by Trivedi and Goel (1986) and total nitrogen content was estimated by Kjeldhal digestion method (Fleck, 1967) in soil samples.

2.5 Cyanobacterial Analyses

Collected samples were mixed throughly, of which 0.5 g was suspended in 10 ml of liquid Chu10 and incubated under standard growth conditions. After 2 weeks, 0.5 ml of the suspension was poured over the Petri plates containing solid Chu10 and allowed to grow (Stanier et al., 1971). Cultures were grown and maintained in a culture chamber set at 28±1°C and illuminated with 40-W daylight fluorescent lamps, giving an average light intensity of 70 $\mu Em^{-2}s^{-1}$ on the surface of the vessels with a 14:10 light: dark rhythm. Cyanobacterial colonies developed on the culture plates were transferred to conical flask with Chu10 medium. Cyanobacterial specimens were identified using the publications of (Geitler, 1932) and (Desikachary, 1959).

3. RESULTS

3.1 Soil pH

The pH values ranged from 6.60–8.40. The soil pH in D and K block was higher than A, B, E and L block as shown in Table 13.1. However, differences in the soil pH values of the different sampling locations were not observed to be statistically significant.

Table 13.1: pH of soil samples collected from different blocks of Keoladeo National Park

S. No.	Sampling Sites	pH of soil samples		
		1st Sampling	2nd Sampling	3rd Sampling
1.	A Block	6.6	7.0	7.1
2.	B Block	6.7	7.2	7.2
3.	D Block	8.4	7.9	7.3
4.	E Block	7.7	7.3	7.3
5.	K Block	8.1	8.0	7.1
6.	L Block	7.3	7.4	7.3

3.2 Levels of Organic Carbon in Soil Samples of Different Blocks

It is evident from the data of Table 13.2 that the concentration of organic carbon ranged from 30 mg/g to 128 mg/g in the soil samples collected from different sampling sites (blocks) during August 2006 (1st sampling). However, the concentration of organic carbon ranged from 41 mg/g ton 132 mg/g during April 2007 (2nd sampling). While, during Nov. 2007 (3rd sampling) the concentration of organic carbon ranged found from 32 mg/g to 130 mg/g. The data of Table 13.2 indicate that highest concentration of organic carbon in A-block and the lowest concentration of organic carbon in D-block were recorded.

Table 13.2: Organic carbon levels in soil samples collected from different blocks of Keoladeo National Park

S. No.	Sampling Sites	Organic carbon (mg/g)		
		1st Sampling	2nd Sampling	3rd Sampling
1.	A Block	128	132	130
2.	B Block	62	71	66
3.	D Block	30	41	32
4.	E Block	51	60	55
5.	K Block	52	63	55
6.	L Block	49	55	51

3.3 Levels of Total Nitrogen Soil Samples of Different Blocks

It is evident from the data of Table 13.3 that total nitrogen level ranged from 0.323 to 0.542 percent in the soil samples collected from different sampling sites (blocks) during August 2006 (Ist sampling). However, the concentration of total nitrogen ranged from 0.256 to 0.723 percent in the soil samples collected from different sampling sites (blocks) during April 2007 (2nd sampling). While, the concentration of total nitrogen ranged from 0.168 to 0.828 percent in the soil samples collected from different sampling sites during November 2007. During the first sampling, the highest (0.542 percent) and the lowest (0.323 percent) concentration of nitrogen was recorded in A- and L- Blocks, respectively. In contrast, during the second and third sampling, the highest (0.723 and 0.828 percent) and the lowest (0.256 and 0.168 percent) levels of nitrogen were recorded in A- and B-Blocks, respectively.

Table 13.3: Total nitrogen levels in soil samples collected from different blocks of Keoladeo National Park

S. No.	Sampling Sites	Total Nitrogen (%)		
		1st Sampling	2nd Sampling	3rd Sampling
1.	A Block	0.542	0.723	0.828
2.	B Block	0.378	0.256	0.168
3.	D Block	0.523	0.612	0.758
4.	E Block	0.311	0.325	0.378
5.	K Block	0.451	0.675	0.655
6.	L Block	0.323	0.400	0.423

3.4 Levels of Inorganic Phosphate in Soil Samples of Different Blocks

It is evident from the data of Table 4 that the concentration of inorganic phosphate ranged from 0.8 to 2.8 μggm^{-1} in the soil samples collected from different sampling sites (blocks) during August 2006 (1st sampling). The concentration of phosphorus ranged from 1.3 to 3.1 μggm^{-1} in the soil samples collected from different sampling sites (blocks) during April 2007 (2nd sampling). However, the concentration of phosphorus ranged 1.0 to 2.9 μggm^{-1} during November 2007 (3rd sampling).

Table 13.4: Phosphorus levels in soil samples collected from different blocks of Keoladeo National Park

S. No.	Sampling Sites	Concentration of inorganic phosphate (μggm^{-1})		
		1st Sampling	2nd Sampling	3rd Sampling
1.	A Block	0.8	1.3	1.0
2.	B Block	1.7	2.1	1.9
3.	D Block	2.6	2.9	2.7
4.	E Block	2.8	3.1	2.9
5.	K Block	2.0	2.6	2.2
6.	L Block	1.5	1.9	1.7

3.5 Levels of Calcium in Soil Samples of Different Blocks

It is evident from the data of Table 13.5 that the concentration of calcium ranged from 361.52 mg/100 g to 822.12 mg/100 g in the soil samples collected from different sampling sites (blocks) during August 2006 (1st sampling). The concentration of calcium ranged from

144.29 mg/100 g to 490.58 mg/100 g in the soil samples collected from different sampling sites (blocks) during April 2007 (2nd sampling). However, the concentration of calcium ranged from 116.59 mg/100 g to 373.23 mg/100 g during November 2007 (3rd sampling).

Table 13.5: Calcium levels in soil samples collected from different blocks of Keoladeo National Park

S. No.	Sampling Sites	Concentration of Ca2+ (mg/100g)		
		1st Sampling	2nd Sampling	3rd Sampling
1.	A Block	822.12	490.58	373.23
2.	B Block	383.16	296.59	234.53
3.	D Block	361.52	144.29	116.59
4.	E Block	538.67	388.77	251.34
5.	K Block	649.20	291.78	154.44
6.	L Block	516.23	248.49	134.21

3.6 Levels of Potassium in Soil Samples of Different Blocks

It is evident from the data of Table 13.6 that potassium concentration ranged from 1320 ppm/g to 3645 ppm/g in the soil samples collected from different sampling sites (blocks) during August 2006 (1st sampling). However, the concentration of potassium ranged from 530 ppm/g to 3100 ppm/g in the soil samples collected from different sampling sites (blocks) during April 2007 (2nd sampling). The concentration of potassium ranged from 2100 ppm/g to 3200 ppm/g in the soil samples during November 2007 (3rd sampling). During first sampling, the highest (3645 ppm/g) and lowest (1320 ppm/g) concentration of potassium was recorded in K-Block and A Block, respectively. In contrast, during second sampling, the highest (3100 ppm/g) and lowest (530 ppm/g) concentration of potassium was recorded in A-Block and K-Block, respectively. However, during the third sampling, the highest (3200 ppm/g) and lowest (2100 ppm/g) concentration of potassium in soil was recorded in E-Block and K-Block, respectively. However, the concentration of sodium was increasing in the soil collected from all the blocks from previous sampling.

Table13.6: Potassium levels in soil samples collected from different blocks of Keoladeo National Park

S. No.	Sampling Sites	K Concentration (ppm/g)		
		1st Sampling	2nd Sampling	3rd Sampling
1.	A Block	1320	3100	3125
2.	B Block	2200	2210	2230
3.	D Block	2820	1580	2321
4.	E Block	3600	2487.5	3200
5.	K Block	3645	530	2100
6.	L Block	2980	2750	3000

3.7 Levels of Sodium in Soil Samples of Different Blocks

It is evident from the data of Table 13.7 that sodium concentration ranged from 27.5 ppm/g to 137.50 ppm/g in the soil samples collected from different sampling sites (blocks) during August 2006 (1st sampling). However, the concentration of sodium ranged from 100 ppm/g to 160 ppm/g in the soil samples collected from different sampling sites (blocks) during April 2007 (2nd sampling). The concentration of sodium 105 ppm/g to 162.3 ppm/g were recorded in the soil samples collected from different sampling sites during November 2007. During the first sampling, the highest (137.50 ppm/g) and lowest (27.5 ppm/g) concentration of sodium was recorded in D-Block and A-Block, respectively. In contrast, during the second and third sampling, the highest (160 ppm/g and 162.3 ppm/g) and lowest (100 ppm/g and 105 ppm/g) concentration of sodium was recorded in D-Block and L-Block, respectively.

Table 13.7: Sodium levels in soil samples collected from different blocks of Keoladeo National Park

S. No.	Sampling Sites	Na Concentration (ppm/g)		
		1st Sampling	2nd Sampling	3rd Sampling
1.	A Block	27.5	105	108
2.	B Block	110	132.5	136
3.	D Block	137.50	160	162.3
4.	E Block	102.50	142.5	145
5.	K Block	82.5	110	112.4
6.	L Block	65.0	100	105

3.8 Variations in Cyanobacteria Types in Different Blocks during 1st Sampling

It is evident from the data of Table 13.8 that *Anabaena* sp. and *Nostoc* sp. were found to be the most common algae present in almost all blocks or sampling sites. Whereas, *Microcystis* sp. and *Spirulina* sp. were the moderately occurring microalgae. *Aphanothece* sp., *Lyngbya* sp.,*Oscillatoria* sp. and *Tolypothrix* sp. were isolated from K-Block, E-Block, A-Block and L-Block respectively. Over all the soil sample of K and E blocks of the park supported large number of cyanobacterial species.

Table 13.8: Occurrence of some genera of Cyanobacteria during the 1st Sampling in different blocks of Keoladeo National Park, Bharatpur

S. No.	Name of isolates	Sampling Station					
		K-Block	D- Block	A- Block	L- Block	B- Block	E- Block
1.	*Anabaena* sp.	+	+	+	+	+	+
2.	*Aphanothece* sp.	+	–	–	–	–	–

3.	*Lyngbya* sp.	–	–	–	–	–	+
4.	*Microcystis* sp.	+	+	–	+	–	+
5.	*Nostoc* sp.	+	+	+	+	+	+
6.	*Oscillatoria* sp.	–	–	+	–	–	–
7.	*Spirulina* sp.	+	+	+	–	+	+
8.	*Tolypothrix* sp.	–	–	–	+	–	–
Total number of genera		05	04	04	04	03	05

3.9 Variations in Cyanobacteria Types in Different Blocks during 2nd Sampling

It is evident from the data of Table 13.9 that *Anabaena* sp., *Nostoc* sp., *Oscillatoria* sp., *Spirulina* sp. and *Tolypothrix* sp. were found to be the most common algae present in almost all blocks or sampling sites. Whereas, *Calothrix* sp. and *Lyngbya* sp. were the moderately occurring microalgae. *Aphanothece* sp., *Aulosira* sp., *Cylindrospermum* sp. and *Microcystis* sp. were isolated from some defined blocks only. *Scytonema* sp. was isolated from L-Block and E-Block only. Over all the soil sample of K-Block and D-Block of the park supported large number of Cyanobacterial species.

Table 13.9: Occurrence of genera of Cyanobacteria during 2nd Sampling in different blocks of Keoladeo National Park, Bharatpur

S. No.	Name of isolates	Sampling Station					
		K-Block	D-Block	A-Block	L-Block	B-Block	E-Block
1.	*Anabaena* sp.	+	+	+	+	+	+
2.	*Aphanothece* sp.	+	–	+	–	+	–
3.	*Aulosira* sp.	–	+	+	+	–	–
4.	*Calothrix* sp.	+	–	+	+	+	–
5.	*Cylindrospermum* sp.	+	+	–	+	–	–
6.	*Lyngbya* sp.	+	+	–	–	+	+
7.	*Microcystis* sp.	+	–	–	+	–	+
8.	*Nostoc* sp.	+	+	+	+	+	+
9.	*Oscillatoria* sp.	+	+	+	–	+	+
10.	*Spirulina* sp.	+	+	+	–	+	+
11.	*Scytonema* sp.	–	–	–	+	–	+
12.	*Tolypothrix* sp.	+	+	–	+	+	+
Total number of genera		10	09	08	08	08	08

3.10 Variations in Cyanobacteria Types in Different Blocks during 3rd Sampling

It is evident from the data Table 13.10 that *Anabaena* sp. and *Nostoc* sp. were found to be the most common algae present in almost all blocks or sampling sites. Whereas, *Lyngbya* sp. and *Spirulina* sp. were the moderately occurring microalgae. *Aphanothece* sp., *Aulosira* sp., *Microcystis* sp., *Oscillatoria* sp., *Spirulina* sp. and *Tolypothrix* sp. were isolated from some defined blocks only. *Calothrix* sp. and *Scytonema* sp. were isolated from L Block only. Over all the soil sample of L and E blocks of the park supported large number of Cyanobacterial sps.

Table 13.10: Occurrence of some genera of Cyanobacteria during 3rd Sampling in different blocks of Keoladeo National Park, Bharatpur

S. No.	Name of isolates	Sampling Station					
		K-Block	D-Block	A-Block	L-Block	B-Block	E-Block
1.	*Anabaena* sp.	+	–	+	+	+	+
2.	*Aphanothece* sp.	–	–	+	–	+	–
3.	*Aulosira* sp.	–	+	–	–	–	+
4.	*Calothrix* sp.	–	–	–	+	–	–
5.	*Lyngbya* sp.	+	+	–	–	+	+
6.	*Microcystis* sp.	–	–	+	+	–	–
7.	*Nostoc* sp.	+	+	+	+	+	+
8.	*Oscillatoria* sp.	+	–	–	–	–	+
9.	*Spirulina* sp.	–	+	–	+	–	–
10	*Scytonema* sp.	–	–	–	+	–	–
11.	*Tolypothrix* sp.	+	–	–	–	–	+
Total number of genera		05	04	04	06	04	06

4. DISCUSSION

Wetlands are attributed to several important ecological functions. They are supposed to serve as natural sewage and wastewater treatment plants and sustain an immensely rich biodiversity. In general, wetlands are fast degrading as a result of increase in anthropogenic activity and altered a management practice that warrants continuous monitoring of their flora, fauna, edaphic, aquatic and other ecological conditions.

A number of studies and various physico-chemical and biological aspects of wetlands were done in India (Hutchinson, 1937; Gonzalves and Joshi, 1946; Sarup, 1961; George, 1964; Khan and Siddiqui, 1970; Greenberg et al.,1992, Singh et al., 1982; Mahajan, 1980; Handoo and Kaul, 1982; Kulshreshtha and Gopal, 1982; Seshavatharam and Chandramohan, 1982; Adoni and Saini, 1984; Vyas, 1984; Khatri, 1984; Ramlingam and Jayaraman, 1985). Bhunya and Mohanty (1990) have also studied the various physico-chemical properties in relation to faunistic composition of Ansupa Lake, Orissa. Various physico-chemical features of Kawar Lake, a natural wetland in Bihar has been discussed in detail by Ahmed and Singh (1990). Other important contributions include those by Seshavatharam (1990), in Kolleru Lake, Andhra Pradesh and Seshavatharam et al. (1990) in Lake Kodakaria, near Visakhapatnam. Physico-chemical properties of water in KNP was studied in detail by Ali and Vijayan (1986; 1987), Mittal et al. (1990) and Vijayan (1991). Whereas Akram (1992) studied the physico-chemical environment of Wular Lake, Kashmir, Abbasi (1997), however, conducted studies on the ecology of wetlands of India.

Soil quality can be assessed by numerous soil variables (Mausbach and Seybold, 1998). In addition to physico-chemical variables, biotic variables also govern soil quality (Pankhurst et al., 1997). Soil variables often show different patterns of response to the same impact (Jha et al., 1992), reflecting the multidimensional quality of soil health (van Straalen, 2002). These soil variables are associated with different soil-related variables, including crop yields. Variations in soil-related variables can be empirically predicted from datasets consisting of multiple soil variables. Therefore, analysis of multiple soil biotic variables can provide comprehensive and multidimensional information (Sena et al., 2008).

The pH values indicate that the soil pH of KNP was neutral to slightly alkaline. This soil pH favors the plants to take up nutrients and also supports the binding of nutrients into forms that are available. These results are in accordance with the findings of Mulongoy (2003). When soil pH of all the blocks during all the sampling seasons was compared, the difference was found to be statistically nonsignificant at $P<0.05$. But when soil pH of D-Block was compared during sampling season 1 and 3 and sampling season 2 and 3, the difference was found to be statistically significant at $P<0.01$.

Soil properties and soil environmental conditions are the important factors influencing the concentration of soil organic carbon (Tao et al., 2000b). The levels of soil organic carbon increased with increasing to its utilization as a microbial substrate. The soluble fraction of organic matter is the main energy substrate of soil and should therefore be utilized preferentially (Marschner and Bredow, 2002). As the soil temperature increases, soil microbial activity is enhanced leading to accelerated decomposition of organic matter and release of organic carbons (that is CO_2) into the atmosphere. This consequently causes an increase in soil organic carbon (Marschner and Bredow, 2002; Wang et al., 2008). When organic carbon of the soils collected from all the blocks during all the sampling seasons was compared, the difference was found to be statistically nonsignificant at $P<0.05$. But when organic carbon of the soils collected from D- and K-Blocks, and B- and E-Blocks during sampling season 1 and 2 was compared, it was found to be statistically significant at $P<0.001$ and $P<0.01$, respectively. However, when organic carbon of the soils collected from D and K block during sampling season 2 and 3 was compared, it was found to be statistically significant at $P<0.01$.

It is well established that nitrogen (N) is the macronutrient often limiting the growth of plants in soil (Vitousek, 1982; Vitousek and Farrington, 1997; Michopoulos et al., 2008). It is evident that nitrification increases the nitrogen level in the soil at any time (Elliott and McCalla, 1972). These results along with the reports of Elliott and McCalla (1972) and

Mielke et al., (1970) suggest that the rapid increase in total nitrogen in the soils during all the sampling seasons may be due to high rate of nitrification probably taking place in this field. Nitrification would account for the fact that nitrogen was high in the field. It has been shown that crop management practices can either enhance or diminish the quantities of soil organic carbon and soil total nitrogen together (Bauer and Black, 1981; Voroney et al., 1981; Campbell and Souster, 1982; Odell et al., 1984; Mann, 1986; Darmody and Peck, 1997; Paustian et al., 1997; Buyanovsky and Wagner, 1998; Dick et al., 1998; Potter et al., 1998; Knops and Tilman, 2000; Kucharik et al., 2001). One hypothesized goal of sustainable agriculture is to increase soil organic carbon and soil total nitrogen, or to maintain these quantities close to native levels (Odell et al., 1984). When total nitrogen of the soils collected from all the blocks during all the sampling seasons was compared, the difference was found to be statistically significant at $P<0.001$. But when total nitrogen of the soils collected from K- and L-Blocks during sampling season 2 and 3 were compared, it was found to be statistically significant at $P<0.01$. However, when total nitrogen of the soils collected from E-Block during sampling season 1 and 2 were compared, it was found to be statistically non-significant.

Compared to other macronutrients, such as sulfur and calcium, the concentration of phosphorus in the soil is very low, generally ranging from 0.001 to 1 mgL^{-1} in infertile to heavily fertilized soils. Plant roots absorb phosphorus dissolved in the soil solution, mainly as HPO_4^{2-} and $H_2PO_4^-$ ions, but some soluble organic phosphorus compounds are also taken up. When inorganic phosphate of all the blocks during all the sampling seasons was compared, the difference was found to be statistically non-significant at $P<0.05$. But when inorganic phosphate of D block was compared with inorganic phosphate of A, E and K-Block during sampling season 1 and 2, the difference was found to be statistically significant at $P<0.05$. It is also evident from the data that the concentration of the inorganic phosphate was low and phosphorus in the form of diamonium phosphate (DAP) and triple super phosphate (TSP) can be supplemented to the soil as recommended by Gary and Stephen (2005).

Calcium is present in adequate amounts in most of the soils. It is a component of several primary and secondary minerals in the soil, which are essentially insoluble for agricultural considerations. These materials are the original sources of the soluble or available forms of calcium. Calcium is also present in relatively soluble forms, as a divalent cation (Ca^{++}) adsorbed to the soil colloidal complex. The ionic form of calcium is considered to be available to the crop plants (Narwal et al., 1993). When calcium of the soils collected from all the blocks during all the sampling seasons was compared, the difference was found to be statistically significant at $P<0.001$.

Potassium is a key essential plant nutrient. Many plant physiologists consider the importance of potassium second only to nitrogen as for as plant growth is concerned. Soil contains 20,000 pop of total potassium. However, all the potassium present in soil is unavailable to the plants. Potassium is present in the form of micas, feldspars and clay minerals and plants cannot use these insoluble forms of potassium. In the present study highest potassium concentration was found in sampling season 2, indicating that the soil was rich with potassium there and no further supplementation of potassium fertilizers to the soil is needed. It is evident that potassium concentration was increasing in the soil collected from all the blocks during previous sampling. When potassium of the soils collected from all the blocks during all the sampling season was compared, the difference was found to be statistically significant at $P<0.001$. But when potassium of the soils collected from B-Block during sampling season 1 and 2 was compared, it was found to be statistically significant at $P<0.01$.

Since, higher concentration of sodium is harmful for the plant growth (Shainberg and Letey, 1984), it seems that whereas the soil of A block was better for the growth of the plants, the soil of D-Block was, however, not suitable for the growth of plant. Most plants do not grow well on the soils that contain salts. Salt causes a reduction in the rate and amount of water that the plant roots can take up from the soil. Also, some salts are toxic to the plants when present in high concentration. When sodium of the soils collected from all the blocks during all the sampling seasons was compared, the difference was found to be statistically non-significant at $P<0.05$. But when sodium of the soils collected from A-Block and E- and L-Blocks during sampling season 1 and 3 were compared, it was found to be statistically significant at $P<0.001$ and $P<0.001$, respectively. However, when sodium of the soils collected from A-Block, E- and L-Blocks, and D- and K-Blocks during sampling season 1 and 2 was compared, it was found to be statistically significant at $P<0.001$ $P<0.01$ and $P<0.05$, respectively.

Role of cyanobacteria in improving soil health and fertility is well known. To exploit the full potential of cyanobacteria in management of soil quality and fertility, coordinated efforts between laboratory and field level research is needed. Cyanobacterial isolates could be more effective in such applications as they are pre-acclimatised to the existing environmental conditions. Hence, a region-specific biodiversity study is important for deriving optimum benefits from indigenous strains. Knowledge of cyanobacterial diversity of a region may help in selecting appropriate cyanobacterial inocula to be applied as biofertilizer consortia in crop fields as well as help in finding strains with other biotechnological potentials. The nature of soil algal communities is the result of the complex influence of vegetation type, soil properties, and climatic conditions (Quesada et al., 1995; 1998). Our previous study indicated that water content (humidity) plays a an important role in the distribution and diversity of soil algae (Lin and Wu, 2013). In the present study, some N-fixing species, including those of *Anabaena* and *Nostoc* were found in the wetland, together with higher amounts of the non N-fixing species, such as *Lyngbya, Microcystis,* and *Oscillatoria.* Presumably, the occurrence of N-fixing species in KNP could be an indication of N deficiency.

Wetland has intensive anthropogenic activity that affects the physico-chemical environment of soils and thus lower the diversity of soil algae. The cyanobacterial species frequently occurred in the studied wetland are of *Anabaena, Lyngbya, Microcystis, Nostoc, Oscillatoria* and *Spirulina.* These species are cosmopolitan and widespread in various types of soils in literatures (Song et al., 2005; Begum et al., 2008; Khaybullina et al., 2010). In contrast, species of cyanobacteria such as *Aphanothece, Cylindrospermum, Scytonema* and *Tolypothrix* occurred only rarely. These species are known to be susceptible to pesticides, disturbance (such as tillage), and heavy metal pollution (Zancan et al., 2006).

5. CONCLUSION

Wetland ecosystem absorbs and recycles essential nutrients, treat sewage and cleanses waste. Trees and forest soil purify water as it flows through forest ecosystems. In preventing soils from being washed away, forests also prevent the harmful situation of rivers and reservoir that may arise from erosion and landslides. Some 130 billion metric tons of organic waste is processed every year by earth's decomposing organisms. Many industrial wastes including detergents, oils, acids and paper are also detoxified and decomposed by the activities of living things.

Microbes make nutrients in the soil available to plants in a form which plants can use. Microbes improve soil structure by the humus they create while digesting organic matter. Microbes help in fixing atmospheric nitrogen. A loopful of soil and water contains zillions of microbes. For instance, an acre of soil might contain 130 pounds each of algae and protozoa, 890 pounds of insects, nearly 900 pounds of earthworm and 2,000 pounds each of bacteria and fungi.

Microbes play an important role in the functioning of wetland ecosystem. Wetlands have been attributed several ecological functions and their sustainability is of great importance as they contain rich biodiversity. Increase in natural and anthrogenic activities is, however, resulting in degradation of wetlands and as a consequence the microorganism are continuously depleting from the these habitats.

Thus, over all it can be concluded that the present study on the physico-chemical and cyanobacterial diversity of KNP wetland in Bharatpur (Rajasthan) was designed to provide baseline data on which the potential of this vast unexploited wetland can be maximized for sustainable cyanobacterial culture. This is the first study investigating the cyanobacterial diversity of the KNP, which has been proven to be an important ecological niche for industrially important cyanobacteria. Culturing cyanobacteria gives us an edge to have more and more cultured cyanobacteria with their taxonomic status from any sort of such environment. Therefore, the strategy to culture microorganisms from any environmental samples and then designing a methodology that delineates biodiversity increases the prospect of having microbial resource in hand which can be further used for other purposes at a later stage.

Acknowledgements

Financial assistance provided by Ministry of Environments and Forests, New Delhi, India Grant No. (F.No. J-22012/44/2005-CS (W) is highly acknowledged.

References

Abbasi SA (1997). Wetlands of India: Ecology and Threats. Discovery Publishing House, New Delhi 2: 33–64.

Adams DG (2000). Cyanobacterial phylogeny and development: Questions and challenges. In: YV Brun, Shimkets LJ (eds) Prokaryotic Development, ASM Press, Washington DC, pp. 51–81.

Adoni AD, and Saini RS (1984). Eutrophication of certain wetland ecosystems in Kerala. Paper presented at the seminar on Wetland Ecology and Management at Keoladeo National Park, Bharatpur (Feb. 23–25).

Ahmed SH, and Singh AK (1990). Hydrobiological features of Kawar Lake, Bihar. Paper presented at the seminar on Wetland Ecology and Management at Keoladeo National Park, Bharatpur (Feb. 23–25).

Akram SMM (1992). Physico-chemical environment of Wular Lake, Kashmir. Dissertation submitted to the University of Kashmir (unpublished). 1–232.

Ali S, and Vijayan VS (1986). Keoladeo National Park Ecology Study: Summary Report 1980-1985.: U.S. Fish and Wildlife Service, Department of Environment and Government of India—Bombay: Bombay Natural History Society, Mumbai, p. 185.

Ali S, and Vijayan VS (1987). Keoladeo National Park - Ecology Study. Annual report 1986. Bombay Natural History Society, Mumbai. p. 110.

Azeez PA, Nadarajan NR, and Prusty BAK (2007). Macrophyte decomposition and changes in water quality. In: Singh KK, Juwarkar A, Singh AK (eds) Environmental degradation and protection., MD Publication: New Delhi. pp. 115–156.

Bauer A, and Black AL (1981). Soil carbon, nitrogen and bulk density comparisons in two cropland tillage systems after 25 years and in virgin grassland. Soil Sci Soc Am J, 45: 1166–1170.

Begum ZNT, Mandal R, and Amin FB (2008). Quantification and nitrogen fixation of cyanobacteria in rice field soils of Bangladesh. Bangladesh J Bot 37: 183–188.

Bhunya SP, and Mohanty BS (1990). Physico-chemical studies in relation to the faunistic composition of Anuspa Lake, Orissa. Paper presented at the seminar on Wetland Ecology and Management at Keoladeo National Park, Bharatpur (Feb. 23–25).

Buyanovsky GA, and Wagner GH (1998). Changing role of cultivated land in the global carbon cycle. Biol Fertil Soils 27: 242–245.

Campbell C A, and Souster W (1982). Loss of organic matter and potentially mineralizable nitrogen from Saskatchewan soils due to cropping. Can J Soil Sci 62: 651–656.

Choudhary KK, Singh SS, and Mishra AK (2007). Nitrogen fixing cyanobacteria and their potential applications. In: Gupta RK, Pandey VD (eds) Advances in Applied Phycology. Daya Publishing House, New Delhi, pp. 142–154.

Darmody RG, and Peck TR (1997). Soil organic carbon changes through time at the University of Illinois Morrow plots. In: Paul EA, Paustian K, Elliott ET, Cole CV (eds) Soil Organic Matter in Temperate Agroecosystems: Long-term Experiments in North America. CRC Press, Boca Raton, Florida.

Datta NP, Khera MS, and Saini TR (1962). A rapid colorimetric procedure for the determination of the organic carbon in soils. J Indian Soc Soil Sci 10: 67–74.

Desikachary TV (1959). *Cyanophyta*, Indian Council of Agricultural Research, New Delhi, p. 686.

Dick WA, Blevins RL, Frye WW, Peters SE, Christenson DR, Pierce FJ, and Vitosh ML (1998). Impacts of agricultural management practices on C sequestration in forest-derived soils of the eastern Corn Belt. Soil Tillage Res 47: 235–244.

Doll EC, and Lucas RE (1973). Testing soil for potassium, calcium and magnesium. In: Walsh LM, Beaton JD (eds) Soil testing and plant analysis, Madison, WI. pp. 133–152.

Edwards AWA (1990). Wetlands in southern Nigeria" in Nigerian Wetlands T.V. Akpata and D.U.U. Okoli, Eds. Ibadan: *Emmi Press Samanda*. pp. 27–34.

Elliott LF, and McCalla TM (1972). Soil atmosphere composition beneath a beef feedlot and a cropped field. Soil Sci Soc Amer Proc 36: 68–70.

Eshiett ET (1992). Physicochemical, morphological and mineralogical characteristics of selected humid region profiles in Southeastern Nigeria. In: Kimble JM (ed) proceedings of the Eighth International Soil Correlation Meeting on Characterization, Classification and Utilization of Wet Soils. USA. Soil Conservation Services, pp. 100–105.

Evans RD, and Johansen JR (1999). Microbiotic crusts and ecosystem processes. Crit Rev Plant Sci 18: 183–225.

Fiske CH, and Subba Row Y (1925). Colorimetric determination of phosphorous. J Biol Chem 66: 375–400.

Fleck A (1967). The determination of organic nitrogen. Proc Assoc Clin Biochem 4: 212.

Gary WV, and Stephen JD (2005). Environmental chemistry, a global perspective. NY: Oxford University Press.

Geitler L (1932). Cyanophyceae. In: L Rabenhorst's Kryptogamen Flora, Akademische Verlagsgesellschaft, Leipzig, pp. 1196.

George MR (1964). Limnological investigations on pond plankto, macrofauna and chemical constituents of water and their bearing on fish production. Ph.D. thesis, University of Calcutta.

Gonzalves E, and Joshi D B (1946). Fresh water Agae near Bombay, the seasonal succession of the algae in a tank at Bandra. J Bombay Nat Hist Soc 46(1): 154–70.

Gopal B, Turner RE, Wetzer RG, and Whigham DF (eds) (1982). Wetlands ecology and management. National Institute of Ecology and International Scientific Publications, New Delhi, India. pp. 11-16.

Greenberg EA, Clesseri S L, Eaton AD (1992). Standard methods for the examination of water and waste water. 18th edn., American Public Health Association.

Hackl E, Pfeffer M, Donat C, Bachmann G, and Zechmeister- Boltenstern S (2005). Composition of the microbial communities in the mineral soil under different types of natural forest. Soil Biol Biochem 37: 661–671.

Handoo JK, and Kaul V (1982). Phytosociology and standing crop in four typical wetlands of Kashmir. In: Gopal B, Turner E, Wetzel RG, Whigam DF (eds) Wetlands Ecology and Management. Indian Institute of Ecology and International Scientific Publications.

He JZ, Xu ZH, and Hughes JF (2006). Molecular bacterial diversity of a forest soil under residue management regimes in subtropical Australia. FEMS Microbiol Ecol 55: 38–47.

Herdman M, Janvier M, Rippka R, and Stanier RY (1979). Genome size of cyanobacteria. J Gen Microbiol 111: 73–85.

Hoffmann L (1989). Algae of terrestrial habitats. Bot Rev 55: 77–105.

Holland J (1977). The Chemistry of the Atmosphere and Oceans, John Wiley & Sons, New York.

Hu CX, Zhang DL, and Liu YD (2004). Research progress on algae of the microbial crusts in arid and semiarid regions. Prog Nat Sci 14: 289–295.

Hutchinson GE (1937). Limnological studies in Indian Tibet. Int Rev Hydrobiol 35: 134–177.

Jha DK, Sharma GD, Mishra RR (1992). Soil microbial population numbers and enzyme activities in relation to altitude and forest degradation. Soil Biol Biochem 24: 761–767.

Johnson LF, Curl EA (1972). Methods for research on the ecology of soil-borne plant pathogens. Minneapolis, Burges Publishing Company.

Kaushik BD (1994). Algalization of rice in salt-affected soils. Annual Agricultural Research, 14: 105–106.

Khan AA, and Siddiqui Q (1970). Diurnal variations in the pond moat at Aligarh. J Inland Fish Soc Ind 2: 146–54.

Khatri TC (1984). Seasonal variation in primary production in relation to some limnological features in Lakhotia Lake. Proc Indian Acad Sci (Anim Sci) 93(7): 697–702.

Khaybullina LS, Gaysina LA, Johansen JR, and Krautová M (2010). Examination of the terrestrial algae of the Great Smoky Mountains National Park, USA. Fottea 10: 201–215.

Knops JMH, and Tilman D (2000). Dynamics of soil nitrogen and carbon accumulation for 61 years after agricultural abandonment. Ecology 81: 88–98.

Kucharik CJ, Brye KR, Norman JM, Foley JA, Gower ST, and Bundy LG (2001). Measurements and modeling of carbon and nitrogen cycling in agroecosystems of southern Wisconsin: Potential for SOC sequestration during the next 50 years. Ecosystems 4: 237–258.

Kulshreshtha M, and Gopal M (1982). Decomposition of freshwater wetlands vegetation, Parts 1 and 2. In: Gopal B, Turner E, Wetzel RG, Whigam DF (eds) Wetlands Ecology Ecology and Management. Indian Institute of Ecology and International Scientific Publications.

Laverman AM, Mbraster M, Roling WFM, and van Verseveld HW (2005). Bacterial community structure and metabolic profiles in a forest soil exhibiting spatially variable net nitrate production. Soil Biol Biochem 37: 1581–1588.

Lin CS, and Wu JT (2013). Environmental factors affecting the diversity and abundance of soil photomicrobes in arid lands of subtropical Taiwan. Geomicrobiol J doi:10.1080/01490451.2013.828135.

Loffelhandt W, and Bohnert H J (1977). Molecular biology of cyanelles. In The Molecular Biology of Cyanobacteria (ed Bryant DA), Kluwer Academic Publishers, Dordrecht, The Netherlands, pp. 65–89.

Mahajan CL (1980). Effect of human activities on the structure and functioning of fresh water ecosystem of Ghana Bird Sanctuary, Bharatpur, Rajasthan. A preliminary report on the Man and Biosphere Project. Department of Science and Technology, Government of India.

Mann LK (1986). Changes in carbon storage after cultivation. Soil Sci 142: 279–288.

Marschner B, and Bredow A (2002). Temperature effects on release and ecologically relevant properties of dissolved organic carbon in sterilized and biologically active soil samples. Soil Biol Biochem 34: 459–466.

Mathur VB, Sinha PR, and Mishra M (2005). Keoladeo National Park World Heritage Site. Technical Report No. 5, UNESCO-IUCN-Wild Life Institute of India, Dehradun, India.

Mausbach MJ, and Seybold CA (1998). Assessment of soil quality; in Soil quality and agricultural sustainability (ed.) R Lal (Chelsea: Ann Arbor Press). pp. 33–43.

May RM (1992). How many species inhabit the Earth? Sci Amer 267(4): 42–48.

Metting B (1981). The systematics and ecology of soil algae. Bot Rev 47: 195–312.

Michopoulos P, Baloutsos G, and Economou A (2008). Nitrogen cycling in a mature mountainous beech forest. Silva Fennica 42: 5–17.

Mielke LN, Ellis JR, Swanson NP, Lorimor JC, and McCalla TM (1970). Groundwater quality and fluctuation in a shallow unconfined aquifer under a level feed lot. pp. 31–40.

Minelli A (1993). Biological Systematics. The State-of-the-art. Chapman & Hall, London. p. 387.

Mishra U, and Pabbi S (2004). Cyanobacteria: A potential biofertilizer for rice. Resonance 9: 6–10.

Mittal DD, Vijayan VS, and Azeez PA (1990). Physico-chemical properties of water at Keoladeo National Park, Bharatpur. Paper presented at the seminar on Wetland Ecology and Management at Keoladeo National Park, Bharatpur (Feb. 23–25).

Mulongoy K, and Merkx R (2003). Soil Organic matter dynamics and sustainability of tropical agricultures. A Wiley sayce Co-publications, New York.

Narwal RP, Gupta AP, Singh A, and Karwasra SPS (1993). Composition of some city waste waters and their effect on soil characteristics. Annals of Biology 9: 239–245.

Odell RT, Melsted SW, and Walker WM (1984). Changes in organic carbon and nitrogen of Morrow plot soils under different treatments, Soil Sci 137: 160–171.

Pankhurst CE (1997). Biodiversity of soil organisms as an indicator of soil health. In Biological indicators of Soil Health. Pankhurst CE, Doube B M, Gupta VVSR (eds) Wallingford, Oxon, UK: CABI International. pp. 297–324.

Paustian K, Collins HP, and Paul EA (1997). Management controls on soil carbon. In: Paul E A, K Paustian, ET Elliott, CV Cole. Soil Organic Matter in Temperate Agroecosystems: Long-term Experiments in North America. CRC Press, Boca Raton, FL.

Pimm S L, Russell GJ, Gittleman L, and Brooks TM (1995). The future of biodiversity. Science 269: 347–350.

Piper CS (1966). Soil and Plant analysis. Hans Publishers, Bombay.

Potter KN, Torbert HA, Jones OR, Matocha JE, Morrison JEJ, and Unger PW (1998). Distribution and amount of soil organic C in long-term management systems in Texas. Soil Tillage Res 47: 309–321.

Prasanna R, Jaiswal P, Nayak S, Sood A, and Kaushik BD (2009). Cyanobacterial diversity in the rhizospere of rice and its ecological significance. Indian Journal of Microbiology 49: 89–97.

Quesada A, Nieva M, Leganés F, Ucha A, Martín M, Prosperi C, and Fernández-Valiente E (1998). Acclimation of Cyanophytal communities in rice fields and response of nitrogenase activity to light regime. Microbial Ecol 35: 147–155.

Quesada A, Sánchez-Maeso E, and Fernández-Valiente E (1995). Seasonal variation of chemical properties of rice field soil from Valencia. Spain Commun Soil Sci Plant Anal 26: 1–19.

Rai LC, Tyagi B, Rai PK, and Mallick N (1998). Inter-active effects of UV-B and heavy metal (Cu and Pb) on nitrogen and phosphorus metabolism of a N_2 fixing cyanobacterium *Anabaena doliolum*. Environ Exp Biol 39: 221–223.

Rai AN (1990). Cyanobacteria in symbiosis. Handbook of Symbiotic Cyanobacteria (ed. Rai AN), CRC Press, Boca Raton.

Ramalingam K, and Jayaraman (1985). A note on the hydroecological features of a pond on a plankton bloom day. Geobios News Report 4: 92–94.

Rippka R (1972). Photoheterotrophy and chemoheterotrophy among unicellular blue-green algae. Arch Microbiol, 87: 93–98.

Roger PA, and Reynaud PA (1982). Free-living blue-green algae in tropical soils. Martinus Nijhoff Publisher, La Hague.

Sarup S (1961). Hydrophytes of Bharatpur–A preliminary study, Univ, Rajaputana studies, Bio Sci 4: 1–10.

Sena MG, Almeida-Vara E, and Duarte JC (2008). Eco-friendly new products from enzymatically modified industrial lignins. Industrial Crops and Products 27(2): 189–195.

Seshavatharam V (1990). Limnological botany of Lake Koeru: a review. Paper presented at the seminar on Wetland Ecology and Management at Keoladeo National Park, Bharatpur. (Feb. 23–25).

Seshavatharam V, and Chandramohan P (1982). Limnological studies on Kondakarla lake–a freshwater body near Visakhapatnam. Technical Report of U.G.C. Research Scheme. Andhra University, Waltair.

Seshavatharam V, Murthy KSN, and Venu P (1990). Ecological studies on Lake Kondakarla: a small freshwater wetland near Visakhapatnam. Paper presented at the seminar on Wetland Ecology and Management at Keoladeo National Park, Bharatpur. (Feb. 23–25).

Shainberg I, and Letey J (1984). Response of soils to sodic and saline conditions. Hilgardia. 61: 21–57.

Sharma S, and Praveen B (2002). Management Plan: Keoladeo National Park, Bharatpur, Plan Period 2002-2006, Department of Forests and Wildlife, Rahasthan, India.

Singh RN (1961). Role of Blue-green Algae in Nitrogen Economy of Indian Agriculture. Indian Council of Agricultural Research. New Delhi, 175.

Singh SP, Pant MC, Sharma PC, and Purohit R (1982). Limnology of shallow water zones of lakes in Kumaon Himalayas (India). In: Gopal B, Turner RK and Wetzel RG, and Whigam DF (eds) Wetlands Ecology and Management. National Institute of Ecology and International Scientific Publications, India, 34-49.

Song T, Martensson L, Eriksson T, Zheng W, and Rasmessen U (2005). Biodiversity and seasonal variation of the cyanobacterial assemblage in rice paddy field in Fujian, China. FEMS Microbiol Ecol 54: 131–140.

Stanier RY, Kunisawa R, Mandel M, and Cohen-Bazire G (1971). Purification and properties of multi-cellular blue-green algae (order Chlorococcales). Bacteriol Rev 35: 171–205.

Tao S, Lin B, Liu X, and Cao J (2000b). Release kinetics of water soluble organic carbon (WSOC) from river sediment and wetland Soil. Water, Air, Soil Poll 118: 407–418.

Thajuddin N, and Subramanian G (1992). Survey of cyanobacterial flora of the southern east coast of India. Bot Mar 35: 305–311.

Thajuddin N, and Subramanian G (2005). Cyanobacterial biodiversity and potential applications in biotechnology. Current Science, 89: 47–57.

Trivedi RK, and Goel PK (1986). Chemical and biological methods for water pollution studies Karad. Environmental Publication pp. 1–251.

van Straalen NM (2002). Assessment of soil contamination—a functional perspective; Biodegradation 13: 41–52.

Venkataraman GS (1981). Blue-green algae: a possible remedy to nitrogen scarcity. Current Science 50: 253–256.

Vijayan VS (1991). KNP Ecology study (1980-1990). Final report. Bombay Natural History Society, Bombay.

Vitousek PM (1982). Nutrient cycling and nutrient use efficiency. American Naturalist 119: 553–572.

Vitousek PM, and Farrington H (1997). Nutrient limitation and soil development: experimental test of a biogeochemical theory. Biogeochemistry 37: 63–75.

Voroney RP, Van Veen JA, and Paul EA (1981). Organic C dynamics in grassland soils. 2. Model validation and simulation of the long-term effects of cultivation and rainfall erosion. Can J Soil Sci 61: 211–224.

Vyas LN (1984). Ecology and conservation of Lakes in and around Udaipur, Rajasthan. Proc 2nd An Workshop on MAB Projects. New Delhi. pp. 184–186.

Wang W, Yang Y, Chen G, Guo J, and Qian W (2008). Profile distribution and seasonal variation of soil dissolved organic carbon in natural *Castanopsis fabric* forest in subtropical China. Chinese J Ecol 27: 924–928 (in Chinese).

Whitton B A, and Potts M (2000). Soils and rice fields. In: Whitton BA, Potts M (eds), The ecology of cyanobacteria, Kluwer Academic Publishers, Dordrecht, The Netherlands, pp. 233–255.

Wilson LT (2006). Cyanobacteria: A Potential Nitrogen Source in Rice Fields. Texas Rice 6: 9–10.

Wilson EO (1988). The Current State of Biological Diversity. In: Wilson EO and Peter FM (eds) , Biodiversity. National Academy Press, New York: 3-18.

Wilson EO (1992). The diversity of life. Harvard University Press, Cambridge. 243–280.

Wyatt JT, and Silvey JHG (1969). Nitrogen fixation by Gleocapsa. Science 165(3896): 908–909.

Zancan S, Trevisan R, and Paoletti MG (2006). Soil algae composition under different agro-ecosystems in North-Eastern Italy. Agricult Ecosyst Environ 112: 1–12.

Zenova GM, Shtina EA, Dedysh SN, Glagoleva OB, Likhacheva AA, and Gracheva TA (1995). Ecological relations of algae in biocenoses. Mikrobiologiya 64: 121–133.

14

Competent Soil Microorganisms for Agricultural Sustainability: New Dimension

ASHOK K. RATHOURE
GM-Consultancy, Ecosystem Resource Management Pvt. Ltd.
Ashoka Pavilion, Opp. Kapadia Health Club, New Civil Road
Surat–395007 (Gujarat) India
E-mail: *asokumr@gmail.com*

1. INTRODUCTION

Microbes such as fungi, bacteria and viruses play a vital role in sustainability of agriculture. Habitually, farmers think of microbes as pests that are destructive to their crops or animals, but many microbes are beneficial. Soil microbes especially bacteria and fungi are essential for decomposing organic matter and recycling old plant material. Some soil bacteria and fungi form relationships with plant roots that provide important nutrients like nitrogen or phosphorus. Fungi can colonize upper parts of plants and provide many benefits, including drought tolerance, heat tolerance, resistance to insects and resistance to plant diseases. Some viruses that benefit to hosts and most probably do not harm their hosts. Microbes in terrestrial environments are important catalysts of global carbon and nitrogen cycles, including the production and consumption of greenhouse gases in soil. Some microbes produce greenhouse gases carbon dioxide (CO_2) and nitrous oxide (N_2O) while decomposing organic matter in soil. Others consume methane (CH_4) from the atmosphere, thus helping to mitigate climate change. The importance of each of these processes is influenced by human activities and impacts the warming potential of Earth's atmosphere. Using contemporary genomic techniques, scientists analyzed DNA extracted directly from soil to assess which microbes were present and their metabolic potential. Communities of microbes were compared from three very different land use types, i.e., row-crop agriculture (corn, soybean and wheat rotations), old fields that had been abandoned from agriculture for decades and deciduous forests. Since these treatments had been in place for 20–50 years, scientists were able to measure the long-term impacts of land management. It is suggested that managing lands to conserve or restore the diversity of methane-consuming bacteria could help mitigate climate change. Likewise, land management also seems to greatly affect the diversity of microbes involved in nitrous oxide production; fertilized soils produce more nitrous oxide and have a different genetic makeup. Agriculture distinctly influences the diversity of bacteria responsible for the production and consumption of two of the three major naturally-occurring greenhouse gases. The uniqueness of microorganisms and their unpredictable nature and biosynthetic capabilities, given a specific set of environmental and cultural conditions, has made them likely candidates for solving particularly difficult problems in the life sciences and other fields as well. The various ways in which microorganisms have been used over the past 50 years to advance medical technology, human and animal health, food processing, food safety and quality, genetic engineering, environmental protection, agricultural biotechnology and more effective treatment of agricultural and municipal wastes provide most impressive record of achievement (Jangid et al., 2011).

The soil quality is the key to sustainable agriculture. Understanding has shown that the transition from conventional agriculture to nature farming or organic farming can involve certain risks, such as initially lower yields and increased pest problems. Once through the transition period, which might take several years, most farmers find their new farming systems to be stable, productive, manageable and profitable without pesticides. Microorganisms are effective only when they are presented with suitable and optimum conditions for metabolizing their substrates Including available water, oxygen, pH and temperature of their environment. Meanwhile, the various types of microbial cultures and inoculants available in the market today have rapidly increased because of these new technologies. Since microorganisms are useful in eliminating problems associated with the use of chemical fertilizers and pesticides, they are now widely applied in nature farming and organic agriculture (Higa, 1991; Parr et al., 1994).

Environmental pollution, caused by soil erosion and the associated transport of sediment, chemical fertilizers and pesticides to surface and groundwater, also improper treatment of human and animal wastes has caused serious environmental and social problems throughout the world. Often engineers have attempted to solve these problems using established chemical and physical methods. However, they have usually found that such problems cannot be solved without using microbial methods and technologies in coordination with agricultural production (Reganold et al., 1990). The beneficial microorganisms are those that can fix atmospheric nitrogen, decompose organic wastes and residues, detoxify pesticides, suppress plant diseases and soil-borne pathogens, enhance nutrient cycling and produce bioactive compounds, such as vitamins, hormones and enzymes that stimulate plant growth. Harmful microorganisms are those that can induce plant diseases, stimulate soil-borne pathogens, immobilize nutrients and produce toxic and putrescent substances that adversely affect plant growth and health (Prasad et al., 2015).

2. FUNCTIONS OF SOIL MICROORGANISMS

The soil microorganisms (actinomycetes, fungi, bacteria and cyanobacteria) so called soil microbial biomass are the agents of transformation of soil organic matter, nutrients and of most key soil processes. Their activities are influenced by soil physico-chemical and ecological interactions. Many actinomycetes produce antibiotic compounds that kill other microorganisms, e.g., streptomycin is produced by growing soil actinomycetes in pure culture. Bacteria (kingdom Monera) can be either autotrophic or heterotrophic and can be both aerobic and anaerobic. They have evolved mechanisms to adapt to life in the most extreme environments. Due to their diversity they play many roles in the soil and are involved in all the organic transactions. Cyanobacteria (formerly blue-green algae) belong to soil microflora and are either autotrophic or heterotrophic. They are also prokaryotic and among the most ancient of organisms. Many genera of cyanobacteria can fix atmospheric nitrogen, e.g., *Anabaena* sp. and are especially plentiful in rice paddies and other wetland soils, where they fix large quantities of nitrogen. Soil bacteria are the primary digestive system of the soil. Their activity is responsible for almost 90 percent of all biological and chemical actions. The followings are the key function of beneficial soil microorganisms:

- Transforming raw elements from one chemical form to another. Important nutrients in the soil are released by microbial activity are nitrogen, phosphorus, sulphur, iron and others.
- Breaking down soil organic matter into a form useful to plants. This increases soil fertility by making nutrients available.
- Degradation of pesticides and other chemicals found in the soil.
- Suppression of soil-borne pathogen that causes diseases. The pathogens themselves are part of this group, but are highly outnumbered by beneficial microbes.
- Fixation of atmospheric nitrogen.
- Decomposition of organic wastes and residues.
- Recycling and increased availability of plant nutrients.
- Production of antibiotics and other bioactive compounds.
- Production of simple organic molecules for plant uptake.
- Complexation of heavy metals to limit plant uptake.
- Solubilization of insoluble nutrient sources.
- Production of polysaccharides to improve soil aggregation.

3. ARBUSCULAR MYCORRHIZAL FUNGI

The symbiotic association of plants and mycorrhizal fungi, i.e., arbuscular mycorrhizal fungi, (AMF) has long been recognized for the benefits it provides with nutrient transport and uptake. However, there is considerable uncertainty about the functional benefits in intensive agricultural systems (Goltapeh et al., 2008; Gupta et al., 2011). Even though

AM symbiosis is widespread, the symbiotic functions of AMF species differ according to specific AMF isolates, host plants and soil properties. AMF associations are generally considered diffuse and non-specific because multiple species colonization linking together two or more plants is not uncommon (Selosse Marc-Andre et al., 2006; Smith et al., 2008). Identification of specific phylotypes of AMF and their relationship with soil properties is a crucial step to fully exploit the benefits from AMF (Selosse Marc-Andre et al., 2006). New knowledge on the microbial interactions in the mycosphere has the potential to enhance our ability to manipulate plant-mycorrhizal associations. A diverse array of bacteria, e.g., *Pseudomonas, Bacillus*, actinomycetes and fungi, e.g., *Aspergillus, Penicillium* species are capable of solubilizing and mineralizing the unavailable forms of phosphorus in soils for plants and the benefits from their use as inoculants are increasingly being recognized (Gyaneshwar et al., 2002; Richardson et al., 2011). The key functions of AMF include the following:

3.1 Drought Resistance

AMF have long been implicated in improved host water relations by influencing everything from stomatal conductance, hydraulic conductance, and leaf and root hydration (Auge, 2001).

3.2 Mineralization of Organic Nutrients

Recently, AMF have been shown to acquire N (Hodge et al., 2001) and P (Tarafdar and Marschener, 1994) directly from decomposing sources, which was thought only to occur in saprobic fungi.

3.3 Seedling Establishment

The role of mycelial networks is thought to equalize resource allocation among neighboring plants, and a potential benefit may be that AM seedlings have access to more resources through the common mycelial network than do non-AM plants during the crucial establishment phase (van der Heijden et al., 2004).

3.4 Nutrient Uptake

AMF mycelia have been shown to increase uptake of many nutrients including P, N, S, B, Cu, K, Zn, Ca, Mg, Na, Mn, Fe, Al, and Si, along with trace elements (Clark and Zeto, 2000). In some cases, AMF may be responsible for acquiring 100 per cent of host nutrients, e.g., P (Smith et al., 2004).

3.5 Increased Soil Stability

AMF greatly increases soil stability and weathering, possibly due to exudation of putative AMF protein glomalin, which is found in very high levels in AM soils (Rillig, 2004).

3.6 Increased Pollination

AMF plants may also be more attractive to pollinators, due to their altered architecture and improved nutrient status (Wolfe et al., 2005).

3.7 Increased Herbivore Tolerance

While AMF plants may be more appealing to herbivores, they have shown to increase tolerance to herbivory through increased compensatory response (Kula et al., 2005).

3.8 Pathogen Resistance

While AMF have yet to be developed for biological control, they appear to reduce disease symptoms for some plant diseases, such as root rot, wilt, yellowing disease, and damping off (Whipps, 2004).

3.9 Heavy Metal Tolerance

AMF have shown to have differential sensitivities to soil metal toxicity, and may buffer hosts from toxic exposure, either through sequestration in the mycelium or by metabolizing it (Gaur and Adholyea, 2004).

4. NITROGEN FIXING BACTERIA

Symbiotic associations between legume plants and root-nodulating bacteria belong to the genera *Rhizobium, Bradyrhizobium, Ensifer* and *Mesorhizobium*. Nitrogen fixation is one of the essential beneficial biological processes for the economic and environmental sustainability of agriculture worldwide. Finding efficient rhizobia for the wide variety of legumes that are cultivated around the world and developing efficient management of symbioses in the field to realize approx 25 kg of N_{2} fixation for each tonne of legume dry matter are the major challenges for the future. Diazotrophy, the ability to fix atmospheric nitrogen catalyzed by the enzyme nitrogenase, is distributed among diverse groups of bacteria and archaea (Reed et al., 2011). Free living N_2-fixing bacteria, e.g., *Azospirillum* sp., *Azotobacter* sp., *Acetobacter diazotrophicus, Herbaspirillum* spp., *Bacillus* sp., *Azoarcus* sp. are found in the rhizosphere and rhizoplane environments of cereal crops. Recent evidence not only identified new genera of N_2 fixing bacteria and archaea in natural and managed ecosystems but also indicated significant edaphic and environmental groupings in genetic diversity and functionality (Buckley et al., 2007). Non-rhizobial N_2-fixing bacteria can grow as endophytes in a number of grasses, e.g., *Pseudomonas* species were the most dominant group of *nif*H carrying bacteria found in the rhizosphere of perennial native grasses (Gupta et al., 2011). The *nif*H gene is present in a number of non-Frankia actinobacteria like *Agromyces, Microbacterium, Corynebacterium* and *Micromonospora*.

5. PLANT GROWTH-PROMOTING RHIZOBACTERIA (PGPR)

The environmental benefits of using PGPR as microbial inoculants have been causing the reduction in the use of agricultural chemicals due to its ecofriendly nature (Egamberdiyeva and Hoflich, 2004). Free-living as well as symbiotic PGPR can improve plant nutrition and growth, plant competitiveness and responses to external stress factors (Mantelin and Touraine, 2004). Mishra et al. (2010) reported that the PGPR were efficient for seed germination and plant growth of *Cicer arietinum* under salinity and can be used as biofertilizer. Bacteria colonising roots that elicit shoot and root growth, referred generally as PGPR, are recommended (marketed) for improving plant growth and disease control both in agriculture and horticulture. PGPR can promote root and shoot growth either by producing plant hormones or secondary metabolites, controlling diseases, induction of systemic resistance or through changing physicochemical interactions with plants (Shrivastava et al., 2014; Prasad et al., 2015). Bacterial inoculants applied as biofertilisers are also being explored to alleviate stress from abiotic factors, such as drought and salinity.

PGPR can promote the growth of plants using direct and indirect mechanisms. Direct mechanisms include lowering the production levels of ethylene through synthesis of 1-aminocylopropane-1-carboxylate (ACC) deaminase in plants (Reed and Glick, 2005; Safronova et al., 2006; Saleem et al., 2007); providing bioavailable phosphorus for plant uptake and atmospheric nitrogen fixation (Patten and Glick, 1996) for plant use; sequestering trace elements like iron using siderophores (Glick 1995), e.g., *Kluyvera ascorbata* SUD 165 that has the ability to synthesize the enzyme ACC deaminase protected *Brassica juncea* and *Brassica campestris* against Ni, Pb and Zn toxicity (Burd et al., 1998) by reducing the stress caused by high ethylene level; and production of plant hormones like gibberellins, IAA, cytokinins and auxins (Glick et al., 1999). Root elongation of *Brassica napus* has also been shown to be stimulated by IAA synthesized by PGPR (Sheng and Xia, 2006) as well as non-identified rhizobacteria on the *B. juncea* roots (Belimov et al., 2005). PGPR that indirectly enhance plant growth by suppressing phytopathogens do so by a variety of mechanisms. These include the ability to produce siderophores that chelate iron, making it unavailable to pathogens; synthesize antifungal metabolites, such as antibiotics, fungal cell wall lysing enzymes, or hydrogen cyanide, which suppress the growth of fungal pathogens; successfully compete with pathogens for nutrients or specific niches on the root; and induce systemic resistance (Bloemberg and Lugtenberg, 2001; Glick, 1995; Persello-Cartieaux et al., 2003; Martınez-Viveros et al., 2010).

6. ENDOPHYTIC MICROORGANISMS

Colonization traits usually relates to the bacterial traits involved in the entire plant-colonization process. In the interactive colonization processes, communication between the plant and microbe has a key role. Bacterial root colonization often starts with the recognition of specific compounds in the root exudates by the bacteria. These compounds probably also have major roles in below-ground community interactions. The plants simultaneously communicate with communalistic, mutuality, symbiotic and pathogenic microorganisms via compounds exuded by their roots (DeWeert et al., 2002; Bais et al.,

2004; 2006). However, it has been suggested that plants can communicate to specifically attract microorganisms for their own ecological and evolutionary benefit (Compant et al., 2005). Owing to the complexity of the plant–microbe interactions in soil, it is extremely difficult to understand the detailed mechanisms involved in these putative selection processes. Rhizobium–plant interaction indicates the existence of highly evolved species specific communication systems (Bais et al., 2006) or from plant–*Pseudomonas* associations, in which two distinct plants have attracted specific minority strains of the *Pseudomonas* sp. involved, rather than the whole *Pseudomonas* community (Lemanceau et al., 1995). Much like the bacteria selected in the rhizosphere, particular endosphere bacteria might also be selected to establish residence inside plants. Bacterial traits required for effective root colonization are subject to phase variation, a regulatory process for DNA rearrangements orchestrated by site-specific recombinase. Endophytic population sizes are dependent on and positively correlated with, plant developmental stage, progressively increasing from the seedling stage onwards and reaching a maximum (Van Overbeek et al., 2008). The endophytic microorganisms can play a vital role in agricultural sustainability, such as growth promoter, insect control, biopesticide, stress tolerance, etc.

6.1 Endophytes as Growth Promoter

There are numbers of mechanisms by which bacteria may promote plant growth and health which includes the production of plant growth phytohormones like auxins, cytokinins, gibberellins and ethylene, solubilised insoluble phosphate, produced HCN and siderophore and can fixed atmospheric N_2. In symbiotic relationships, the microorganism helps the plant with nutrient assimilation or contributes biochemical activities that the plant lacks; microbes also confer a degree of protection against plant diseases. The plants supply competitive advantage to the corresponding microbes. Volatile substances, such as 2-3 butanediol and aceotin produced by bacteria seem to be a newly discovered mechanism responsible for plant-growth promotion. Endophytes produce adenine ribosides that stimulate growth and mitigate browning of pine tissues (Smith and Read, 2008; Ryu et al., 2003; Pirttila et al., 2004).

6.2 Endophytes for Insect Control

The endophytic fungus *Phomopsis oblonga* was reported to protected elm trees against the beetle *Physocnemum brevilineum*. The capacity of endophytic fungus to repel insects, induce weight loss, growth and development reduction and even to increase pest death rate shown that the mode of action based on the capability to render the plant unpalatable to several types of pests like aphids, grasshoppers, beetles, etc. (Carroll, 1988). Alkaloids from *N. lolii* and *L. perenne* are capable of altering insect behaviour. Several of these alkaloids were added to the diet of adult individuals of Coleoptera *Heteronychus arator*. Peramine, lolitrem B, lysergol-type alkaloids, festuclavine and lisergic acid showed no effects on the insect. Ergovine showed moderate effects whereas ergotamine, ergovaline from the ergot-type alkaloid family seem to be responsible for the plant resistance (Ball et al., 1997). Miles et al., (1998) showed that endophytic isolates of *Neotyphodium* sp. produce

N-formilonine and a paxiline nalogous in the host *Echinopogon ovatus.* Dougherty et al., (1998) showed *Haematobia irritans* larvae of horn fly (cattle ectoparasite) killed when cattle manure was amended with seed extracts containing lolines from plants infected with *N. coenophialum.*

6.3 Endophyte as Biopesticidal

Biopesticides, agent of biological origin, may be viruses, bacteria, pheromones, plant or animal compounds. The outstanding feature of biopesticides is environment friendly, easy biodegradability, faster rate of product development, low research expenditure, do not disturb natural biocenosises also application of biopreparation lowers the chemical loading to the environment, does not deteriorate the soil fertility, natural ways of decomposition in nature, non-toxicity for warm-blooded organisms and overcoming the anti-environment chemical pesticide. *Burkholderia brasilensis* is an endophyte of roots, stems and leaves of sugarcane plant while *Burkholderia tropicalis* is confined to its roots and stems, with effective biopesticide properties. Many cultivated and wild type plants have been investigated for endophytic fungal metabolites that include guanidine and pyrrolizidine alkaloids, indole derivatives, sesquiterpenes, isocoumarin derivatives (Kumar et al., 2008).

6.4 Tolerance to Environmental Stress

Endophytic microorganisms actively respond to various biotic and abiotic stress factors which hampers the overall agricultural scenario. Biotic stress tolerance mechanisms, such as biocontrol of phytopathogens in the root zone through production of antifungal or antibacterial agents, pathogen antagonism, siderophore production aiding plant nutrition by iron chelation, phosphate solubilisation and induction of systematic acquired host resistance (Johansson et al., 2004). Endophytes from potato plants showed antagonistic activity against fungi and also inhibited bacterial pathogens belonging to the genera *Erwinia* and *Xanthomonas.* Some of the endophytic isolates produced in particular, various bacteria and fungi especially of the genera *Peudomonas, Bacillus* and *Trichoderma* produce a range of metabolites against other phytopathogenic fungi (Raaijmkers et al., 2002). Production of antibiotics sensitive to plant pathogens, such as alkaloids, terpenoids, aromatic compounds, polypeptides and secretion of enzymes that has the capacity to hydrolyse compounds like cellulose, hemicellulose, chitin, proteins, etc.

7. CYANOBACTERIA

The soils in desert and semi-arid regions have poor physical properties, low fertility and water deficiency. Cyanobacterial application to the organically poor semi-arid soil played a significant role in improving the status of carbon, nitrogen and other nutrients in the soil (Nisha et al., 2007). Organic matter content of dry land soils is chemically and biologically less stable and tends to decrease rapidly in arid regions thus leading to poor organic matter contents. Poor soil structure usually associated with low organic carbon, compaction, salinity and sodicity results in reduced aeration and rates of water infiltration, hence more soil erodability and the reduced number as well as biodiversity of microflora ultimately has the adverse impact on plant growth and productivity. Flaibani et al., (1989) reported that exopolysaccharides from cyanobacteria also contribute to reclamation and improvement of desert soils. It is reported that cyanobacteria, as carbon and nitrogen fixers, can contribute to the improvement of soil nutrient status in arid soils (Jeffries et al., 1992; Lange et al., 1994). Cyanobacterial sheaths and EPS also play a major role in water retention due to their hygroscopic nature (Decho 1990). The soil microflora contributes to increased water holding capacity of soils (Verrecchia et al., 1995). Belnap and Gillette (1997) and Campbell (1979); Belnap and Harper (1995) reported that formation of algal crust by *Microcoleus vaginatus* and *M. chthonoplastes* in extremely dry climates, make significant contributions in stability of soils. It is reported that cyanobacteria produce extracellular polymeric substances (EPS) that help them to overcome conditions of water stress and also bind soil particles (Mazor et al., 1996). Marathe (1972) and Falchini et al., (1996) demonstrated that cyanobacteria play a crucial role in bioamelioration of soils and enhancing crop yield not only aggregation initiation, but also through protection of soil porosity due to reduction of the damaging effect of water addition. The diazotrophic cyanobacteria contribute substantially to the fertility of the soil especially under tropical paddy field conditions. Some native cyanobacterial strains have been reported to play an important role in improving soil aggregation, water holding capacity and soil aeration in paddy and several other agriculture fields (Rogers and Burns, 1994; DeCaire et al., 1997; Hegde et al., 1999).

Better soil aggregation in cyanobacterial biofertilizer treated soils may be attributed to polysaccharides produced by the blue green algae (Ahmed et al., 2000). The native strains of cyanobacteria in semi-arid soils showed remarkable potential for improving structural stability, nutrient status and productivity of the soil due to their inherent tolerance capacity under limited soil moisture condition (Manchanda and Kaushik, 2000). Cyanobacteria produced organic matter, which increased soil organic C as well as water stable soil aggregates (Malam Issa et al., 2001). Cyanobacterial inoculations in disturbed soils have been reported to restore the population of carbon and nitrogen cycle microorganisms (Acea et al., 2003). Cyanobacteria seem to exert a mechanical effect on soil particles by the superficial network of the trichomes/filaments as well as enmeshing with soil particles at depth. Increase in total sugar content in soils having a cyanobacterial mat has been found to correspond to the enhanced soil aggregation process. The cyanobacterial consortium used as biofertilizer also produces abundant EPS about 25 per cent of their total biomass (Nisha et al., 2007). Algal biomass and activity in soil crust are concentrated in the upper soil layers and EPS have gluing effect on soil particles leading to accelerate soil aggregation. EPS produced by the cyanobacteria also seem to promote the activity of soil microflora as indicated by high soil enzymatic activities. Such microbial flora may produce more EPS, thus further amplifying the effect. Cyanobacteria play an important role in fixing C and N in soil and have been considered important for desert ecosystem. Removal of cyanobacteria from the arid sites reduces productivity and increases erosion by exposing unprotected subsurface soils to wind and water.

8. MICROORGANISMS FOR BIOREMEDIATION

Several chemicals are released into the soil ecosystem either as a method of disposal or as a consequence of the technology of their utilization. In particular, the application of pesticides, many of which are toxic or contain toxic contaminants, is central to the high yields in modern agriculture. With the advances in biotechnology, bioremediation has become one of the major developing research areas of environmental restoration, utilizing microorganisms to reduce the concentration/toxicity of various chemical pollutants, such as heavy metals, dyes pesticides, etc. Biological treatment especially using cyanobacteria, for treatment of water bodies has manifold advantages over conventional methods, as cyanobacteria are cosmopolitan in environment and known to accumulate high levels of metal/pollutant; therefore, the process involved is relatively cheap and environment friendly. Systems are Cu, Zn, Ni, Co, Pb, Cr, Cd, etc. (Kaushik et al., 1999) Among the photoautotrophs, cyanobacteria are relatively more tolerant to heavy metals (Fiore and Trevors, 1994). The uptake of metal ions, such as Cu, Pb, Zn, Ni, Cd, Cr, etc. has been reported in some of the efficient cyanobacteria are *Spirulina platensis, Oscillatoria anguistissima, Microcystis* sp., *Synechococcus* sp. (Verma and Singh, 1995; Rai et al., 1998; Pradhan and Rai, 2000; Yee et al., 2004).

The remediation of soils contaminated with heavy metals can be performed using chemical, physical and biological techniques. The metal and organic pollutants can be removed by the microbial flora. *Bacillus* sp. was efficient to remove the Au, Cd, Cr, Fe, Mn, Ni, Pb, U and Zn. It was recorded that *Bacillus* sp. can efficiently remove the metal pollutants from the waste or industrial effluents (Brierley and Brierley, 1993; Philip et al., 2000; Gunasekaran et al., 2003). *Pseudomonas* sp. is also reported to absorb the metals such as Cu, Cr, Cd, Pb, Ni, U and Zn from soil by various investigators (Kapley et al., 1999; Sar and D'Souza, 2001; Cybulski et al., 2003; Tarangini, 2009). Ilhan et al., (2004) observed that *Staphylococcus saprophyticus* reduced Cr, Cu and Pb ions. The isolate was adsorbed 100 percent Pb ions at different pH range of 3–5 and at 27°C temperature. *Corynebacterium* sp. and *Flavobacterium* sp. were reported for removal of organic contaminants and halogenated hydrocarbons, phenoxyacetates and aromatic hydrocarbons (Jogdand, 1995). *Corynebacterium glutamicum* was able to reduce the thorium and uranium (Tarangini, 2009). The biosorption of heavy metal Cd and Cu by *Flavobacterium* sp. was reported by Rajbanshi (2008) whereas, biosorption of Cu, Fe and Zn by *Corynebacterium* sp. was reported by Odokuma (2009). Many investigators (Friis and Myers-Keith, 1986; Mattuschka et al., 1993; Jogdand, 1995; Tarangini, 2009) reported that *Streptomyces* sp. can be used to remove the metal ions of Ag, Cd, Cu, Cr, Ni, Pb, U and Zn from the waste. *Streptomyces noursei* was able to reduce the metal ions in order Pb>Ag> Cu>Cr (Mattuschka et al., 1993). *Aspergillus niger* was reported to remove Ag, Au, Cd, Cu, Th, U and Zn (Townsley et al., 1986; Kuyucak and Volesky, 1988; Khalid et al., 1993; Guibal et al., 1995; Gunasekaran et al., 2003).

9. PLANT MICROBE INTERACTIONS

It is important to recognize that plants exist in intimate association with microorganisms, some of which cause plant disease while others protect against disease. Identifying, understanding and utilizing microorganisms or microbial products to control plant disease and enhance crop production are integral parts of sustainable agriculture. Biological control has the potential to control crop diseases while causing no or minimal detrimental environmental impact. Some of the benefits of utilizing microorganisms include:

- reduced dependence on chemical pesticides, important because of the increasing restrictions on chemical usage due to environmental and public concerns;
- lack of development of pathogen resistance to biological control organisms, important due to the observed increase in resistance to many chemical controls;
- improvement of soil and enhancement of agricultural sustainability;
- selective action against specific groups of pathogens and not against beneficial organisms; and
- biodegradability of microbial pesticides and the by-products of their manufacture.

The soil-borne plant pathogens result in economic losses for farmers. For root diseases of mature crops, there are a few effective and economical post-plant strategies for control. The use of chemical pesticides to control soil-borne pathogens has caused significant changes in air and water quality, altered natural ecosystems resulting in direct and indirect effects on wildlife and caused human health problems, e.g., methyl bromide, a fumigant used to control soil-borne diseases, has become notorious in recent years for depleting the ozone layer and changing the climate of our planet. The future of sustainable agriculture will increasingly rely on the integration of biotechnology with traditional agricultural practices. Although genetic engineering promises enhanced yields and disease resistance, it is also important to recognize that plants exist in intimate associations with microorganisms, some of which cause plant disease while others protect against disease.

The probiotics, new concept of managing plant health through the manipulation of probiotic organisms associated with plants, has gained interest recently (Picard et al., 2008). Plant-specific stimulation of specific microbial groups in their rhizosphere suggests that plants may have evolved to strategically stimulate and support particular microbial groups capable of producing antibiotics as a defense against diseases caused by soil-borne pathogens (Weller et al., 2002). Soil bacteria belonging to the genus *Pseudomonas* are ubiquitous in most soils and have been linked to wide-ranging processes including plant growth promotion and inhibition, disease control, nutrient cycling, nitrogen fixation and bioremediation. Their ability to respond quickly to changes in physical, chemical, carbon and nutritional conditions in soil has been linked to their functional significance in agricultural ecosystems (Hornick, 1992; Shrivastava et al., 2014). Antibiosis is the most commonly suggested trait responsible for their activity against plant pathogens and a number of antimicrobial compounds have been identified, e.g., 2,4-diacetylphloroglucinol (2,4-DAPG), phenazines (PHZ), pyrrolnitrin (PRN), pyoluteorin (PLT), hydrogen cyanide (HCN) and biosurfactant antibiotics (Picard et al., 2008). Conventional biochemistry-based characterisation of the chemicals is now complemented with molecular techniques, e.g., metabolomics and transcriptomics, to unravel the mechanisms of production, interactions with pathogens and plants, genotypic and phenotypic diversity of organisms capable of producing similar compounds and determine their activity in natural soil environments.

10. BIOLOGICAL CONTROL ORGANISMS

The bacteria, fungi and actinobacteria can act as biocontrol agents against root diseases (Whipps, 2001). A number of bacterial and fungal inoculant formulations are available commercially to control diseases in agricultural and horticultural crops (Westerdijk, 2000). The success of biocontrol inoculants depends upon the ability to:

a. lengthen the period, during which a threshold population density is sustained in the rhizosphere;
b. maintain the adequate populations needed to provide effective biological control; and
c. increase the magnitude of disease control provided by introduced rhizobacteria.

The cyanobacteria can also be used as biocontrol agents based on previous research. In modern agriculture practices, the conventional control methods of plant diseases have not been found quite efficient due to survival of the reproductive structures of pathogens in the soil. At the same time, fungicides and other artificial chemical agents inhibit the growth and development of crop plants (Nyporko et al., 2002). Various antagonistic microorganisms have been identified for almost every life cycle stage of several plant pathogenic microbes and cyanobacteria are the efficient antagonistic agents against several pathogenic microbes (Yuen et al., 1994). Various investigators have proved that cyanobacteria can be efficiently used as biocontrol agent (DeCaire et al., 1990; DeMule et al., 1991; Teuscher et al., 1992; Zulpa et al., 2003; Tassara et al., 2008). Biological control is an attractive approach for the control of soil-borne diseases (Cook, 2007). Actinobacterial endophytes can colonize plants without disrupting the normal endophytic populations, can produce antifungal antibiotics and plant growth hormones, and can also induce systemic disease resistance in plants (Conn et al., 2008).

Advantages of a biological approach to disease control include a lack of environmental damage, reduced human health risks, lack of resistance development in the pathogen, selectivity in mode of action, lack of activity against most beneficial microorganisms and improved soil conditions and agricultural sustainability. Biocontrol inoculants are generally tested for their antibiosis potential due to antagonism, hyper-parasitism, competition and predation by indigenous organisms; however, organisms that induce a systemic resistance to diseases and pests have the greatest potential to succeed under field conditions (Kloepper et al., 2004). Some soils can suppress the severity of disease even in the presence of a pathogen, host plant and favorable climatic conditions for the disease. There are a number of examples, both in Australia and worldwide, where agricultural soils have become suppressive to soil-borne pathogens (Weller et al., 2002; Cook, 2007; Gupta et al., 2011).

11. CONCLUSION

Soil microbes especially bacteria and fungi are essential for decomposing organic matter and recycling old plant material. Microbes in terrestrial environments are important catalysts of global carbon and nitrogen cycles, including the production and consumption of greenhouse gases in soil. Some microbes produce the greenhouse gases carbon dioxide (CO_2) and nitrous oxide (N_2O) while decomposing organic matter in soil. Others consume methane (CH_4) from the atmosphere, thus help to mitigate climate change. The importance of each of these processes is influenced by human activities and impacts the warming potential of the earth's atmosphere. The soil quality is the key to a sustainable agriculture. The symbiotic association of plants and mycorrhizal fungi, i.e., arbuscular mycorrhizal fungi, (AMF) has long been recognized for the benefits it provides with nutrient transport and uptake. Nitrogen fixation is one of the essential beneficial biological processes for the economic and environmental sustainability of agriculture worldwide. Free-living as well as symbiotic PGPR can improve plant nutrition and growth, plant competitiveness and responses to external stress factors. PGPR can promote the growth of plants using direct and indirect mechanisms. The endophytic microorganisms can play a vital role in agricultural sustainability such as growth promoter, insect control, biopesticide, stress tolerance, etc. Cyanobacteria seem to exert a mechanical effect on soil particles by the superficial network of the trichomes/filaments as well as enmeshing with soil particles at depth. The cyanobacterial consortium used as biofertilizer also produces abundant EPS. With the advances in biotechnology, bioremediation has become one of the major developing research area of environmental restoration, utilizing microorganisms to reduce the concentration/toxicity of various chemical pollutants, such as heavy metals, dyes pesticides, etc. It is important to recognize that plants exist in intimate associations with microorganisms, some of which cause plant disease while others protect against disease. Identifying, understanding and utilizing microorganisms or microbial products to control plant disease and enhance crop production are integral parts of sustainable agriculture. Biological control has the potential to control crop diseases while causing no or minimal detrimental environmental impact. The competent soil microorganisms can be used in various ways to sustain our agriculture practices.

Acknowledgements

Author is thankful to Kanchan Prabha Rathoure (Mrs.) for technical and for her valuable suggestions to make it more effective.

References

Acea MJ, Prieto Fernandez A, and Diz Cid N (2003). Cyanobacterial inoculation of heated soils: effect on microorganisms of C and N cycles and on chemical composition in soil surface. Soil Biol Biochem 35: 513–524.

Ahmed D, El Gamal MS, Ammar UM, and Abd El Raouf TM (2000). Role of some cyanobacteria in enhancement of soil characteristics. Egypt J Plycol 1: 99–106.

Auge RM (2001). Water relations, drought and vesicular-arbuscular mycorrhizal symbiosis. Mycorrhiza 11: 3–42.

Bais HP, Park SW, Weir TL, Callaway RM, and Vivanco JM (2004). How plants communicate using the underground information superhighway. Trends Plant Sci 9: 26–32.

Bais HP, Weir TL, Perry LG, Gilroy S, and Vivanco JM (2006). The role of root exudates in rhizosphere interactions with plants and other organisms. Annu Rev Plant Biol 57: 233–266.

Ball OJP, Miles CO, and Prestidge RA (1997). Ergopeptide alkaloids and Neotyphodium lolli-mediated resistance in perennial ryegrass agains adult *Heteronynchus arator* (Colcoptera: scarabaeidae). J Econ Entomol 90: 1382–1391.

Belimov AA, Hontzeas N, Safronova VI, Demchinskaya SV, Piluzza G, Bullitta S, and Glick BR (2005). Cadmium-tolerant plant growth-promoting bacteria associated with the roots of Indian mustard (*Brassica juncea* L Czern). Soil Biol Biochem 37: 241–250.

Belnap, J, and Gillette DA (1997). Disturbance of biological soil crust: impacts on potential wind erodability of sandy desert soils in Utah. USA Land Degrad Develop 8: 355–362.

Belnap J, and Harper KT (1995). Influence of cryptobiotic soil crust on elemental content of tissue of two desert seed plants. Arid Soil Res Rehabilit 9: 107–115.

Bloemberg GV, and Lugtenberg BJJ (2001). Molecular basis of plant growth promotion and biocontrol by rhizobacteria. Curr Opin Plant Biol 4: 343–350.

Brierley CL, and Brierley JA (1993). Immobilization of biomass for industrial application of biosorption In: Torma AE, Apel ML, Brierley CL (eds) Biohydrometallurgical Technologies, The Minerals, Metals & Materials Society: Warrendale, PA 2: 35–44.

Buckley, DH V Huangyutitham, SF Hsu, and TA Nelson (2007). Stable isotope probing with 15N2 reveals novel non-cultivated diazotrophs in soil. Appl Environ Microbiol 73: 3196–3204.

Burd GI, Dixon DG, and Glick BR (1998). A plant growth-promoting bacterium that decreases nickel toxicity in seedlings. Appl Environ Microbiol 64: 3663–3668.

Campbell SE (1979). Soil stabilization by a prokaryotic desert crust implications for precambrian land biota. Origin Life 9: 335–348.

Carroll G (1988). Fungal endophytes in stems and leaves-from latent pathogen to mutualistic symbiont. Ecology 69: 2–9.

Clark RB, and Zeto SK (2000). Mineral acquisition by arbuscular mycorrhizal plants. J Plant Nutr 23: 867–902.

Compant S, Duffy B, Nowak J, Clement C, and Barka EA (2005). Use of plant growth-promoting bacteria for biocontrol of plant diseases: principles, mechanisms of action, and future prospects. Appl Environ Microb 71: 4951–4959.

Conn V, Walker A, and Franco C (2008). Endophytic actinobacteria induce defense pathways in *Arabidopsis thaliana*. Mol Plant Microbe Interact 21: 208–218.

Cook, RJ (2007). Toward cropping systems that enhance productivity and sustainability. Proc Natl Acad Sci 103: 18389–18394.

Cybulski Z, Dzuirla E, Kaczorek E, and Olszanowski A (2003). The influence of emulsifiers on hydrocarbon biodegradation by Pseudomonadacea and Bacillacea strains. Spill Sci Technol B 8: 503–507.

De Weert S, Hans V, Mulders HM, Irene K, Nico H, Guido VB, Jos V, René De M, and Ben JJL (2002). Flagella-driven chemotaxis towards exudates components is an important trait for tomato root colonization by *Pseudomonas fluorescens*. Mol Plant Microbe Interact 15: 1173–1180.

DeCaire, GZ, De Cano, MS, De Mule, MC, Palma, RM, and Colombo K (1997). Exopolysaccharide of *Nostoc muscorum* (cyanobacteria) in the aggregation of soil particles. J Appl Phycol 4: 249–253.

DeCaire GZ, De Cano, MS, De Mule, MCZ, and De Halperin DR (1990). Antimycotic products from the cyanobacterium *Nostoc muscorum* against *Rhizoctonia solani*. Phyton 51: 1–4.

Decho AW (1990). Microbial exopolymer secretions in ocean environments: their role(s) in food webs and marine processes. Oceanogr Mar Biol Ann Rev 28: 73–153.

DeMule MCZ, De Caire GZ, De Cano MS, and De Halperin DR, (1991). Bioactive compounds from *Nostoc muscorum* (Cianobacterias). Cytobios 66: 169–172.

Dougherty CT, Knapp FW, Bush LP, Maul JE, and Van Willigen J (1998). Mortality of horn fly (Diptera: Muscidae) larvae in bovine dung supplemented with loline alkaloids from tall fescue. J Med Entomol 35: 798–803.

Egamberdiyeva D, and Hoflich G (2004). Effect of plant growth-promoting bacteria on growth and nutrient uptake of cotton and pea in a semi-arid region of Uzbekistan. J Arid Environ 56: 293–301.

Falchini L, Sparvoli E, and Tomaselli L (1996). Effect of Nostoc (cyanobacteria) inoculation on the structure and stability of clay soils. Biol Fertil Soils 23: 246–252.

Fiore MF, and Trevors JT (1994). Cell composition and metal tolerance in cyanobacteria. Biometals 7: 83–103.

Flaibani A, Olsen Y, and Painter TJ (1989). Polysaccharides in desert reclamation: composition of exocellular proteoglycan complexes produced by filamentous blue-green and unicellular green edaphic algae. Carbohydrate Res 190: 235–248.

Friis N, and Myers-Keith P (1986). Biosorption of uranium and lead by *Streptomyces longwoodensis*. Biotechnol Bioeng 28: 21–28.

Gaur A, and Adholeya A (2004). Prospects of arbuscular mycorrhizal fungi in phytoremediation of heavy metal contaminated soils. Curr Sci 86: 528–34.

Glick BR (1995). The enhancement of plant growth by free-living bacteria. Can J Microbiol 41: 109–117.

Glick BR, Patten CL, Holguin G, and Penrose DM (1999). Biochemical and genetic mechanisms used by plant growth promoting bacteria. Imperial College Press, London, UK, pp. 86–179.

Goltapeh EM, Danesh YR, Prasad R, and Varma A (2008). Mycorrhizal Fungi: what we know and what should we know? In: Varma A (ed), Mycorrhiza 3rd edn. Springer-Verlag Berlin Heidelberg, 3–27.

Guibal E, Roulph C, and Le Cloirce P (1995). Infrared spectroscopic study of uranyl biosorption by fungal biomass and materials of biological origin. Environ Sci Technol 29: 2496–2503.

Gunasekaran P, Muthukrishnan J, and Rajendran P (2003). Microbes in Heavy Metal Remediation. Indian J Exp Biol. 41: 935–944.

Gupta VVSR, Rovira AD, and Roget DK (2011). Principles and management of soil biological factors for sustainable rainfed farming systems In: Tow P (ed) Rainfed farming systems. Springer Science and Business Media. pp. 149–184.

Gyaneshwar P, Kumar GN, Parekh LJ, and Poole PS (2002). Role of soil microorganisms in improving P nutrition of plants. Plant and Soil 245: 83–93.

Hegde DM, Dwivedi BS, and Babu SNS (1999). Biofertilizers for cereal production in India-a review. Indian J Agric Sci 69: 73–83.

Higa T (1991). Effective microorganisms: A biotechnology for mankind In: Parr JF, Hornick SB, Whitman CE (ed) Proceedings of the First International Conference on Kyusei Nature Farming US Department of Agriculture, Washington, DC, USA pp. 8–14.

Hodge A, Campbell CD, and Fitter AH (2001). An arbuscular mycorrhizal fungus accelerates decomposition and acquires nitrogen directly from an organic source. Nature 413: 297–99.

Hornick SB (1992). Factors affecting the nutritional quality of crops. Amer J Alternative Agric 7: 63–68.

Ilhan S, Nourbakhsh MN, Kilicarslan S, and Ozdag H (2004). Removal of chromium, lead and copper ions from industrial wastewaters by *Staphylococcus saprophyticus*. Turk Electronic J Biotech 2: 50–57.

Jangid K, Williams MA, Franzluebbers AJ, Schmidt TM, Coleman DC, and Whitman WB (2011). Land-use history has a stronger impact on soil microbial community composition than above-ground vegetation and soil properties. Soil Biol Biochem 43: 2184–2193.

Jeffries, DL, Klopatek, JM, Link, SO, and Bolton Jr H (1992). Acetylene reduction by cryptogamic crusts from a blackbrush community as related to resaturation and dehydration. Soil Biol Biochem 24: 1101–1105.

Jogdand SN (1995). Environmental Biotechnology. 1st edn., Himalaya Publishing House, Bombay India, pp. 104–120.

Johansson JF, Paul LR, and Finlay RD (2004). Microbial interactions in the mycorrhizosphere and their significance for sustainable agriculture. FEMS Microbiol Ecol 48(1): 1–13.

Kapley A, Purohit HJ, Chhatre S, Shanker R, and Chakrabarti T (1999). Osmotolerance and hydrocarbon degradation by a genetically engineered microbial consortium. Biores Tech 67: 241–245.

Kaushik S, Sahu BK, Lawania RK, and Tiwari RK (1999). Occurrence of heavy metals in lentic water of Gwalior region. Pollut Res 18: 137–140.

Khalid AM, Ashfaq SR, Bhatti TM, Anwar MA, Shemsi AM, and Akhtar K (1993). The uptake of microbially leached uranium by immobilized microbial biomass. In: Torma AE, Apel ML, Brier-ley CL (eds) Biohydrometallurgical Technologies. The Minerals, Metals & Materials Society: Warrendale PA 2: 299–308.

Kloepper JW, Ryu CM, and Zhang S (2004). Induced systemic resistance and production of plant growth by *Bacillus* spp. Phytopathology 94: 1259–1266.

Kula AR, Harnett DC, and Wilson GWT (2005). Effects of mycorrhizal symbiosis on tallgrass prairie plant–herbivore interactions. Ecol Lett 8: 61–69.

Kumar MS, Kirubanandan S, Sripriya R, and Sehgal PK (2008). Triphala promotes healing of infected full-thickness dermal wound. J Surg Res 144: 94–101.

Kuyucak N, and Volesky B (1988). Biosorbents for recovery of metals from industrial solutions. Biotechnol Left 10(2): 137–142.

Lange OL, Meyer A, Zellner H, and Heber U (1994). Photosynthesis and water relations of lichen soil crusts: field measurements in the coastal fog zone of the Namib Desert. Funct Ecol 8: 253–264.

Lemanceau P, Corberand T, Gardan L, Latour X, Laguerre G, Boeufgras J, and Alabouvette C (1995). Effect of two Plant Species, Flax (*Linum usitatissinum* L) and Tomato (*Lycopersicon esculentum* Mill), on the diversity of soilborne populations of Fluorescent Pseudomonads. Appl Environ Microb 61: 1004–1012.

Malam Issa O, LeBissonnais Y, Defarge C, and Trichet J (2001). Role of a cyanobacterial cover on structural stability of sandy soils in the sahelian part of western Niger. Geoderma 101: 15–30.

Manchanda, H, and Kaushik A (2000). Algal flora of aridisols of Rohtak and salt tolerance of the indigenous cyanobacteria. Trop Ecol 41: 217–223.

Mantelin S, and Touraine B (2004). Plant growth-promoting bacteria and nitrate availability: Impacts on root development and nitrate uptake. J Exp Bot 55: 27–34.

Marathe KV (1972). Role of some blue-green algae in soil aggregation In: 1st International Symposium on Taxonomy and Biology of Blue-Green Algae, Madras, India, pp. 328–331.

Martínez-Viveros O, Jorquera MA, Crowley DE, Gajardo G, and Mora ML (2010). Mechanisms and practical considerations involved in plant growth promotion by rhizobacteria. J Soil Sci Plant Nutr 10(3): 293–319.

Mattuschka B, Junghaus K, and Straube G (1993). Biosorption of metals by waste biomass In: Torma AE, Apel ML, Brierley CL (eds) Biohydrometallurgical Technologies. The Minerals, Metals & Materials Society: Warrendale, PA 2: 125–132.

Mazor G, Kidron GJ, Vonshak A, and Abeliovich A (1996). The role of cyanobacterial exopolysaccharides in structuring microbial crusts. FEMS Microbiol Ecol 21: 121–130.

Miles CO, diMena ME, Jacobs SWL, Garthwaite I, Lane GA, Prestidge RA, Marshal SL, Wilkinson HH, Schardl CL, Ball OJP, and Latch CM (1998). Endophytic fungi in indigenous australasian grasses associated with toxicity to livestock. Appl Environ Microb 64: 601–606.

Mishra M, Kumar U, Mishra PK, and Prakash V (2010). Efficiency of plant growth promoting rhizobacteria for the enhancement of *Cicer arietinum* L. growth and germination under salinity. Adv Biol Res 4(2): 92–96.

Nisha R, Kaushik A, and Kaushik CP (2007). Effect of indigenous cyanobacterial application on structural stability and productivity of an organically poor semi-arid soil. Geoderma 138: 49–56.

Nyporko AY, Yemets AI, Klimkina LA, and Blume YB (2002). Sensitivity of *Eleusine indica* callus to Trifluralin and Amiprophosmethyl in correlation to the binding of these compounds to tubulin. Russ J Plant Physiol 49: 413–418.

Odokuma LO (2009). Effect of culture age and biomass concentration on heavy metal uptake by three axenic bacterial cultures. Advances in Natural and Applied Sciences 3(3):339–349.

Parr JF, Hornick SB, and Kaufman DD (1994). Use of microbial Inoculants and organic fertilizers in agricultural production In: Proceedings of the International Seminar on the Use of Microbial and Organic Fertilizers in Agricultural Production Published by the Food and Fertilizer Technology Center, Taipei, Taiwan.

Patten CL, Glick BR (1996). Bacterial biosynthesis of Indole-3-acetic acid. Can J Microbiol 68:3795–3801.

Persello-Cartieaux F, Nussaume L, and Robaglia C (2003). Tales from the underground: molecular plant-rhizobacteria interactions. Plant Cell and Environ 26: 189–199.

Philip L, Iyengar L, and Venkobacher L, (2000). Site of interaction of copper on *Bacillus polymyxa*, Water Air Soil Pollution, 119: 11–21.

Picard C, and Bosco M (2008). Genotypic and phenotypic diversity in populations of plant-probiotic *Pseudomonas* spp. colonizing roots. Naturwissenschaften 95: 1–16.

Pirttilä AM, Joensuu P, Pospiech H, Jalonen J, and Hohtola A (2004). Bud endophytes of Scots pine produce adenine derivatives and other compounds that affect morphology and mitigate browning of callus cultures. Physiol Plant 121: 305–312.

Pradhan S, and Rai LC (2000). Optimization of flowrate, initial metal ion concentration and biomass density for maximum removal of Cu2+ by immobilized Microcystis. World J Microbiol Biotechnol 16(50): 579–584.

Prasad R, Kumar M, and Varma A (2015). Role of PGPR in soil fertility and plant health. In: Egamberdieva D, Shrivastava S, Varma A (eds) Plant Growth-Promoting Rhizobacteria and Medicinal Plants. Springer International Publishing Switzerland 247–260.

Raaijmakers JM, Vlami M, and de Souza JT (2002). Antibiotic production by bacterial biocontrol agents Antonie van Leeuwenhoek 81: 537–547.

Rai LC, Singh S, and Pradhan S (1998). Biotechnological potential of naturally occurring and laboratory grown Microcystis in biosorption of Ni^{2+} and Cd^{2+}. Curr Sci 74: 461–463.

Rajbanshi A (2008). Study on heavy metal resistant bacteria in Guheswori sewage treatment plant. Our Nature 6: 52–57.

Reed M, and Glick B (2005). Growth of canola (*Brassica napus*) in the presence of plant growth-promoting bacteria and either copper or polycyclic aromatic hydrocarbons. Can J Microbiol 51: 1061–1069.

Reed SC, Cleveland CC, and Townsend AR (2011). Functional ecology of free-living nitrogen fixation: A contemporary perspective. Annu Rev Ecol Evol Syst 42: 489–512.

Reganold JP, Papendick RI, and Parr JF (1990). Sustainable Agriculture. Sci Am. 262(6): 112–120.

Richardson A, and Simpson RJ (2011). Soil microorganisms mediating phosphorus availability. Plant Physiol 156: 989–996.

Rillig MC (2004). Arbuscular mycorrhizae terrestrial ecosystem processes Ecol Lett 7: 740–54.

Rogers SL, and Burns RG (1994). Changes in aggregate stability, nutrient status, indigenous microbial populations, and seedling emergence, following inoculation of soil with *Nostoc muscorum*. Biol Fertil Soils 18: 209–215.

Ryu CM, Farag MA, Hu C-H, Reddy MS, Wei H-X, Pare PW, and Kloepper JW (2003). Bacterial volatiles promote growth in Arabidopsis. Proc Natl Acad Sci 100:4927–4932.

Safronova V, Stepanok V, Engqvist G, Alekseyev Y, and Belimov A (2006). Root associated bacteria containing 1-aminocyclopropane-1-carboxylate deaminase improve growth and nutrient uptake by pea genotypes cultivated in cadmium supplemented soil. Biol Fertil Soils 42: 267–272.

Saleem M, Arshad M, Hussain S, and Bhatti AS (2007). Perspective of plant growth promoting rhizobacteria (PGPR) containing ACC deaminase in stress agriculture. J Ind Microbiol Biotechnol 34: 635–648.

Sar P, and D'Souza SF (2001). Biosorptive uranium uptake by Pseudomonas strain: Characterization and equilibrium studies. J Chem Tech Biot 76: 1286–1294.

Selosse Marc-Andre Richard F, He X, and Simard S (2006). Mycorrhizal networks: des liaisons dangereuses? Trends Ecol Evol 21: 621–628.

Sheng XF, and Xia JJ (2006). Improvement of rape (*Brassica napus*) plant growth and cadmium uptake by cadmium-resistant bacteria. Chemosphere 64:1036–1042.

Shrivastava S, Prasad R, and Varma A (2014). Anatomy of root from eyes of a microbiologist. In: Morte A, Varma A (eds) Root Engineering. Springer-Verlag Berlin Heidelberg 40: 3–22.

Smith S, Smith F, and Jakobsen I (2004). Functional diversity in arbuscular mycorrhizal (AM) symbiosis: the contribution of the mycorrhizal P uptake pathway is not correlated with mycorrhizal responses growth or total uptake. New Phytol 162: 511–24.

Smith SE, and Read DJ (2008). Mycorrhizal symbiosis. 3rd edn Academic Press London, UK.

Tarafdar JC, and Marschner H (1994). Efficiency of VAM hyphae in utilization of organic phosphorus by wheat plants. Soil Sci Plant Nutr 40: 593–600.

Tarangini K (2009). Biosorption of Heavy Metals using Individual and Mixed cultures of *Pseudomonas aeruginosa* and *Bacillus subtilis* M Tech Thesis, Department of Chemical Engineering, National Institute of Technology, Rourkela India.

Tassara, C, Zaccaro, MC, Storni, MM, Palma, M, and Zulpa, G, (2008). Biological control of lettuce white mold with cyanobacteria. Int J Agric Biol 10: 487–492.

Teuscher E, Lindequist U, and Mundt S (1992). Cyanobakterien, Quellen biogener Wirkstoffe Pharm Ztg Wiss 137: 57–69.

Townsley CC, Ross IS, and Atkins AS (1986). Biorecovery of metallic residues from various industrial effluents using filamentous fungi. In: Lawrence RW, Branion RMR, Ebner HG (eds) Fundamental and Applied Biohydrometallurgy, Elsevier: Amsterdam, pp. 279–289.

van der Heijden MGA, Wiemken A, and Sanders IR (2004). Arbuscular mycorrhizal fungi as support systems for seedling establishment in grassland. Ecol Lett 7: 293–303.

Van Overbeek L, and van Elsas JD (2008). Effects of plant genotype and growth stage on the structure of bacterial communities associated with potato (*Solanum tuberosum* L). FEMS Microbiol Ecol 64: 283–296.

Verma SK, and Singh SP (1995). Multiple metal resistance in the cyanobacterium *Nostoc muscorum*. Bull Environ Contam 54: 614–619.

Verrecchia, E, Yair, A, Kidron, GJ, and Verrecchia K (1995). Physical properties of the psammophile cryptogamic crust and their consequences to the water regime of sandy soils, north-western Negev desert. Israel J Arid Environ 29: 427–437.

Weller D, Raaijmakers JM, Mcspadden Gardener BB, and Thomashow LS (2002). Microbial populations responsible for specific soil suppressiveness to plant pathogens. Annu Rev Phytopathol 40: 309–348.

Westerdijk CE (2000). Biological control of leaf rot in lettuce. Bull, Volleglondsgroenteteelt pp. 6–8.

Whipps JM (2004). Prospects and limitations for mycorrhizas in biocontrol of root pathogens. Can J Bot 82: 1198–227.

Whipps JM (2001). Microbial interactions and biocontrol in the rhizosphere. J Exp Bot 52:487–511.

Wolfe B, Husband B, and Klironomos JN (2005). Effects of a belowground mutualism on an aboveground mutualism. Ecol Lett 8: 218–23.

Yee N, Benning LG, Phoenix VR, and Ferris FG (2004). Characterization of metal cyanobacteria sorption reactions: a combined macroscopic and infrared spectroscopic investigation Environ Sci Technol 38: 775–782.

Yuen GY, Craig LM, Kerr ED, and Steadman JR (1994). Influences of antagonist population levels, blossom development stage and canopy temperature on the inhibition of the Sclerotinia sclerotiorum on dry edible bean by *Erwinia herbicola*. Phytopathology 84: 495–501.

Zulpa G, Zaccaro MC, Boccazzi F, Parada JL, and Storni M (2003). Bioactivity of intra and extracellular substances from cyanobacteria and lactic acid bacteria on wood blue stain fungi. Biol Control 27: 345–348.

15

Mutant of Cyanobacteria Proved to be Better Biofertilizer

ILA DUBEY[1*] AND SURENDRA SINGH[2]
[1] Department of Biotechnology, Indian Institute of Technology, Roorkee, India
[2] Centre of Advanced Study in Botany, BHU, Varanasi, India
*Email: *iladvivedi@gmail.com*

1. INTRODUCTION

Pollution and contamination of soil is one of the major concerns in today's world. Use of chemical fertilizers and pesticides has caused tremendous harm to the environment. Solution to this problem is use of environment-friendly biofertilizer; Biofertilizers are organisms that enrich the nutrient quality of soil (El-Habbasha et al., 2007; Yosefi et al., 2011). Biofertilizer are natural fertilizers defined as living fertilizers mainly inoculant of microorganisms (bacteria, algae, fungi) which promotes plant growth by increasing nutrient supply (Vessey, 2003). There are different types of microorganisms which are used as biofertilizers. Some are capable of nitrogen fixation such as *Azotobacter*, blue green algae, *Rhizobium* and *Azospirillum*. Blue-green algae, such as cyanobacteria are diazotrophic (nitrogen fixing) oxygenic, photosynthetic prokaryotes which have long been recognized as having potential as biofertilizer. One of the biotechnological aspects receiving considerable attention as biofertilizer is especially by using diazotrophic cyanobacteria is the photobiological production of ammonia. Cyanobacteria are Gram-negative bacteria, which have been defined as "microorganisms that harbour, within a typically prokaryotic cell, a photosynthetic apparatus similar in structure and function to that located in the chloroplast of phototrophic eukaryotes" (Stanier, 1977). Their presence can be found in the fossil record of early and middle Precambrian period approximately 2×10^9 years ago (Schopf and Walter, 1982). Morphologically, they can occur as branched or unbranched filaments or as single cells (Rippka et al., 1979). Cyanobacteria possess a wide range of morphological diversity, ranging from a unicellular to a branched filamentous organization and can differentiate heterocysts for N_2-fixation, akinites for surviving adverse conditions (cold and desiccation) and hormogonia for taxis. This structural-functional plasticity confers great versatility, enabling cyanobacteria to adopt and inhabit a wide range of environmental niches (Gallon, 1992; Tandeau de Marsac and

Houmard, 1993; Bryant, 1994; Bergman, et al., 1997). For over 150 years, cyanobacteria were considered a special group of algae called the blue-green algae (Rippka, 1988). As a result, phycologists working under the provisions of the International Code of Botanical Nomenclature (Botanical Code) developed their classification. Under this code, they were known by a variety of names, such as Myxophyceae, Phycochromophyceae, Cyanophyceae and Schizophyceae (Rippka, 1988). The name "cyanobacteria" was first proposed by stanier (Rippka et al., 1979) in recognition of their prokaryotic affinities. For example, in addition to the absence of the intracellular membranous organelles, the cyanobacterial DNA, like that of the bacteria, is not enclosed by a nuclear membrane (Rippka, 1988).

2. CYANOBACTERIA AS BIOFERTILIZER

Cyanobacteria play an important role in maintenance of soil fertility, as a natural biofertilizer (Song et al., 2005). The research on N_2-fixing property of cyanobacteria as a source of fertilizers started five decades ago. The pioneering research was carried out by first international R&D collaboration (Burlew, 1953; Oswald and Golueke, 1960) at the University of California Berkeley. However, till date several other R&D projects are ongoing, regarding biofertilizer production with nitrogen-fixing cyanobacteria with genetic modifications, to increase microalgae productivities (Benemann et al., 2003). The primary reason for such genetic modifications is to provide ability to cyanobacteria so as to sustain extreme conditions. In saline habitats, high sodium chloride content affects growth and development of plants. Cyanobacteria can decrease soil salinity (Saadatnia and Riahi, 2009). Cyanobacteria, because of phototrophy (Calvin cycle) and diazotrophy (dinitrogen fixation) like properties proved to be better biofertilizer especially in *Nostoc* and *Anabaena* strains (Wolk, 1996).

In India, considerable progress has been made in the development of cyanobacteria based biofertilizer technology. It has also been demonstrated that this technology can be a powerful means of enriching the soil fertility and improving rice crop yields. However, the technology needs to be improved further for better exploitation under sustainable agriculture systems (Mishra and Pabbi, 2004). Thus, it is imperative to understand potential use of cyanobacteria as a N_2-fixing biofertilizer with a perspective to improve productivity by imparting ability to resist soil toxicity, viz., sodium azide (Stewart, 1980; Hader et al., 1987; Kumar and Kaushik, 1992; Sinha and Kumar, 1992; Kumar et al., 1996).

Large number of trials conducted at the state and district levels under the DST sponsored All India Coordinated project on cyanobacteria during 1976–1984 have revealed certain important advances of using cyanobacterial biofertilizer in rice production systems.

- In the absence of chemical fertilizer, the application of cyanobacteria can provide 20–30 kg of biologically fixed N ha^{-1}.
- By supplementing the chemical nitrogen fertilizer with cyanobacteria, the farmer could save to the extent of 20–30 kg N ha^{-1} without affecting the crop yield.

- Even with fully recommended levels of chemical nitrogen, complementation with cyanobacteria has been found to be ecologically and agronomically beneficial.

Nitrogen is the most limiting nutrient of bioproductivity. The success of modern agriculture depends on the availability of fixed nitrogen. Present source of fixed nitrogen is fossil fuel-based chemical nitrogen fertilizer. In view of the limited availability of fossil fuels, chemical fertilizers would soon become impracticable. It has been emphasized that an improvement in the efficiency of biological nitrogen fixation and conferring this ability to other systems, such as cereal crops would significantly reduce the dependence on nitrogen fertilizers. Two principal approaches to the problem have been suggested (Merrick and Dixon, 1984). The first one involves manipulation of existing association between nitrogen-fixing bacteria crop plants and the second involves the transferring of nitrogen fixing (*nif*) genes into the plant genome. Cyanobacteria form both heterocyst and nitrogenase in the aerobic growth medium devoid of fixed source of nitrogen. Time course study of the kinetics of heterocyst development and nitrogenase activity in *A. cylindrica* (Stewart, 1973) has clearly demonstrated heterocyst differentiation preceding nitrogenase activity. The available physiological and biochemical evidences also support the view that heterocyst is the site of nitrogenase activity and synthesis under aerobic conditions (Fleming and Haselkorn, 1973).

3. SALT STRESS ON CYANOBACTERIA

Osmotic misbalance in the ecosystem is due to high salinity (salt concentration above normal concentration level). High salinity in the environment leads to serious problem of low productivity rate by deteriorating the soil quality. In case of cyanobacteria, salt stress above its tolerant capacities can affect its growth and metabolism.When studied the stress effect of sodium chloride (NaCl) on wild and mutant strain of *Anabaena variabilis* reveals the fact that under NaCl stress, both the strains take up more nitrate and nitrite to counteract the deleterious effect of NaCl on N_2 fixing machinery (Chauhan et al., 1999 a,b). However, due to NaCl induced inhibition of nitrogen metabolizing enzyme (NR) the wild type strain could not assimilate nitrate and hence could not grow and survive under such conditions. Since, NaCl resistant mutant strain having impaired GS showed higher uptake of nitrate and nitrite as compared to its wild type counterpart, a role of reduced Na^+ influx and increased Na^+ efflux in the regulation of nitrate and nitrite uptake and NaCl tolerance was suggested in the NaCl resistant mutant strain of *A. variabilis* (Chauhan et al., 2001). Shows that NaCl-resistant ($NaCl^r$) mutant strain of *A. variabilis* having impaired GS (Glutamine Synthetase) activity also took up nitrate and nitrite more rapidly as compared to its wild type counterpart in the presence and absence of NaCl, thus suggesting that Het– Fix–mutant strain of *A. variabilis* lacking the ability to produce heterocyst and nitrogenase, also exhibited low level of nitrate and nitrite uptake as compared to its wild type counterpart (Singh et al., 2001).

Although transduction of signals of salt stress in cyanobacteria is not yet known, but it is studied that target of signal transduction could be either the DNA, leading to the transcriptional induction of stress relevant genes, or at translational level resulting in the activity of stress specific proteins.

For salt acclimation process for possible post-translational process are described (Varela and Mager, 1996; Toroser et al., 1998; Toroser et al., 1997; Li et al., 1998)

- Due to conformational changes in key enzymes, there is alteration in endogenous concentration of Na^+. This is thought to be the most effective pathway.
- Osmotic signal transduction and enzyme activation by protein phosphorylation or dephosphorylation as reported in bacteria and plants.
- Salinity induced enzyme activation by phosphorylation or dephosphorylation with controlled interaction with a regulatory protein.
- Proteolytic processing of native enzyme.

Apart from DNA damage, protein damage is another outcome of stress proteins.

Cyanobacteria provide distinct advantages over higher plants and plant based systems as experimental materials. Cyanobacteria forms substantial fraction of the blooms in several important types of habitats. Thus, to conserve natural sources and their use in best possible manner, the activity of cyanobacteria can be further deliberated. Further, cyanobacteria have practical importance in several respects. There is ample evidence that nitrogen-fixing blue-green algae make a major contribution in fertility of paddy fields. Farmers do not use fertilizer in several southeastern states of India and eastern countries of world. It is apparent that cyanobacteria permits moderate rice harvest to be gathered. However, absence of cyanobacteria results in poor harvest. It is certain that humankind survives largely because of nitrogen fixation by cyanobacteria. In addition, cyanobacterial nitrogen fixation is advantageous over nitrogenous fertilizers owing to economic benefits. The economic benefits by use of cyanobacteria are due to *in situ* nitrogen fixation that reduces transportation costs. Thus, research with the objective of improving effectiveness of cyanobacterial nitrogen fixation under sodium azide (NaN_3) stress is of considerable economic importance. The present study will be significant because of the reasons stated as under:

- Sodium azide resistant mutant strain having derepressed nitrogenase, impaired GS and producing ammonia extracellularly would be helpful in increasing the soil fertility of degraded soils.
- The information obtained would help in understanding the genetic basis of NaN_3 tolerance in cyanobacteria in general and in *A. cylindrica* in particular.

4. SODIUM AZIDE EFFECT

The nitrogen fixing ability of several genera has implications for the maintenance of the fertility of natural and cultivated ecosystems and many trials have been conducted to increase rice yield by cyanobacterial inoculation (Mishra and Pabbi, 2004). However, natural strains of cyanobacteria are incapable of growth vis-a-vis release limited or no ammonia under high sodium azide (NaN_3) concentration. NaN_3 is a colourless, odourless, crystalline solid (salt-like) or solution. It is highly toxic in nature. Azide is an alternative substrate for nitrogenase and is reduced to ammonia and elemental nitrogen by the enzyme (Schollhorn and Burris, 1967). It is also a general inhibitor of growth and respiration (Hardy and Knight, 1967). Thus, it is well known for its use in pesticide formulation. The first known and recorded use of NaN_3 as a pesticide was in the 1920 (Richards, 2006). NaN_3 is also used a pesticide in rice fields along with carbofuran, propanil, bifenex. Literature revealed that application of pesticides inhibits nitrification. Further, NaN_3 has ability to retard nitrification even after 60 days of incubation (Turner, 1979).

Thus, an azide resistant cyanobacterial strain that has ability to form active nitrogenase in different medium can improve N_2-fixation. The capacities to fix nitrogen in the presence of combined nitrogen and to tolerate azide stress and release most of the fixed nitrogen as ammonia are of particular importance. It is theorized that mutants of N_2-fixing organisms derepressed for nitrogenase in fixed inorganic nitrogen medium could easily be selected by looking for azide-resistance on the medium. Derepressed nitrogenase would detoxify azide thus rendering the system resistant to azide. In the absence of either a natural or ammonia liberating mutant strain, treatment of N_2-fixing cyanobacterial cells with GS inhibitors seem to be the preferred way for induced ammonia liberation (Singh and Singh, 1978). This chapter provides an account to understand the development of an azide resistant mutant strain of a diazotrophic cyanobacterium *A. cylindrica* which can withstand NaN_3 stress, fix nitrogen at higher rate and subsequently release higher levels of fixed nitrogen as ammonia extracellularly.

5. MUTANT DEVELOPMENT

One of the technological aspects receiving considerable attention, especially by using diazotrophic cyanobacteria, is the photobiological production of ammonia. However, free-living diazotrophic cyanobacteria fix dinitrogen sufficient for their own needs and do not generally liberate significant amounts of ammonia into their external environments. Attempts are being made to induce ammonia production either by the addition of GS inhibitor or glutamate analogues, such as L-methionine-DL-sulfoximine1 (MSX), 5-hydroxylysine, and phosphinothricin or by producing ammonia liberating mutant strains. By using MSX, continuous photo-production of ammonia by diazotrophic cyanobacteria *A. variabilis* has been reported (Chauhan et al., 1999b). In the absence of either a natural or ammonia liberating mutant strain, treatment of N_2 fixing cyanobacterial cells with GS inhibitors seem to be the preferred way for induced ammonia liberation. However, prolonged exposure to MSX has been reported to be toxic for cyanobacteria, causing nitrogen starvation and inhibition in growth and protein synthesis. Therefore, chemical mutation of strain to develop salt tolerant mutant strain could be better solution of this problem.

By using chemical mutagens, such as N-methyl-N-nitro-N-nitrosoguanidine (NTG), several cyanobacterial mutant strains resistant to MSX or defective in GS have been isolated and shown to liberate ammonia in the external medium. However, these mutant strains are slow growing and cannot withstand the environmental extremes like NaCl stress. In this perspective, isolation of salt tolerant mutant (that can withstand high salt concentration, fix nitrogen at higher rate and subsequently release higher levels of fixed nitrogen as ammonia) may help in reclamation of degraded soils. The present investigation is an attempt to develop and characterize a salt resistant mutant strain of diazotrophic cyanobacteria. The developed mutant may exhibit derepressed nitrogenase, impaired GS and produced ammonia extracellularly. Exponentially grown cells of cyanobacterial strain were harvested by centrifugation and resuspended in 10mM HEPES/NaOH (pH 7.0). N-methyl-N-nitrosoguanidine (NTG, 250μg ml^{-1}) was added to the cultures (Chauhan et al., 2001). The cultures were incubated at 30°C for 4 hours in light with shaking. The cultures were washed three times with BG-11 medium and resuspended in same medium. Mutagenized cells were grown photoautotrophically for several generations and transferred into liquid BG-11 medium containing 5 mM KNO_3 with 20 mg/100 ml sodium azide. The strain thus isolated showed more tolerance against sodium azide and designated as sodium azide resistant mutant.

6. SODIUM AZIDE TOLERANT MUTANT STRAIN VERSUS WILD STRAIN

After mutation both, the wild type strain and the mutant strain grown properly for analysing and comparing changes occur due to mutation and resistance development. Wild type and mutant strain of *Anabaena cylindrica* were axenically grown in BG-11 medium (Rippka, et al., 1979) without nitrogen source. Cultures were incubated in a culture room at 25±1°C and illuminated with day-light fluorescent tubes having photon fluence rate of 50 μmol m^2 s^{-1} at surface of the vessels. Unless otherwise stated, all the experiments were carried out with mid-log phase cultures. The medium was buffered to pH 7.5 with 10 mM Hepes/NaOH.

In terms of nitrogen fixation changes noticed are described below:

6.1 Percent Survival

Percent survival of wild type *A. cylindrica* and its mutant counterpart was determined by growing cells of both the strains in liquid BG-11 medium containing different concentrations of NaN_3 (0, 10, 20, 30, 40, 50 μg ml^{-1}). Growth (in terms of the chl *a*) of both the strains was recorded after 7 days and expressed in terms of percent survival against NaN_3-untreated cells as control.

In order to select a mid-inhibitory survival concentration of NaN_3 for wild type *A. cylindrica* and its mutant counterparts and also to see its impact on survival of both the strains, both wild type and its mutant counterparts were exposed to various concentrations of NaN_3 (10, 20, 30, 40 and 50 μg ml^{-1}) for 7 days. After that growth of both the strains was recorded (in terms of chl *a*) and expressed in terms of percent survival. NaN_3 at 10 μg ml^{-1} concentration hardly had any significant effect on the survival of both wild type and its mutant counterpart. Whereas, the survival of mutant strain was found to be slightly more at 20 μg ml^{-1} NaN_3, this concentration of NaN_3, however, supported only 50 percent survival of its wild type counterparts. Thus, this concentration of NaN_3 was chosen and employed in all further experiments.

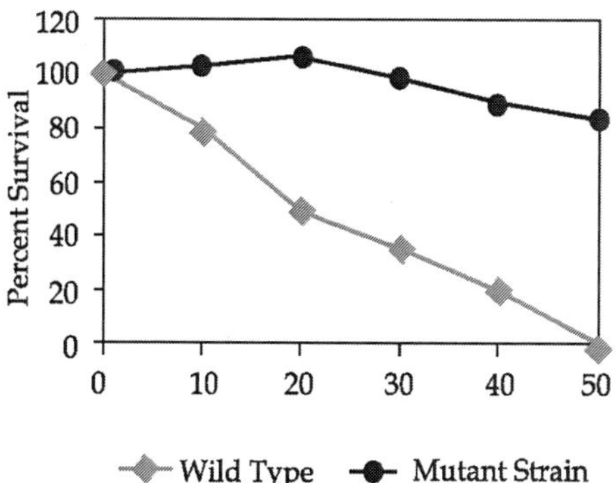

Fig. 15.1 Growth of wild type *Anabaena cylindrica* and its NaN^r_3 mutant counterpart (in terms of percent survival) in response to NaN_3

6.2 Ammonia Production

For assay of extracellular ammonia production, the cells of wild type *A. cylindrica* and its NaN_3 mutant counterpart were placed in 100 ml conical flasks containing 20 ml N_2-medium and were incubated in a culture room at 25°C and illuminated in light (with daylight fluorescent tubes having the photon fluence rate of 50 μmol $m^{-2}s^{-1}$). Samples were harvested at the end of 24 h and ammonia in the cell-free supernatants was estimated by hypochlorite method (Solorzano, 1969).

For assay of extracellular ammonia production, the cells of wild type *A. cylindrica* and its NaN_3 mutant counterpart were placed in 100 ml conical flasks containing 20 ml N_2-medium and were incubated in a culture room at 25°C and illuminated in light (with daylight fluorescent tubes having the photon fluence rate of 50 μmol m^{-2} s^{-1}). Samples were harvested at the end of 24 h and ammonia in the cell-free supernatants was estimated by hypochlorite method (Solorzano, 1969). The rate of ammonia production was expressed as μM ammonia produced mg^{-1} protein h^{-1}.

It is evident from the data of Table 15.1 that NaN_3 mutant strain of *A. cylindrica* produced nearly twofold higher ammonia as compared to its wild type counterpart in the absence of NaN_3. Interestingly, NaN_3 mutant strain of *A. cylindrica* produced about 100 times more ammonia than its wild type counterpart in the presence of NaN_3. In contrast, wild type *A. cylindrica* produced negligible amount of ammonia in response to NaN_3. The negligible ammonia production by wild type *A. cylindrica* in the presence of NaN_3 and very high ammonia production by NaN_3 mutant strain suggests that NaN_3 mutant strain of *A. cylindrica* does not have any inhibitory effect of NaN_3 on ammonia production mechanism. Since, NaN_3 mutant strain of *A. cylindrica* produced ammonia efficiently even in the presence of NaN_3 and can be used as a better biofertilizer. This higher production of ammonia, even in presence of NaN_3 in case of NaN^r_3 mutant strain, could be supported by results obtained from detection of nitrogenase and GS activity.

Table 15.1: Photoproduction of ammonia by wild type *Anabaena cylindrica* and its NaN_3^r mutant strain in response to NaN_3

Conditions	Ammonia production ($\mu M\ NH_4^+\ mg^{-1}$ protein h^{-1})
Wild type in the absence of NaN_3	96
Wild type in the presence of NaN_3	2.06
NaN_3^r Mutant strain in the absence of NaN_3	196
NaN_3^r Mutant strain in the presence of NaN_3	200

The data in each column are means of three independent experiments with four replicates each. The maximum variation from the mean value was less than 5 percent.

6.3 Nitrogenase Activity and Glutamine Synthetase (Transferase) Assay

Nitrogenase activity in whole cells was assayed by using acetylene reduction technique (Stewart et al., 1967). Glutamine synthetase (GS) transferase activity was assayed following the method of (Sampaio et al., 1979) by quantifying the amount of γ-glutamyl hydroxamate produced in the following transfer reaction catalyzed by GS:

$$\text{Glutamate} + NH_4^+ + \text{ATP} \xrightarrow{\text{Divalent cation}} \text{Glutamine} + \text{Pi} + \text{ADP}$$

$$\text{Glutamine} + \text{hydroxylamine} \xrightarrow[\text{ADP/Arsenate}]{\text{GS}} \gamma - \text{glutamyl hydroxamate} + NH_3$$

One unit of the enzyme in this reaction has been defined as the amount of GS required to catalyze the synthesis of 1 μmol γ-glutamyl hydroxamate min^{-1} under standard transfer assay conditions.

To study the effect of NaN_3 on nitrogenase activity (in terms of nmol C_2H_4 μg^{-1} chl*a* h^{-1}) of wild type *A. cylindrica* and its NaN_3 mutant counterpart, the enzyme activity in both the strains was assayed in response to various concentrations of NaN_3 (10, 20, 30, 40 and 50 μg ml^{-1}) and the data are presented in Figs. 15.1 and 15.2. It is evident from the data that nitrogenase activity of NaN_3 mutant strain was found to be higher than its wild type counterpart in the absence of NaN_3. Nitrogenase activity of NaN_3 mutant strain of *A. cylindrica* increased with increase in NaN_3 concentration up to 20 μg ml^{-1} Nitrogenase activity of wild type *A. cylindrica* increased slightly at 10 μg ml^{-1} of NaN_3; further increase in NaN_3 concentrations (20-50 μg ml^{-1}), however, resulted in a significant decrease in nitrogenase activity (Fig. 15.2). Nitrogenase activity of NaN_3^r mutant strain of *A. cylindrica* increased with increase in NaN_3 concentration up to 20 μg ml^{-1} (Fig. 15.3). However, further increment in NaN_3 concentrations has resulted in a significant decline in the nitrogenase activity of NaN_3^r mutant strain.

This shows the resistant nature of NaN_3^r mutant strain for NaN_3. The most likely reason for NaN_3 induced nitrogenase activity in NaN_3 resistant mutant strain appears to be due to enhanced availability of cellular nitrogen reserved for fresh protein synthesis and hence for nitrogenase as well as for growth. These results are in accordance with the previous studies (Peterson et al., 1976; Thomas et al., 1977; Peterson and Wolk 1978; Bagchi et al., 1986; Pandey and Singh 1984). Further, GS comparison in Figs. 15.4 and 15.5, shows that NaN_3 mutant strain having reduced GS activity produced ammonia in the external medium in the presence as well as absence of NaN_3. This suggests that higher nitrogenase activity accompanied by low GS activity in NaN_3 mutant strain could be the cause of liberated ammonia. However, wild-type strain having normal GS activity did not liberate ammonia. These results further suggest that metabolic linkage exists between GS (ammonia assimilation or release) and nitrogenase in NaN_3 mutant strain.

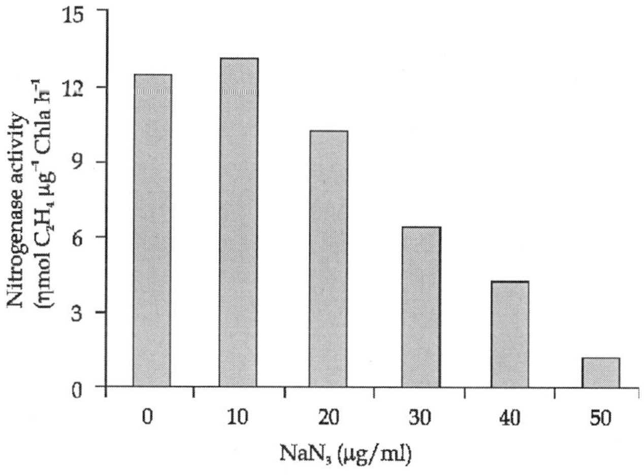

Fig. 15.2 Nitrogenase activity of wild type *Anabaena cylindrica* in response to various concentrations of NaN_3

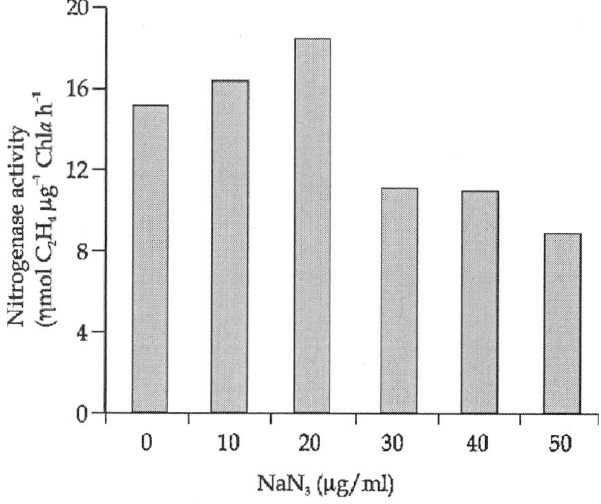

Fig. 15.3 Nitrogenase activity of NaN_3 mutant strain of *Anabaena cylindrica* in response to various concentrations of NaN_3

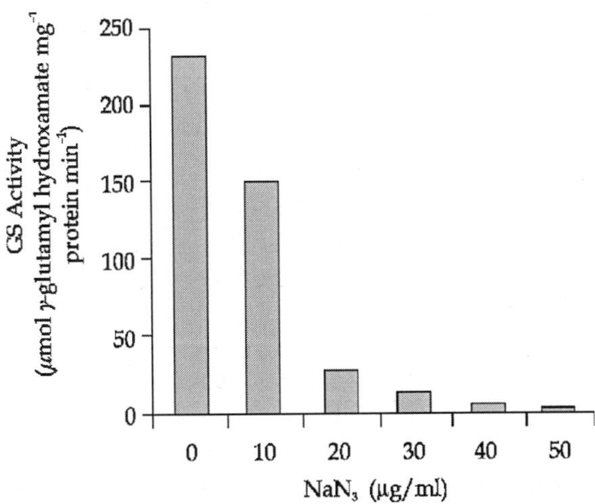

Fig. 15.4 Glutamine synthetase activity of wild type *Anabaena cylindrica* in response to various concentration of NaN_3

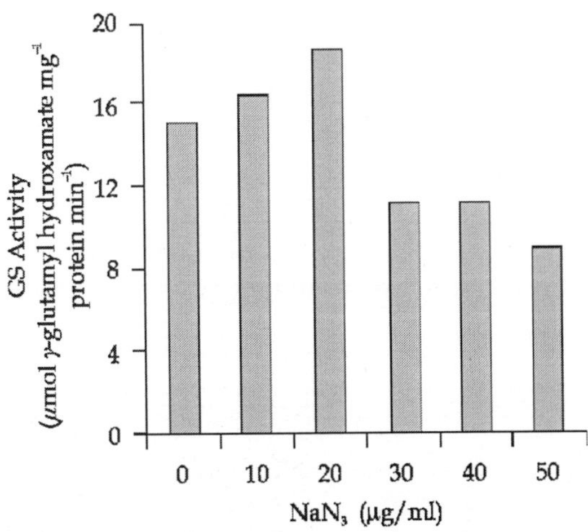

Fig. 15.5 Glutamine synthetase activity of NaN_3 mutant strain of *Anabaena cylindrica* in response to various concentration of NaN_3

6.4 SDS-PAGE

In order to see the effect of NaN_3 on protein profile of wild type *A. cylindrica* and its mutant counterpart, the cells were grown with carried out by using the gradient (12-25%) gel. Rest of the procedure was same as described (Laemmli, 1970).

It clearly indicates (Fig. 15.6) that in presence of 10 μg ml^{-1} NaN_3, a protein having molecular weight of 36 kDa was expressed and synthesized in the wild type *A. cylindrica* (Lane 3, Fig. 15.5). However, this disappeared with increase in NaN_3 concentration from 20 and 30 μg ml^{-1} (Lanes 4, 5; Fig. 15.5). Thus, NaN_3 induced protein profiling for wild type *A. cylindrica* suggests degradation of proteins in response to NaN_3. These results can be correlated with the results of percent survival, where the growth of wild type *A. cylindrica* reduced to half after 20 µg ml^{-1} NaN_3 concentrations. It implies that with an increase in NaN_3 concentration, the proteins of wild type *A. cylindrica* are degraded resulting in complete loss of vital proteins and ultimate death of the wild type *A. cylindrica* at 50 µg ml^{-1} NaN_3. NaN_3 induced protein profiling for NaN_3^r mutant strain of *A. cylindrica* suggests that in the absence of NaN_3, the NaN_3^r mutant strain expressed and synthesized the proteins having molecular weights of 18, 20, 42 and 50 kDa (Lane 6, Fig. 15.6). The exact functions of these proteins are yet to be explored, however; they may be developed after mutation in the wild type *A. cylindrica* following treatment with NTG. However, under NaN_3 stress of 10 µg ml^{-1}, some of the above-mentioned expressed and synthesized proteins disappeared (Lane 7, Fig. 15.6). Moreover, a protein of 36 kDa appeared with maximum intensity in the NaN_3^r mutant strain cultured with 20 µg ml^{-1} NaN_3 (Lane 8, Fig. 15.6). This can be correlated with the growth of NaN_3^r mutant counterpart of *A. cylindrica* exhibiting maximum growth at 20 µg ml^{-1} NaN_3. This further suggests that this protein is expressed and synthesized in order to provide resistance to the NaN_3^r mutant strain against NaN_3. The decrease in growth of NaN_3^r mutant strain of *A. cylindrica* after 20 µg ml^{-1} NaN_3 is also supported by NaN_3 induced protein profiling of NaN_3^r mutant strain showing disappearance of 36 kDa protein expression at this particular concentration of NaN_3 (Lane 9, Fig. 15.6). Thus, exposure of the cells to NaN_3 resulted in qualitative and quantitative regulation of individual proteins as these are the proteins required by the organism to tolerate the NaN_3 up to certain concentration and also to fix nitrogen even under such stressful condition.

A strain that continuously releases ammonia can be a better biofertilizer (Singh, 1992; Singh et al., 1995; Chauhan et al., 1999c). Thus, NaN_3 mutant strain of *A. cylindrica* isolated in this study shall be more useful in this respect. Persistence of GS activity in NaN_3 mutant strain even in the presence of NaN_3 suggests that GS system in NaN_3 mutant strain is more resistant to NaN_3 than that of its wild type counterpart. These results further suggest that although NaN_3 mutant strain is impaired in GS activity, yet, this impairment did not cause any inhibition in growth and survival of this strain (Singh et al., 1983; Spiller et al., 1986; Kerby et al., 1986; Singh, 1993). The synthesis of specific set of proteins in response to NaN_3 in case of NaN_3^r mutant strain of *A. cylindrica* proved its resistance to NaN_3 and also indicated that these proteins help in the maintenance of vital cellular functions in NaN_3 mutant strains up to certain concentration of NaN_3.

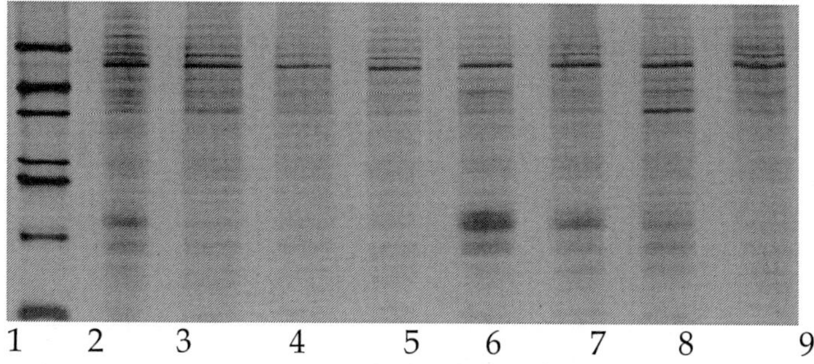

Fig. 15.6 Cellular protein synthesis in wild type *Anabaena cylindrica* and its NaN_3^r mutant counterpart in response to NaN_3

Lane 1: Marker proteins (Mol. wt.: 14, 20, 24, 29, 36, 45 and 66 kDa)
Lane 2: Wild type *A. cylindrica* in absence of NaN_3
Lane 3: Wild type *A. cylindrica* in presence of 10 μg ml^{-1} NaN_3
Lane 4: Wild type *A. cylindrica* in presence of 20 μg ml^{-1} NaN_3
Lane 5: Wild type *A. cylindrica* in presence of 30 μg ml^{-1} NaN_3
Lane 6: NaN_3^r mutant strain in absence of NaN_3
Lane 7: NaN_3^r mutant strain in presence of 10 μg $ml^{-1}$$NaN_3$
Lane 8: NaN_3^r mutant strain in presence of 20 μg $ml^{-1}$$NaN_3$
Lane 9: NaN_3^r mutant strain in presence of 30 μg $ml^{-1}$$NaN_3$

7. WHICH ONE IS BETTER: BIOFERTILIZER MUTANT OR WILD

Nitrogenase activity of mutant strain was found to be higher than its wild type counterpart in the absence of NaN_3. Nitrogenase activity of wild type strain decrease with increase in NaN_3 concentration (20 μg), whereas at this concentration nitrogenase activity of NaN_3 resistant mutant strain increases significantly. NR activities of wild type *A. cylindrica* and its mutant counterparts increased with time in the absence of NaN_3. NR activity of mutant strain of *A. cylindrica* stimulated significantly in the presence of NaN_3. NaN_3 induced impairment of nitrogen metabolism, primary reasons for impaired growth of wild type *A. cylindrica* in the presence of NaN_3. Metabolic linkage exists between GS (ammonia assimilation or release) and nitrogenase in mutant strain higher nitrogenase activity accompanied by low GS activity in mutant strain (cause of ammonia liberation and release). Mutant counterpart of *A. cylindrica* took up maximum ammonia both in the presence as well as in the absence of NaN_3. NaN_3 induced inhibition of nitrogen metabolizing enzymes is a reason for non-survival of wild type *A. cylindrica* (not assimilate nitrogen sources and hence could not grow). Increase in NaN_3 concentration with wild type *A. cylindrica* results in degradation of proteins and ultimate death at higher concentration. The synthesis of specific set of proteins in response to NaN_3 in case of mutant strain of *A. cylindrica* proved its resistance to NaN_3 and also indicated that these proteins help in the maintenance of vital cellular functions in mutant strains up to certain concentration.

8. CONCLUSION AND FUTURE PROSPECTS

Cyanobacteria are probably one of the most important groups of organisms. This is due to their global importance as they populate many environments. In Asia and the Far East, N_2-fixing cyanobacteria have been used as source of biofertilizer in rice-fields since early (De, 1939; Watanabe, 1956; Singh, 1961). However, for successful exploitation of the cyanobacterial biofertilizer, the basic data regarding their nitrogen metabolism are necessarily required. In the rice fields, the nitrogen input resulting from the cyanobacterial N_2-fixation is gauged by estimating the rate of N_2-fixation and heterocyst differentiation (Venkataraman, 1975). The response of cyanobacteria to NaN_3 stress will give insight into how and when azide can limit their growth *in situ*. This knowledge will greatly increase our understanding of how NaN_3 influences metabolic processes, such as carbon and nitrogen metabolism in cyanobacteria. The results obtained from this study would help in the better understanding of the process of nitrogen metabolism and the exploitation of cyanobacteria as biofertilizer.

References

Bagchi SN, Singh DT, and Singh HN (1986). Regulation by azide of heterocyst and nitrogenase in azide-resistant mutants of the cyanobacterium, *Nostoc muscorum.* New Phytol 102(1): 51–64.

Benemann JR (2003). Biofixation of CO_2 and greenhouse gas abatement with microalgae: Technology Roadmap. U.S. department of energy and steering committee of the international network on biofixation of CO_2 and greenhouse gas abatement with microalgae. Available at: www.ieaghg.org/docs/01roadmp.pdf.

Bergman B, Gallon JR, Rai AN, and Stal LJ (1997). N_2-fixation by non-heterocystous cyanobacteria. FEMS Microbiol Rev 19: 139–185.

Bryant DA (1994). The Molecular Biology of Cyanobacteria. Kluwer Academic Publishers. Dordrecht, The Netherlands.

Burlew D (1953). Algae culture: From laboratory to pilot plant. Carnegie Inst of Washington.

Chauhan VS, Singh B, Singh S, Bisen PS (2001). Regulation of sodium influx in the NaCl-resistant ($NaCl^r$) mutant strain of the cyanobacterium *Anabaena variabilis.* Curr Microbiol 42: 100–105.

Chauhan V S, Singh S, and Bisen P S (1999a). Development of NaCl resistant mutant strain of the diazotrophic cyanobacterium *Anabaena variabilis* having derepressed nitrogenase and impaired glutamine synthetase activities. Physiol Mol Biol Plants 5: 33–36.

Chauhan VS, Singh S, Pandey PK, and Bisen PS (1999b). Isolation and partial characterization of NaCl tolerant mutant strain of *Anabaena variabilis* with impaired glutamine synthetase activity. J Basic Microbiol 39: 219–226.

Chauhan VS, Singh S, Gour RK and Bisen PS (1999c). Immobilization caused physiological and biochemical changes in NaCl resistant ($NaCl^r$) mutant strain of diazotrophic cyanobacterium *Anabaena variabilis*. Indian J Exp Biol 32: 696–700.

De PK (1939). The role of blue-green algae in nitrogen fixation in rice fields. Proc Roy Soc London B 127: 121–139.

El-Habbasha SF, Hozayn M, and Khalafallah MA (2007). Integration effect betweenphosphorus levels and biofertilizers on quality and quantity yield of faba bean (*Vicia faba* L.) in newly cultivated sandy soils. Research J of Agri and Biolo Sci 3(6): 966–971.

Fleming H, and Haselkorn R (1973). Differentiation in *Nostoc muscorum*: nitrogenase issynthesized in heterocysts. Proc Natn Acad Sci USA. 70 : 2727–2731.

Gallon JR (1992). Reconciling the incompatible N_2- fixation and O_2. New Phytol 122: 571–609.

Häder DP, Worrest RC, and Kumar HD (1987). Aquatic ecosystems. UNCP environmental effects panel report, pp. 39–48.

Hardy RWF, and Knight EJR (1967). ATP dependent reduction of azide and HCN by N_2 fixing enzymes of *Azotobacter vinelandii* and *Clostridium pasteurianum* Biochimica Biochim Biophy Acta 139: 69–90.

Kerby NW, Rowell P, and Stewart WDP (1986). The uptake and metabolism of methylamine by N_2-fixing cyanobacteria. Arch Microbiol 143: 353–358.

Kumar A, Sinha RP, and Hader DP (1996). Effect of UV-B on enzyme of nitrogen metabolism in the cyanobacterium *Nostoc calcicola.* J Plant Physiol 148: 86–91.

Kumar H, and Kaushik D (1992). Standardization of algalization technology for amelioration of salt affected soils. In: Kaushik BD (ed.) Proc Natl symp on cyanobacterial nitrogen fixation., Associate Publishing Company, New Delhi, India 343–352.

Laemmli UK (1970). Cleavage of structural proteins during the assembly of the head of bacteriophage $T_{4.}$˙ Nature (London) 227: 680–685.

Li S, Ault A, Malone CL, Raitt D, Dean S, Johnston LH, Deschenes RJ, and Fassler JS (1998). The yeast histidine protein kinase, Sln 1p, mediates phosphotransfer to two response regulators, Ssklp and Skn7p. EMBO J 17: 6952–6962.

Merrick M, and Dixon R (1984). Why don't plants fix nitrogen? Trends Biotechnol 6: 162–166.

Mishra U, and Pabbi S (2004). Cyanobacteria: A potential biofertilizer for rice. Resonance, Centre for Conservation and Utilization of Blue-Green Algae, Indian Agricultural Research Institute, New Delhi, India.

Oswald WJ, and Golueke CG (1960). The biological conversion of solar energy. Adv Appl Microbiol 11: 223–242.

Pandey KD, and Singh PK (1984). Isolation and characterization of nitrate reductase mutants and regulation of nitrate reductase and nitrogenase in the cyanobacterium *Nostoc muscorum*. Mol Gen Genet 195 (1): 180–185.

Peterson RB, and Burris RH (1976). Properties of heterocysts isolated with colloidal silica. Arch Microbiol 108: 35–40.

Peterson RB, and Wolk CP (1978). High recovery of nitrogenase activity and of ^{55}Felabeled nitrogenase in heterocysts isolated from *Anabaena variabilis*. Proc Natl Acad Sci USA. 75 (12): 6271–6275.

Richards D (2006). Sodium azide based pesticides. 13th Annual International Research Conference on Methyl Bromide Alternatives and Emissions Reduction, American Pacific Corporation.

Rippka R (1988). Recognition and identification of cyanobacteria. In: Parker L, Glazer AN (eds) Methods of Enzymology. Academic Press, San Diego, Vol 167, pp. 28–67.

Rippka R, Deruelles J, Waterbury JB, Herdman M, and Stanier RY (1979). Generic assignments, strain histories and properties of pure cultures of cyanobacteria. J Gen Microbiol 111: 1–61.

Saadatnia H, and Riahi H (2009). Cyanobacteria from paddy-fields in Iran as a biofertilizer in rice plants. Plant Soil Environment 55(5): 207–212.

Sampaio MJM, Rowell P, and Stewart WDP (1979). Purification and some properties of glutamine synthetase from the nitrogen-fixing cyanobacteria *Anabaena cylindrica* and *Nostoc* sp. J Gen Microbiol 111: 181–191.

Schollhorn R, and Burris RH (1967). Reduction of azide by N_2-fixing enzyme. Proc Natl Acad Sci, USA 57: 1317–1323.

Schopf JW, and Walter MR (1982). Origin and evolution of cyanobacteria: the geological evidence. In: Carr NG, Whitton BA (eds) The Biology of Cyanobacteria, Blackwell Scientific Publications, Oxford, 543–564.

Singh B, Chauhan VS, Singh S, and Bisen PS (2001). Isolation and partialcharacterization of Het- Fix-mutant strain of the diazotrophic cyanobacterium *Anabaena variabilis* showing chromatic adaptation. Curr Microbiol 42: 321–327.

Singh HN, and Singh HN (1978). An azide-resistant mutant of the blue-green *alga Nostoc muscorum* producing heterocyst and nitrogenase in the presence of fixed inorganic nitrogen source. Arch Microbiol 119: 197–20l.

Singh HN, Singh RK, and Sharma R (1983b). An L-methionine -D,L-sulfoximine-resistant mutant of the cyanobacterium *Nostoc muscorum* showing inhibitor-resistant γ-glutamyltransferase, defective glutamine synthetase and producing extracellular ammonia during N_2 fixation. FEBS Letters 154 (1): 10–14.

Singh RN (1961). Role of Blue-green Algae in Nitrogen Economy of Indian Agriculture. Indian Council of Agricultural Research, New Delhi, India.

Singh S (1992). Photobiological production of ammonia by an L-mthionine-DLsulfoximne-resistant mutant of cyanobiont Nostoc ANTH. Indian J Exp Biol 30: 284–287.

Singh S (1993). Role of glutamine synthetase, glutamine and NH_4^+ in the regulation of NO2- uptake in the cyanobiont Nostoc ANTH. J Plant Physiol 142: 403–406.

Singh S, Singh BB, and Bisen PS (1995). Role of ammonium assimilation in urea inhibition of nitrogenase activity in cultured cyanobiont *Nostoc* ANTH. Indian J Exp Biol 33: 94–96.

Sinha RP, and Kumar A (1992). Screening of blue-green algae for biofertilizer. In: Proc National Seminar on Organic Farming, Pune, pp. 95–97.

Solorzano L (1969). Determination of ammonia in natural water by the phenol-hypochlorite method. Limnol Oceanogr 14: 799–801.

Song T, Martensson L, Eriksson T, Zheng W, and Rasmessen U (2005). Biodiversity and seasonal variation of the cyanobacterial assemblage in rice paddy field in Fujian, China. FEMS Microbiol Ecol 54: 131–140.

Spiller H, Latorre C, Hassan ME, and Shanmugam KT (1986). Isolation and characterization of nitrogenase, derepreesed mutant strain of cyanobacterium *Anabaena variabilis*. J Bacteriol 165: 412–419.

Stanier RY (1977). The position of the cyanobacteria in the world of phototrophs. Calsberg Res Commun 42: 77–98.

Stewart WDP (1973). Nitrogen fixation by photosynthetic microorganisms. Ann Rev Microbiol 27: 283- 318.

Stewart WDP (1980). Some aspects of structure and function in N_2-fixing cyanobacteria. Ann Rev Microbiol 34: 497–536.

Stewart WDP, Fitzgerald GP, and Burris RH (1967). *In situ* studies on N2-fixation using acetylene reduction technique. Proc Natl Acad Sci, USA 58: 2071–2078.

Tandeau de Marsac N, and Houmard J (1993). Adaptation of cyanobacteria to environmental stimuli: new steps towards molecular mechanisms. FEMS Microbiol Review 104: 119–190.

Thomas J, Meeks JC, Wolk CP, Shaffer PW, and Austin SM (1977). Formation of glutamine from [^{13}N]ammonia, [^{13}N]dinitrogen, and [^{14}C]glutamate by heterocysts isolated from *Anabaena cylindrica*. J Bacteriol 129 (3): 1545–1555.

Toroser D, Athwal GS and Huber SC (1997). Protein phosphorylation as a mechanism for osmotic stress activation of sucrose phosphate synthase in spinach leaves. Plant Physiol 114:947–955.

Toroser D, Athwal GS, and Huber SC (1998). Site specific regulatory interaction between spinach leaf sucrose phosphate synthase and 14-3-3 proteins. FEBS Lett 435:110–114.

Turner FT (1979). Soil nitrification retardation by rice pesticides. Soil Sci Soc Am J 43(5): 955–957.

Varela JCS and Mager WH (1996). Response of *Saccharomyses cerevisiae* to change in external osmolarity. Microbiology 142: 721–731.

Venkatraman GS (1975). The role of blue-green algae in tropical rice cultivation. In: Stewart WDP (ed) Nitrogen-Fixation by Free-Living Microorganisms. Cambridge Univ Press, London, pp. 207–218.

Vessey JK (2003.) Plant growth promoting rhizobacteria as biofertilizers. Plant Soil 255: 571–586.

Watanabe A (1956). The effect of atmospheric nitrogen-fixing blue-green algae on yield of rice. Bot Magazine 69: 530–535.

Wolk CP (1996). Heterocyst formation. Annu Rev Genet 30:59–78.

Yosefi K, Galavi M, Ramrodi M, and Mousavi SR (2011). Effect of bio-phosphateand chemical phosphorus fertilizer accompanied with micronutrient foliar application on growth, yield and yield components of maize (Single Cross 704). Aust J of Crop Sci 5(2) 175–180.

16

Reclamation of Saline Soil Using Vermicomposting for Sustainable Agriculture

Alok Bharadwaj[1*], Surendra Singh[2] and Ashok Agarwal[3]
[1]Department of Biotechnology, Microbiology & Immunology
GLA University, Mathura, Uttar Pradesh, India
[2]Department of Advanced Botany, Banaras Hindu University
Varanasi, Uttar Pradesh, India
[3]Department of Botany, B.S.A. (PG) College
Mathura, Uttar Pradesh, India
*Email: *alok.bhardwaj@gla.ac.in*

1. INTRODUCTION

Soil is the outermost covering of the earth's crust, which consists of loosely arranged layers of materials composed of inorganic and organic constituents in different stages of organization. It is a natural medium in which plants grow, multiply and die thus providing a perennial source of organic matter, which could be recycled for plant nutrition. It also provides the physical support for the anchorage of root system and also serves as a reservoir of air, water and nutrients, which are essentially required for growth of the plants. The portion of earth beneath the soil is known as bedrock and it does not contribute directly to the growth of plants (Subba Rao, 1986; Giri et al., 2005).

Reclamation and management of degraded soils have gained global attention in view of the increasing needs of growing population, rapid industrialization and deteriorating environmental quality. In north India, particularly in states like Uttar Pradesh, a large area of land (1.3 mha) is suffering from the problem of high salinity or alkalinity and is lying either unutilized or partially utilized. Such land constitutes potential land resources for meeting future needs. Intensive R&D efforts are, however, required to manage these soils for sustainable development, for prosperity of the society and for environmental conservation.

District Mathura, the chosen study site is having a vast area of land affected with salinity (nearly 5718 ha), as in major part of the area, the water used for irrigation and other purposes is saline in nature. Considering the potentiality of organic matter amendments in reclaiming and managing saline/usar soils, the present investigation has been undertaken to determine the efficacy of vermicompost in the reclamation of saline/usar soils, via altering the physico-chemical and microbiological characteristics of such soils. It has been stated that organic materials together with inorganic amendments are found to be more cost effective, hasten the reclamation process and increase the yield (Dargan et al., 1982). Therefore, their use should be encouraged.

2. MATERIALS AND METHODS

The present study includes survey of selected sites, soil sampling, characterization of chosen sites, influence of organic amendments in maintaining physico-chemical and microbiological properties of salt affected/usar soils. The details of sites, geographical and climatic information of the location along with experimental set-up are given below.

2.1 Survey of Selected Sites

Soil samples were collected from the six chosen sites of Raya village (Table 16.1) situated in Mant tehsil of Mathura district. The area is situated in the saline tract of Mathura district. Mathura district lies between 27° 14′ and 27° 58′ north latitude and 77° 17′ and 78° 12′ east longitude with an area of 3329.4 square kilometres. It is situated on Bombay-Delhi main Railway lines (WR, CR and NCR). It is linked with rest of the country with network of rails and roads. The district is bounded on the northwest by Gurgaon district of Haryana, northeast by Aligarh and for a short distance (13 km), in the east by Etah District.

Table 16.1: Names of villages from where soil samples collected

S. No.	Name of Villages	Soil Sampling Site
1	Deewana-Kalan	S1
2	Kherari	S2
3	Kumha	S3
4	Veerahana	S4
5	Sonai	S5
6	Pirasua	S6

2.2 Collection of Soil Samples

Soil samples were collected up to 0–6 cm depth from the surface with the help of sterilized iron borer, following the method of Johnson and Curl (1972), during the month of April 2005. Simultaneously, small amount (10 g approx) of soil from each site was transferred to small tin containers which were previously weighed to a constant weight, for determination of soil moisture. For the collection of soil samples, six villages of Raya block were chosen and each was numbered as below. Soil samples from each site were collected in triplicate and were mixed together to make a composite sample for detailed physico-chemical and microbiological studies.

2.3 Doses of Vermicompost Amendment and Preparation of Samples

To evaluate the role of organic enrichment on physico-chemical and microbiological attributes of saline alkali soils, vermicompost was used. Freshly collected, air-dried and sieved soil samples, weighed in equal amount, were taken in separate fresh polythene bags. To each polythene bag vermicompost was added separately in the ratio of 5, 10, 15, 20 and 25 percent (w/w), respectively.

After adding the respective doses, the samples were shaken and mixed thoroughly by hand manipulation and finally passed through the sieve for uniform mixing. Moisture status of the amended soil samples was maintained up to 60–70 percent water holding capacity by adding water. The samples were stored at room temperature. Detailed, analyses were made after 10 days of amendment. Natural soil samples without any amendment serving as control were also similarly maintained.

2.4 Physico-Chemical Analysis

The methods for analyses of various physico-chemical characteristics of soil are presented below:

2.4.1 *Soil Temperature*

Soil temperature was recorded at the time of its collection by using soil thermometer. The bulb of the thermometer was buried at a depth of 6 cm in the soil for about half an hour and the temperature were recorded when it became constant.

2.4.2 *Soil Moisture*

The moisture content of the soil was determined by using Oven Dry Method as described by Piper (1966).

2.4.3 *Water Holding Capacity*

Water holding capacity of various soil samples was determined following the method as suggested by Piper (1966).

2.4.4 *Soil pH*

Calomel electrode method was used for determining the soil pH (Piper, 1966). Soil was dispersed in the distilled water in a ratio of 1:5 (w/v) and was shaken vigorously for one hour on an electric shaker. The pH of the soil suspension was recorded by using pH meter.

2.4.5 *Soil Organic Carbon*

Soil organic carbon was estimated following the method of Piper (1966).

2.4.6 *Organic Matter*

Total soil organic matter was calculated by multiplying the organic carbon by a factor 1.724. This factor sometimes referred to as the 'van Beinmelon factor' and is based on the assumption that soil organic matter contains 58 percent carbon (Jackson, 1973).

2.4.7 *Total Nitrogen*

Total nitrogen was estimated following semi micro-Kjeldahl method as suggested by Jackson (1973) and as modified by Iswaran and Marwaha (1980).

2.4.8 *Carbon: Nitrogen (C/N) Ratio*

The carbon to nitrogen ratio was calculated by dividing the percent carbon content by the percent total nitrogen content.

2.4.9 *Electrical Conductivity*

Electrical conductivity is a measure of a water capacity to convey electric current. Electrical conductivity of water is directly proportional to its dissolved mineral matter content. Soil was dispersed in the distilled water in a ratio of 1:5 (w/v) and was shaken vigorously with the help of electric shaker. The conductivity of the soil suspension was recorded by using conductivity meter.

2.4.10 *Exchangeable Calcium and Magnesium*

Air dried soil (7.5 g) was taken in 500 ml beaker to which 40 per cent alcohol (15 ml) was added. This mixture was well shaken and kept for 15 min. The suspension was filtered through Whatman No. 50 filter paper and the soil was washed 4–5 times with 7.5 ml of alcohol 40 percent. Final washing was performed with 7.5 ml of absolute alcohol to dry the soil. The filter paper along with soil was scrapped in a 250 ml beaker and the funnel and filter paper were simultaneously washed with ammonium acetate solution for removing any adhered portion of the soil. The suspension was kept overnight. Finally, the suspension was filtered through Whatman No. 42 filter paper with additional ammonium acetate solution in pre-measured 50 ml volumetric flask. Now the filtrate (50 ml) was taken in a beaker and to this 2 ml NaOH and 200 mg murexide indicator were added. Thus, pink colour appeared. This solution was titrated against EDTA until pink colour changes to purple. The amount of calcium and magnesium were calculated by using the following formula:

$$\text{Ca}\ (\%) = \frac{A \times 400.8 \times V}{\text{v} \times 10000 \times \text{s}}$$

$$\text{Mg}\ (\%) = \frac{(\text{B}-\text{A}) \times 400.8 \times \text{V}}{\text{v} \times 1.645 \times 10000 \times \text{s}}$$

where

A = Volume of EDTA used for calcium determination

B = Volume of EDTA used for calcium + magnesium determination

V = Total volume of extract prepared soil

S = Weight of soil taken

v = Volume of soil extract titrated

2.4.11 *Exchangeable Sodium and Potassium*

Air dried soil (1 g) was taken in a conical flask, to which 25 ml of triacid solution (H_2SO_4: HNO_3: $HClO_4$ = 9:3:1) was added to make a suspension. The suspension was heated continuously on heater till it became transparent. Total volume of solution was made up to 100 ml with the help of distilled water. The solution was filtered through Whatman filter No. 50. The concentration of sodium and potassium in different soil samples were measured by using flame photometer. Flame photometer was calibrated by using standard solutions (10–1000 ppm) of NaCl and KCl prior to use.

2.5 Microbiological Analysis

2.5.1 *Isolation of Soil Mycoflora*

The mycoflora of different soil samples were studied following the soil plate method as described by Warcup (1950). For the isolation of soil mycoflora, Martin's rose Bengal agar medium (Martin, 1950) was used. The fungal isolates thus obtained were subcultured on potato dextrose agar (PDA) slants and the pure cultures were stored in refrigerator at 5±1°C.

2.5.2 *Isolation of Other Microflora*

2.5.2.1 *Actinomycetes*

Dilution method was employed for the isolation and determination of the actinomycetes population per gram of soil. Soil sample (5 g) was taken in 50 ml of sterilized distilled water in a conical flask and shaken on rotatory shaker for 15 min. The supernatant was successively diluted through a series of dilution up to 10^{-6}. One milliliter of each dilution was pipetted out into sterilized Petri dish, and about 20 ml of Thronton's agar medium prior to its solidification were mixed by gentle rotation of Petri dish.

Petri dishes were incubated at 26±1°C for the development of actinomycetes colonies. The colonies developed on Thronton's agar medium were counted after 96 and 168 hr of incubation and the population of actinomycetes per gm of the air dried soil were calculated.

2.5.2.2 *Bacteria*

Bacteria usually prefer a neutral or slightly alkaline medium and many will not grow on the media commonly used for fungi. Nutrient agar is suitable and recommended for the isolation of most kind of bacteria. The dilution method similar to actinomycetes was followed for the determination of bacterial population.

2.5.3 *Enumeration of Microflora*

Fungi, bacteria and actinomycetes were enumerated by following the procedures given by Aneja (1996) and Johnson and Curl (1972).

2.5.3.1 *Enumeration of mycoflora*

Population of soil microfungi was expressed as fungal population g^{-1} soil. It was calculated by finding out the amount of soil transferred to the Petri dish.

The percent frequency, abundance, density and importance value index (IVI) for each fungal species in various soil types were determined following the standard ecological methods outlined by Weaver and Clements (1938) and Tansley (1946).

$$\text{Density} = \frac{\text{Total number of individual of each fungal species}}{\text{Total number of Petri plates studied}}$$

$$\text{Percent frequency} = \frac{\text{Total number of Petri plates in which a species occurred}}{\text{Total number of Petri plates studied}} \times 100$$

$$\text{Abundance} = \frac{\text{Total number of individual of each fungal species}}{\text{Number of Petri plates occupied by that species.}}$$

From these characters, their relative value was computed as follows:

$$\text{Relative density} = \frac{\text{Density of a species}}{\text{Total density of all the species}} \times 100$$

$$\text{Relative frequency} = \frac{\text{Frequency of a species}}{\text{Total frequency of all the species}} \times 100$$

$$\text{Relative abundance} = \frac{\text{Abundance of a species}}{\text{Total abundance of all the species}} \times 100$$

IVI was calculated as follows:

$$\text{IVI} = \text{RD} + \text{RF} + \text{RA}$$

where

RD = Relative density

RF = Relative frequency

RA = Relative abundance

The fungal species with the highest IVI and the next in order was considered dominant and co-dominant members of the community, respectively.

2.5.3.2 *Enumeration of bacteria and actinomycetes*

Fungi, bacteria and actinomycetes were enumerated by following the procedures given by Aneja (1996) and Johnson and Curl (1972). The number of fungi, bacteria and actinomycetes was quantified by their colony forming units (C.F.U.), which can be calculated by using the following formula:

$$\text{Colony forming unit (C.F.U.)} = \frac{\text{Number of colonies x dilution factor}}{\text{Weight of soil (g)}}$$

2.6 Studies on Soil Fungistasis

2.6.1 *Selection of Test Organism*

For assaying the fungistatic activity of various soils, *Fusarium solani* and *Alternaria alternata* were selected as test organisms.

2.6.2 *Assay Methods of Soil Fungistasis*

The following methods were employed to evaluate the fungistatic behavior of unamended and vermicompost amended soil samples.

2.6.2.1 *Direct assay method*

Direct assay method as per the method described by Lingappa and Lockwood (1963).

2.6.2.2 *Agar disc method*

Agar disc method as per the method described by Jackson (1958).

2.6.2.3 *Soil emanation agar method*

Soil emanation agar method as per the method described by Hora and Baker (1970).

2.7 Collection of Vermicompost

Vermicompost, the excreta of earthworm possesses high quantity of different nutrients, such as nitrogen, potassium, phosphorus and C:N ratio, etc. In the present study, prepared vermicompost was collected from Shri Braj Gau Seva Trust, Vrindavan (Mathura) (Figs. 3.2–3.5). Such vermicompost has pH value of 6.5–7.4, organic carbon 27.43–30.3 percent, total nitrogen 1.80–2.0 percent, phosphorus 1.32 to 1.93 percent, potassium 1.28 to 1.50 percent, C:N ratio 14-15.1, calcium 3.0–4.5 percent, magnesium

0.4–0.7 percent and sodium 0.2–0.3 percent. In addition to this, such vermicompost also contains sufficient amount of bacteria, actinomycetes, fungi and growth hormones such as giberellins, auxins and cytokinins that enhance the fertility level of saline/usar soils.

3. RESULTS AND DISCUSSION

In the present study, vermicompost, an organic enricher, was used for the improvement of fertility status of saline soils. Reclamation of barren/alkali soils by addition of organic material has been widely reported. Organic amendments had profound effect on physical condition of the saline/usar soil. In India vermicomposting has been recognized as a promising tool for sustainable agriculture and land development. Vermicomposting is the biodegradation and instabilization of organic residues through the ingestion and egestion by earthworms. Vermicompost is considered to be the best out of organic wastes. On proper handing of the organic wastes, i.e., proper maintenance of certain physical and chemical properties of the organic wastes, it can be converted into vermicompost that acts as a conditioner for the soil health as well as a rich nutrient source for the crops.

In vermicompost unamended control soil, soil temperature was slightly higher (29.05°C), while it declined in the soil amended with vermicompost (28.41°C) (Table 16.2). The maximum decrease in soil temperature was recorded in the soils amended with 20 and 25 percent vermicompost. These results are supported by the findings of the earlier workers (Gautam, 2002; Singh, 2002). Gautam (2002) and Singh (2002) have reported that composted materials like rice husk compost, farm yard manure and other types could bring reduction in soil temperature following applications of various doses of amendments. Decrease in soil temperature following vermicompost amendments could be explained on the basis of the fact that vermicompost contains humus that reacts with problem soil particles coupled with endothermic reaction, resulting into the reduction in soil temperature.

Table 16.2: Physico-chemical characteristics of unamended and different vermicompost amended soils

S. No.	Physico-chemical Parameter	Vermicompost Amended Soil (%)	S_1	S_2	S_3	S_4	S_5	S_6	Average Mean	Pool Mean
1	Temperature	Unamended Soil	30.6	28.8	28.1	28.4	28.6	29.8	29.05	29.05
		5	28.4	28.1	28.4	28.6	28.8	29.4	28.62	
		10	28.6	28.4	28	28.1	28.4	29.6	28.52	
		15	28.1	29.2	28.4	28.4	28.4	29	28.58	28.41
		20	28	28.4	28.1	28	28.2	28.6	28.22	
		25	28.1	28	28	28	28	28.6	28.12	

2	Moisture	Unamended Soil	3.62	2.92	2.7	2.96	2.9	3.12	3.04	3.04
		5	3.84	2.98	3.69	3.22	3.08	3.18	3.33	
		10	4.29	3.4	4	3.68	3.42	3.57	3.73	
		15	4.24	3.66	4.87	3.98	3.84	3.84	4.07	3.96
		20	4.39	4.1	4.24	4.2	3.96	4.32	4.20	
		25	4.84	4.16	4.46	4.61	3.96	4.86	4.48	
3	Water holding Capacity	Unamended Soil	22.4	21.6	22.1	22.8	23.1	22.8	22.47	22.47
		5	23.8	21.8	22.6	23	23.6	23.6	23.07	
		10	24.1	22.6	23.2	23.4	23.6	23.3	23.37	
		15	25	23.2	23.6	23.8	23.8	23.8	23.87	23.85
		20	25.1	23.8	24.4	24	24.2	24.1	24.27	
		25	25.5	24.5	24.6	24.6	24.6	24.4	24.70	
4	pH	Unamended Soil	8.8	8.4	8.5	8.4	8.2	8.5	8.47	8.47
		5	8.6	8.2	8.4	8.3	8.2	8.5	8.37	8.14
		10	8.5	8.1	8.2	8.3	8	8.3	8.23	
		15	8.3	8.1	8.2	8.1	8.1	8.2	8.17	
		20	8.1	8	8	8	8	8.1	8.03	
		25	7.9	7.9	7.8	7.9	7.9	8	7.90	
5	Organic Carbon	Unamended Soil	0.03	0.015	0.02	0.04	0.02	0.01	0.02	0.02
		5	0.045	0.032	0.58	0.105	0.18	0.15	0.18	
		10	0.36	0.26	1.03	0.33	0.36	0.49	0.47	
		15	1.11	0.98	1.11	0.765	0.615	0.7	0.88	0.78
		20	1.29	1.32	1.12	0.945	0.945	0.99	1.10	
		25	1.33	1.36	1.2	1.14	1.26	1.18	1.25	
6	Organic Matter	Unamended Soil	0.051	0.025	0.344	0.068	0.034	0.017	0.09	0.09
		5	0.077	0.055	0.999	0.181	0.31	0.258	0.31	
		10	0.62	0.448	1.775	0.568	0.62	0.844	0.81	
		15	1.91	1.68	1.913	1.318	1.06	1.2	1.51	1.34
		20	2.22	2.27	1.93	1.629	1.629	1.7	1.90	
		25	2.29	2.34	2.068	1.965	2.172	2.034	2.14	
7	Total Nitrogen	Unamended Soil	0.003	0.0014	0.0192	0.004	0.0021	0.001	0.01	0.01
		5	0.004	0.0027	0.0499	0.01	0.017	0.0134	0.02	
		10	0.034	0.022	0.0887	0.031	0.033	0.044	0.04	
		15	0.111	0.084	0.099	0.0691	0.058	0.062	0.08	0.07
		20	0.114	0.113	0.099	0.082	0.082	0.089	0.10	
		25	0.118	0.119	0.103	0.099	0.112	0.101	0.11	

Contd...

S. No	Physico-chemical Parameter	Vermicompost Amended Soil (%)	S_1	S_2	S_3	S_4	S_5	S_6	Average Mean	Pool Mean
8	C:N ratio	Unamended Soil	10	10.71	10.41	10	9.52	10	10.11	10.11
		5	11.25	11.85	11.6	10.5	10.58	11.19	11.16	
		10	10.58	11.81	11.61	10.64	10.9	11.13	11.11	
		15	10	11.66	11.14	11.07	10.6	11.29	10.96	11.22
		20	11.31	11.68	11.21	11.52	11.52	11.12	11.39	
		25	11.27	11.42	11.6	11.51	11.25	11.68	11.46	
9	Electrical Conductivity (dsm^{-1})	Unamended Soil	0.86	0.71	1.08	1.16	0.76	1.18	0.96	0.96
		5	0.59	0.69	1.01	0.92	0.78	1.02	0.84	
		10	0.48	0.53	0.92	0.75	0.68	0.98	0.72	
		15	0.58	0.59	0.91	0.68	0.59	0.79	0.69	0.66
		20	0.39	0.42	0.78	0.64	0.46	0.61	0.55	
		25	0.33	0.39	0.79	0.61	0.39	0.59	0.52	
10	Exchangeable Calcium (%)	Unamended Soil	0.098	0.084	0.089	0.084	0.105	0.089	0.09	0.09
		5	0.137	0.113	0.107	0.101	0.129	0.104	0.12	
		10	0.162	0.128	0.121	0.125	0.140	0.129	0.13	
		15	0.143	0.152	0.126	0.173	0.145	0.134	0.15	0.15
		20	0.176	0.251	0.141	0.174	0.160	0.143	0.17	
		25	0.185	0.240	0.197	0.211	0.171	0.157	0.19	
11	Exchangeable Potassium (mgl^{-1})	Unamended Soil	5.4	8	5.2	6.1	5	6.6	6.05	6.05
		5	5.9	8.4	5.9	6.8	5.4	7.12	6.59	
		10	6.8	9.6	6.6	7.4	6.8	10.62	7.97	
		15	8	10.12	7.21	11.24	7.86	11.16	9.27	9.54
		20	9.12	12.64	9.16	14.1	9.18	13.26	11.24	
		25	9.21	12.8	10.41	14.62	14	14.65	12.62	
12	Exchangeable Sodium (mgl^{-1})	Unamended Soil	27.6	32.1	22.4	26.8	21.4	26.4	26.12	26.12
		5	24.9	30.8	20.2	26	21	24.1	24.50	
		10	34.1	26.1	22.1	26.4	20.4	22.2	25.22	
		15	13.1	19.8	20	21.8	18.1	22.2	19.17	19.68
		20	11.7	15.4	16.5	17.6	16.8	16.4	15.73	
		25	9.2	14.1	14.2	16.8	14.2	14.2	13.78	

13	Exchangeable Magnesium (%)	Unamended Soil	0.03	0.016	0.0087	0.016	0.008	0.01	0.01	0.01
		5	0.028	0.023	0.012	0.018	0.01	0.01	0.02	
		10	0.039	0.031	0.019	0.024	0.019	0.013	0.02	
		15	0.039	0.045	0.025	0.14	0.029	0.016	0.05	0.04
		20	0.046	0.061	0.058	0.038	0.023	0.023	0.04	
		25	0.048	0.046	0.07	0.057	0.031	0.052	0.05	

Soil moisture in vermicompost amended soils showed marginal increase from 3.33 to 4.48 percent over the soil moisture (3.04 percent) of vermicompost unamended control soil. The moisture efficiency of the soil to a large extent depends on soil texture, organic content and characters of colloids (Shukla and Misra, 1993). Water holding capacity also improved from 23.07 to 24.70 percent in the soil amended with vermicompost over the water holding capacity (22.47%) of vermicompost unamended control soil. Vermicompost at its dose of 25 percent showed greater efficiency in improving the soil physical properties irrespective of its lower doses. Goel (2007) has also reported 61.70 and 6.09 percent gain in water holding capacity and moisture content of sodic soil, respectively after addition of vermicompost. Ohi and Ebo (1995) while studying the effect of organic and inorganic amendments on soil physical properties observed that the addition of poultry manure (10 t ha^{-1}), significantly increased the soil organic matter content, total porosity and hydraulic conductivity.

Enrichment of soil with vermicompost amendments reduced the soil pH as compared to vermicompost unamended control saline/usar soil. The average soil pH of vermicompost unamended control soil was 8.47, which was reduced by 3.5 to 4.5 percent after application of different doses of vermicompost. Maximum decrease in soil pH (7.90) was recorded in the soils amended with 25 percent vermicompost. Singh (2006) has also reported a significant reduction in soil pH by using different doses of composted material in saline/usar soils. Gautam (2002) while applying different doses of organic enrichers in saline/alkali soils has also observed similar results. The application of organic matter including vermicompost showed significant reduction in pH of saline/usar soils (Maheswarappa et al., 1999). Sharma et al. (2004) have suggested that soil pH is the key factor influencing the availability of micronutrients like Fe and Mn in the soil.

Decrease in electrical conductivity showed definite pattern in the soils amended with various doses of vermicompost. The decline in electrical conductivity was found to be dose specific as it reduced gradually with the increasing doses of vermicompost. The electrical conductivity was 0.96 dsm^{-1} in vermicompost unamended control soil, however, it decreased from 0.84 to 0.52 dsm^{-1} following application of different doses of vermicompost. These findings are also supported by earlier works (Ajit and Hooda, 2002; Warman and AngLopez, 2002; Elvira et al., 1998).

Improvement in nutritional properties of saline/usar soils was recorded following amendment of the soils with different doses of vermicompost. Organic carbon content was low (0.02%) in vermicompost unamended control soil, however, its value increased to 1.25

percent in the soil amended with 25 percent vermicompost, followed by its subsequent doses. This indicated gain 20.65 percent in organic carbon content in the soil amended with 25 percent dose of vermicompost. However, Goel (2007) recorded gain 26.51 percent in organic carbon content in sodic soils amended with vermicompost. Badale and More (2000) reported that addition of organic sources, such as vermicompost improved the chemical fertility of soil. Application of 75 percent dose level of vermicompost recorded maximum organic carbon. Katyal (1993) has also reported that application of organic manures improves the soil structure and chemical properties, particularly its nutrient status as carbon and nitrogen content, and thus prevents soil degradation. Singh et al., (2000) have reported manifold increase in the status of organic carbon (12 times), nitrogen (3 times) and calcium (2 times) during development of social forestry plantation in sodic wastelands, indicating a good degree of reclamation. Joshi et al. (1990) have reported the effect of green manuring on the improvement of physical conditions of the soil under rice-wheat cropping system.

Enrichment of soil with different organic amendments has a positive effect on nitrogen content. In control soil without any amendment, nitrogen content was low (0.01%) and it reached to its maximum level (0.11%) in the soil amended with 25 percent vermicompost. The increase in nitrogen content was related to the increase in dose of vermicompost amendment. Increment in nitrogen content in the soils amended with 25 percent vermicompost was nearly 11 times higher as compared to vermicompost unamended control soils. Balaji (1994) and Venkatesh (1995) have reported a significant increase in total nitrogen content by the application of 75 percent vermicompost with *Azospirillum* in problem soils. Bharadwaj et al. (2007) while adding vermicompost in the problem soils have also recorded about increase (15.47 percent) in total nitrogen content.

Amendments of soils with various doses of vermicompost resulted in marginal increase in C:N ratio from 11.16 to 11.46. The C:N ratio in vermicompost unamended control soil was 10.11. Bharadwaj et al. (2007) while comparing the effect of compost and vermicompost in saline/usar soils have reported that vermicompost had far better reclamating ability of problem soils as compared to compost. Vermicompost enrichment of the soils also resulted in 48.71 and 176.23 per cent increase in C:N ratio and organic matter, respectively.

Organic matter content of vermicompost unamended control soil was 0.09 percent, however, its levels increased significantly from 0.31 to 2.14 percent in the soils amended with different doses of vermicompost. Masciandaro et al. (2000b) also showed improvement in soil structure by increasing porosity and reducing bulk density of the soil amended with composted biological sludge. Since, a composted municipal solid waste contains more than 50 percent organic matter, its application had positive impact on soil organic matter. Kaviraj and Sharma (2003) have reported that the organic manure enrichment resulted in improvement in the values of moisture content, water holding capacity, organic carbon, total nitrogen organic matter in problem soils.

Exchangeable sodium content in vermicompost unamended control soil was 26.12 mgl^{-1}, however, its levels decreased from 24.50 to 13.78 mgl^{-1} in the soils amended with different doses of vermicompost. These results indicated 6.2 to 47.24 percent reduction in sodium content in the soils amended with different doses of vermicompost. Ansari (2002) has also reported that application of organic amendments decreases the soil pH, and effectively helps in reducing the soil salinity through replacement of the Na^+ by Ca^{++} in saline/sodic soils. Shukla and Mishra (1993) have reported the reduction in exchangeable sodium content under different vegetative covers. Whereas sodium is known for its dispersing action and compacting the soil particles, calcium, however, promotes flocculation (Somani, 1981; Ram and Prasad, 1982). Singh et al., (2004) have reported that among exchangeable cations, a reduction in Na^+ and elevation in Ca^{++}, Mg^{++} and K^+ concentration are indicative of good soil amelioration in afforested sodic soils. These results also support the contention of the findings in the present study.

Bhargava and Abrol (1978) while working with sodic soils treatment reported a high degree of relationship between exchangeable sodium and pH of the soil. In the present study, similar findings were also recorded with the increasing doses of vermicompost amendments. The values of exchangeable sodium decreased along with soil pH.

Exchangeable calcium content in vermicompost unamended control soil was 0.09 percent, however, its level increased (0.12 to 0.19 percent) in the soils amended with various doses of vermicompost. The increase in calcium content ranged from 33.3 to 111 percent in the soils amended with different doses of vermicompost. Robbins (1986) suggested that respiring roots increase the CO_2 concentration in the soil atmosphere, which on dissolution in water produces carbonic acid. Consequently, this acid would increase the calcium carbonate minerals by lowering the pH and dissolving the lime minerals and forming a host of complex calcium ion pairs, thus increasing the soil solution Ca^{++} concentration. Exchangeable calcium was also reported to increase in sodic/saline soils after organic amendments (Goel, 2007; Singh, 2006; Gautam, 2002).

Exchangeable potassium content in vermicompost unamended control soil was 6.05 mgl^{-1}, however, its levels increased from 6.59 to 12.62 mgl^{-1} in the soil amended with various doses of vermicompost. The increase in exchangeable potassium content ranged from 8.92 to 108.59 percent in the soils amended with different doses of vermicompost enrichment. Goel (2007) also suggested greater efficacy of vermicompost enrichment as compared to other soil enricher. Albanell et al. (1988) have also reported the increase in phosphorus and potassium in the soil due to chemical changes during vermicomposting of sheep manure mixed with cotton industrial wastes.

Amendments of the soils with various doses of vermicompost resulted in the improvement of exchangeable magnesium (0.01%) over the soil unamended with vermicompost. Maximum increase in exchangeable magnesium (0.05 percent) was recorded in the soil amended with 25 per cent vermicompost. However, the soil amended with 5 percent vermicompost supported minimum level of exchangeable magnesium (0.02 percent). The increase in exchangeable magnesium content was 100 to 400 percent in the soils amended with different doses of vermicompost over the vermicompost unamended control soils. Garg and Singh (2003) reported a significant reclamation in barren saline/sodic soils under energy plantation established for six years due to increase in magnesium, organic carbon, total nitrogen, exchangeable calcium contents and reduction in soil pH and bulk density.

3.1 Microbiological Analysis of Soil

The data in Table 16.3 and 16.4 show that a total of 43 fungal species were isolated from the vermicompost unamended control soil and from the soil amended with 5, 10, 15, 20 and vermicompost (25 percent). Maximum number of fungal species (23 species) were, however, recorded in vermicompost unamended control soil representing 53.48 percent of the total fungal spectrum. Twenty-two species representing 51.16 percent of the total fungal spectrum were recorded in the soil amended with vermicompost (5 percent). The soils amended with 10 and 15 percent vermicompost supported 19 species each, representing 44.18 percent of the total fungal spectrum. However, 18 and 17 species were isolated from the soils amended with 20 and 25 percent vermicompost, respectively; the percent fungal spectrum was found to be 41.86 percent and 39.53 percent, respectively. The abundance of *Aspergilli* in tropical areas has already been reported (Peyronel, 1956; Waksman, 1917; Gautam, 2002; Singh, 2004, 2006). Present investigation also reveals the dominance of *Aspergilli* in the selected soils. The poor growth of the members of mucorales in the selected soils indicates that the environment in these areas is deserted (Garrett, 1963; Gautam, 2002).

Table 16.3: IVI of Fungal species isolated from the soils amended with control soil, 5 percent and 10 percent doses of vermicompost

S. No.	Name of species	Control soil				5%				10%			
		F	D	A	IVI	F	D	A	IVI	F	D	A	IVI
1	*Absidia butleri*					100	1.33	1.33	14.82				
2	*Acremonium* spp.	66.6	0.66	1	11.17								
3	*Alternaria alternata*	100	1.33	1.33	17.81					100	1	1	15.41
4	*Aspergillus fumigatus*	66.6	0.66	1	11.17	100	1.33	1.33	14.82				
5	*Aspergillus flavus*	100	2.66	2.66	29.37					100	1	1	15.41
6	*Aspergillus glaucus*					100	1.66	1.66	17.04				
7	*Aspergillus niger*$_1$	100	1.66	1.66	20.68	100	4	4	32.8	100	2	2	24.3
8	*Aspergillus niger*$_2$	100	1.33	1.33	17.81	100	3.33	3.33	28.28	100	1	1	15.41
9	*Aspergillus nidulans*$_1$	66.6	1	1.5	14.73	66.6	1.33	2	14.87	100	2	2	24.3
10	*Aspergillus nidulans*$_2$					66.6	1	1.5	12.14	100	1	1	15.41
11	*Aspergillus terreus*					66.6	1	1.5	12.14				
12	*Botrytis cinerea*					66.6	1	1.5	12.14	66.6	1	1.5	15.22
13	*Curvularia geniculate*	66.6	0.66	1	11.17					66.6	0.66	1	11.55
14	*Curvularia lunata*					66.6	0.66	1	9.37				
15	*Cladosporium herbarum*					100	1.33	1.33	14.82				

16	*Chaetomium globosum*					100	1.66	1.66	17.04				
17	*Chaetomium erraticum*									66.6	0.66	1	11.55
18	*Fusarium chlamydosporum*					33.3	0.33	1	6.18				
19	*Fusarium solani*	33.3	0.33	1	7.46					66.6	1	1.5	15.22
20	*Fusarium oxysporum*	100	1	1	14.94					66.6	1	1.5	15.22
21	*Fusarium moniliforme*									66.6	1	1.5	15.22
22	*Fusarium globosum*	66.6	1	1.5	14.73								
23	*Humicola fuscoatra*					100	1	1	12.6				
24	*Myrothecium verrucaria*												
25	*Mucor hiemalis*	100	1.33	1.33	17.81					66.6	1	1.5	15.22
26	*Mucor mucedo*									66.6	1	1.5	15.22
27	*Penicillium chrysogenum*	66.6	0.66	1	11.17	66.6	1	1.5	12.14	100	1.33	1.33	18.34
28	*Penicillium funiculosum*	66.6	0.66	1	11.17								
29	*Penicillium expensum*					66.6	0.66	1	9.37	66.6	1	1.5	15.22
30	*Penicillium citrinum*					100	1	1	12.6				
31	*Penicillium digitatum*												
32	*Phialophora* spp.	33.3	0.33	1	7.46								
33	*Rhizopus nigricans*	33.3	0.33	1	7.46	66.6	0.66	1	9.37				
34	*Rhizopus stolonifera*	66.6	0.66	1	11.17								
35	*Rhizopus oryzae*	33.3	0.33	1	7.46								
36	*Syncephalastrum recemosum*					66.6	0.66	1	9.37				
37	*Stachybotrys atra*	66.6	0.66	1	11.17								
38	*Trichoderma viride*					66.6	1	1.5	12.14				
39	*Trichoderma glaucum*												
40	*Mycellia sterilia*$_1$	66.6	0.66	1	11.17					100	1.33	1.33	18.34
41	*Mycellia sterilia*$_2$	100	1.33	1.33	17.81					100	1	1	16.72
42	*Unidentified*$_1$	33.3	0.33	1	7.46	66.6	0.66	1	9.37	33.3	0.33	1	7.76
43	*Unidentified*$_2$	66.6	0.66	1	11.17	33.3	0.33	1	6.18				

Table 16.4: IVI of Fungal species isolated from the soils amended with 15 percent, 20 percent and 25 percent doses of vermicompost

S. No.	Name of species	15%				20%				25%			
		F	D	A	IVI	F	D	A	IVI	F	D	A	IVI
1	*Absidia butleri*												
2	*Acremonium* spp.	66.6	0.66	1	10.2					33.3	0.33	1	9.07
3	*Alternaria alternate*					66.6	0.66	1	11.75	100	1.33	1.33	21.01

Contd...

4	*Aspergillus fumigatus*	100	3.33	3.33	30.77					100	1.33	1.33	21.01
5	*Aspergillus flavus*					100	1.33	1.33	18.56	100	1.33	1.33	21.01
6	*Aspergillus glaucus*	100	3.33	3.33	30.77	66.6	1	1.5	15.3				
7	*Aspergillus niger* 1	100	3.33	3.33	30.77	100	2.33	2.33	27.16	100	1.66	1.66	24.49
8	*Aspergillus niger* 2	100	2.33	2.33	23.45	100	1.33	1.33	18.56	100	1.33	1.33	21.01
9	*Aspergillus nidulans* 1	100	1.33	1.33	16.12	66.6	0.66	1	11.75				
10	*Aspergillus nidulans* 2	100	1.33	1.33	16.12	66.6	0.66	1	11.75				
11	*Aspergillus terreus*					100	1.33	1.33	18.56	100	1.66	1.66	24.49
12	*Botrytis cinerea*												
13	*Curvularia geniculate*	100	1	1	13.7					66.6	1	1.5	17.63
14	*Curvularia lunata*									100	1	1	17.52
15	*Cladosporium herbarum*												
16	*Chaetomium globosum*	100	1.33	1.33	16.12								
17	*Chaetomium erraticum*												
18	*Fusarium chlamydosporum*												
19	*Fusarium solani*	33.3	0.33	1	6.75	66.6	1	1.5	15.3				
20	*Fusarium oxysporum*	66.6	0.66	1	10.2	66.6	1	1.5	15.3				
21	*Fusarium moniliforme*	100	1.33	1.33	16.12	66.6	1	1.5	15.3				
22	*Fusarium globosum*												
23	*Humicola fuscoatra*	66.6	1	1.5	13.23								
24	*Myrothecium verrucaria*									100	1	1	17.52
25	*Mucor hiemalis*					66.6	1	1.5	15.3	100	1	1	17.52
26	*Mucor mucedo*					66.6	0.66	1	11.75				
27	*Penicillium chrysogenum*	66.6	0.66	1	10.2								
28	*Penicillium funiculosum*	66.6	0.66	1	10.2								
29	*Penicillium expensum*												
30	*Penicillium citrinum*	33.3	0.33	1	6.75								
31	*Penicillium digitatum*					66.6	1	1.5	15.3				
32	*Phialophora* sp.												
33	*Rhizopus nigricans*									66.6	0.66	1	13.26
34	*Rhizopus stolonifera*	100	1	1	13.7					66.6	0.66	1	13.26

35	*Rhizopus oryzae*												
36	*Syncephalastrum recemosum*	66.6	0.66	1	10.2								
37	*Stachybotrys atra*												
38	*Trichoderma viride*												
39	*Trichoderma glaucum*					100	1.33	1.33	18.64				
40	*Mycellia sterilia$_1$*	66.6	0.66	1	10.2	66.6	1.66	2.5	22.3	100	1	1	17.52
41	*Mycellia sterilia$_2$*					66.6	1	1.5	15.3	100	1.33	1.33	21.01
42	*Unidentified$_1$*					100	1.66	1.66	21.49	33.3	0.33	1	9.07
43	*Unidentified$_2$*									66.6	0.66	1	13.26

Out of 43 species isolated from various soils, 2 species (*A. niger$_1$* and *A. niger$_2$*) were found to be common in vermicompost-unamended control as well as in vermicompost amended soils. The wider ecological amplitude shown by these fungal species due to their repeated isolation during vermicompost amendments tends us to believe them as autochthonous species. Garrett (1956) has proposed some of the criteria or operational procedure as a basis for considering a microorganism native to a particular habitat. These include: (i) the repeated isolation of the species from samples of the environment, (ii) its existence at a high population density; and (iii) its ability to tolerate environmental extremes typifying that site.

The microenvironment of the soil is determined by both biotic and abiotic parameters of environment. Greater variation in temperature, moisture, relative humidity and the vegetation cover during diverse climatic conditions are supposed to give rise to a specificity to the soil microenvironment and hence to their fungal flora. Therefore, *A. alternata, F. oxysporum, Phialophora* species, *C. geniculata* were only isolated from the soil sample collected from natural undisturbed area. *F. chlamydosporum* was isolated from the soil amended with 5 percent vermicompost. However, *B. cinerea* was present in the soils amended with 5 and 10 percent vermicompost.

Mucor hiemalis, M. mucedo, F. moniliforme, Trichoderma glaucum were only isolated from the soil amended with 20 percent vermicompost. *R. nigricans* was isolated from vermicompost unamended control soil and from the soil amended with 5 and 25 percent vermicompost. *Stachybotrys atra* and *F. globosum* were isolated only from control/ undisturbed soil. Whereas, *A. alternata* was present in control/undisturbed soil as well as in the soils amended with 10, 20 and 25 percent vermicompost. *Myrothecium verrucaria* was isolated only from the soil amended with 25 percent vermicompost. *T. viride* and *Trichoderma glaucum* were, however, isolated from the soil amended with 5 and 20 percent vermicompost, respectively. *M. sterilia$_1$*, was present in vermicompost unamended control as well as in the soils amended with all the doses of vermicompost except the soil amended with 5 percent vermicompost. The restricted distribution of these fungal species indicates that they can act as valuable indicators in certain specific environment (Cohen, 1949; Saksena, 1955; Tresner et al., 1954).

Tresner et al. (1954); Mallik and Rice (1966); Agarwal (1983) and Singh (2002) have reported that during successional sequence, the fungal population and the members of species increased from pioneer to climax community of soil fungi in the same way as the succession of the vegetation. Present investigation revealed that the fungal community composition vary during different doses of vermicompost amendments. Vermicompost unamended control soil was dominated by *A. flavus, A. niger*$_{1,}$ *A. niger*$_2$ and *A. nidulans*$_1$ with IVI values of 29.37, 20.68, 17.81 and 14.73, respectively. Dominance of *Aspergillus* was also reported by Saksena et al. (1966); Gautam (2002) and Singh (2006).

In the soils amended with vermicompost (5 per cent), the dominant microfungal members were *A. niger*$_1$, *A. niger*$_2$, *A. glaucus* and *A. nidulans*$_1$ having their IVI values of 32.8, 28.28, 17.04 and 14.87, respectively. The soil amended with 10 percent vermicompost was dominated by *A. niger*$_1$, *A. nidulans*$_1$, *A. flavus* and *P. chrysogenum* with their IVI values of 24.30, 24.30, 15.41 and 18.34, respectively.

Whereas, *A. fumigatus A. glaucus, A. niger*$_1$, and *A. niger*$_2$ having IVI values of 30.77, 30.77), 30.77 and 23.45, respectively were dominant in the soil amended with 15 percent vermicompost. The soil amended with vermicompost (20 percent) was dominated by *A. niger*$_1$, *M. sterilia*$_1$, and unidentified$_1$ fungal species having their IVI values of 27.16, 22.30 and 21.49, respectively.

In vermicompost (25 percent) amended soil, *A. niger*$_1$ (IVI = 24.49), *A. terreus* (IVI = 24.49), *A. alternata, A. fumigatus, A. flavus* and *A. terreus* (IVI = 21.10) were found to be the dominant fungi. These findings are in accordance with the finding of Alexander (1971), who assigned the soil microenvironment in itself to be a special microcosm possessing a characteristic microbial community made up of population coexisting and interacting with each other.

Aspergillus, Fusarium, Mucor, Cladosporium and *Trichoderma* were dominantly recorded microfungi in vermicompost. This clearly indicates that the fungal diversity is more in vermicompost (Subler et al., 1998). The digestive epithelium of the simple straight tubular gut of worms is known to secrete the enzymes cellulase, amylase, invertase, protease, phosphatese (Ranganathan and Vinotha, 1998). Earthworms inevitably consume the soil microbes during the ingestion of litter and soil. It has been reported that earthworms necessarily have to feed on microbes, particularly fungi to meet their protein/nitrogen requirement (Ranganathan and Parthasarathi, 2000). Earthworms also play a major role in affecting populations of soil organisms, especially causing changes in the soil microbial community (Coleman, 1985).

It is evident from the data of Tables 16.3 and 16.4 that *A. niger*$_1$ exhibited higher IVI values during all the experiments with different doses of vermicompost amendment. Its isolation from the soil showed a definite trend with respect to its dominance during various type of vermicompost amendments. In vermicompost unamended control soil, its IVI value (20.68) was lower than *A. flavus* (29.37) and it was not dominant microfungus in control or undisturbed saline/usar soil. *A. niger*$_1$ with an IVI value of 32.8 was dominant in vermicompost (5 percent) amended soil. *A. niger*$_1$ along with *A. nidulans*$_1$, having IVI values of 24.30 was also dominant in the soil amended with vermicompost (10 percent). *A. niger*$_1$ along with *A. fumigatus* and *A. glaucus* having IVI values of 30.77 was also found to be dominant in the soil amended with 15 percent vermicompost. *A. niger*$_1$ was also dominant in the soils amended with 20 and 25 percent vermicompost with IVI values of 27.16 and 24.49, respectively.

The presence of *A. niger*$_1$ as the dominant community member in 10, 15, 20 and vermicompost (25 percent) amended soil and co-dominant member in the soil amended with 5 percent vermicompost suggests that the serial succession represents a cyclic series of events with a large system. In the succession (Trensar et al., 1954; Singh, 2002; Singh, 2006) that is analogous with higher plants, fungi do not form discrete communities, but form a series of progressively changing species combinations along with the sequence of succession.

3.2 Microbial Population in Vermicompost Amended Soil

The data in Table 16.5 shows the population dynamics of fungi, bacteria and actinomycetes in saline/usar soil amended with different doses of vermicompost. Effectiveness of vermicompost amendments increased with decreasing level of exchangeable sodium reflecting the increase in microbial activity.

Table 16.5: Population dynamics of fungi, bacteria and actinomycetes in saline soil amended with varying doses of vermicompost

S. No.	Doses of vermicompost amendment (%)	Microbial Population × 10^3/g soil		
		Fungi	Bacteria	Actinomycetes
1	Unamended control soil	3.1	4.9	3.9
2	5	3.1	5	4
3	10	3.4	5.1	4.2
4	15	3.4	5.2	4.3
5	20	3.6	5.3	4.4
6	25	3.9	5.5	4.5

The population of fungi in natural undisturbed soil was found to be 3.1×10^3 g^{-1} soil. The fungal population, however, varied from 3.1×10^3 to 3.9×10^3 g^{-1} soil in the soils amended with 5 and 25 percent vermicompost, respectively. A significant increase in fungal population was recorded in response to the increment in the doses of vermicompost amendment. These results are supported by the findings of earlier workers (Gautam 2002; Gupta and Singh, 2004).

Fungi are sensitive to salinity (Singh, 1950). Bacterial counts were 4.9×10^3 g^{-1} soil in vermicompost unamended control soil, however, their population increased as the doses of vermicompost amendment increased. Bacterial population of 5.0×10^3 and 5.5×10^3 g^{-1} soil was recorded in 5 and 25 percent vermicompost amended soils, respectively. The population of actinomycetes also followed the same trend. However, its population was found to be minimum (3.9×10^3 g^{-1} soil) in vermicompost unamended control soils. The population of actinomycetes also increased with increase in doses of vermicompost amendments and reached to its maximum level 4.5×10^3 g^{-1} soil in the soils amended with

25 percent vermicompost. The increase in fungal, bacterial and actinomycetes population in the soils amended with vermicompost clearly indicates that the vermicompost has the ability to enrich the saline/usar soil with nutrients. Thus, the improvement in physico-chemical attributes of saline/usar soil due to vermicompost enrichment may be directly correlated to the increased microbial population in that soil. Vermicompost contains higher amounts of growth promoting substances, vitamins and enzymes, which in turn increased the microbial (bacterial, fungal and actinomycetes) population in the soil (Gunadi et al., 1999; Masciandaro et al., 2000b).

Some works pertaining to microbiological characterization of saline-sodic soil (James, 1959) and saline/contaminated soils amended with different doses of organic enrichers have also been done (Gupta and Singh, 2004; Singh, 2006). Gupta and Singh (2004) and Singh (2006) has reported that whereas the population of fungi, bacteria and actinomycetes was minimum in control soil, it increased with increasing doses of organic enrichers. The pattern of distribution of microbial population in vermicompost amended soil was found to be in the order of bacteria> actinomycetes>fungi. Gupta and Bajpai (1974) also observed the higher population and total counts of actinomycetes in salt affected soils as compared to normal soil. Some information is also available on the influence of amendment of saline-alkali soil on microbial properties (Mahmood et al., 1969). Higher bacterial population associated with the application of 75 percent vermicompost with *Azospirullum* has also been reported (Kannan et al., 2005).

3.3 Fungistatic Activity

The fungistatic behaviour of natural/vermicompost-unamended soils coupled with 5, 10, 15, 20 and 25 percent vermicompost amended soils was studied. Direct and indirect methods of soil fungistasis were used to evaluate the fungistatic activity in vermicompost unamended control soils as well as the soils amended with different doses of vermicompost. An attempt was also made to compare their efficacy of the vermicompost amendments against the test organisms, *F. solani* and *A. alternata.*

To assess the role of vermicompost amendments on fungistatic behaviour in chosen soils, the soils were amended with different doses of vermicompost. Fresh soil samples (composite samples) were collected, air-dried, sieved and taken in a screw capped bottle. After adding the respective concentrations (5, 10, 15, 20 and 25 percent w/w) of vermicompost, the soil samples were mixed thoroughly by hand manipulation and finally passed through the sieve for uniform mixing. Moisture status of the vermicompost amended soil samples was maintained to 30 percent water holding capacity by adding water. The vermicompost amended soil samples were stored at room temperature and studies on soil fungistasis were conducted after 10 days. The soil moisture was adjusted to 60 percent water holding capacity before use. Natural soil samples without any vermicompost amendment were also similarly maintained to serve as controls. All the fungistasis experiments were repeated at least twice.

After 10 days of vermicompost amendment, varying levels of sensitivity by *F. solani* and *A. alternata* towards soil fungistasis were observed. The soils amended with different doses of vermicompost, however, showed variable percent inhibition in fungistatic activity (Tables 16.6 and 16.7). In general, *F. solani* and *A. alternata* showed variable percent inhibition in different assay methods. The conidia germination in both the test fungi exhibited greater percent inhibition with the increase of the concentration of vermicompost in all the three methods adopted in the present study.

Table 16.6: Fungistatic activity (in terms of % inhibition) of chosen soils amended with vermicompost after 10 days of amendment against *F. solani*

S. No.	Doses of amendments (%)	Test Fungi: *Fusarium solani*		
		SEA	AD	DA
1	Unamended control soil	0	0	68
2	5	2	4	69
3	10	5	8	79
4	15	10	8	79
5	20	12	14	80
6	25	15	14	84

Table 16.7: Fungistatic activity (in terms of % inhibition) of chosen soils amended with vermicompost after 10 days of amendment against *Alternaria alternata*

S.No.	Doses of amendments (%)	Test Fungi: *Alternaria alternata*		
		SEA	AD	DA
1	Unamended control soil	0	0	51
2	5	2	3	53
3	10	5	6	58
4	15	7	6	62
5	20	10	8	68
6	25	10	10	78

(SEA: Soil Emanation Method, AD: Agar Disc Method, DA: Direct Assay Method)

The emanation of inhibitory volatile components seemed to affect the germination of conidia of *F. solani* because in almost all the soils amended with different doses of vermicompost, soil emanation method exhibited greater inhibition over the vermicompost unamended control soils. Maximum percent inhibition was observed in the soil amended with 25 and 15 percent vermicompost. However, the soil amended with 5 percent vermicompost showed only 2 percent inhibition in germination of the conidia. Similar results were also observed with *A. alternata*. However, in *A. alternata* maximum percent inhibition (10 percent each) was recorded in the soils amended with 20 and 25 percent vermicompost.

Similar emanation of inhibitory volatile components was also recorded in the soils amended with in oil cakes (Agarwal, 1983). In this study, an irregular sensitivity of *F. solani* in response to different oil cakes amendments was also recorded. Lewis and Papavizas (1975) and Lockwood (1977) stressed that energy sources and other organic materials added to the soil may reduce or completely annul fungistasis initially. After a week or more, a long-term effect may become evident as manifested by an enhancement of fungistasis.

In the present study, the emanation of volatiles as a cause of inhibition of conidia germination in *F. solani* may be due to higher nutritive content of vermicompost. Mac Millan (1956) has suggested that in the immediate vicinity of a nitrogen rich substrate, a considerable amount of ammonia is released, which had inhibitory effect on the germination of fungal propagules.

It is evident from the data of Tables 16.6 and 16.7 that percent inhibition of conidia germination in *F. solani* and *A. alternata* was found to be almost similar in soil emanation agar method as well as in agar disc method. Whereas, maximum percent inhibition of conidia germination (10%) in *A. alternata* was recorded in the soil amended with 25 percent vermicompost, *F. solani*, however, showed its maximum percent inhibition of conidia germination (14 per cent each) in the soils amended with 20 and 25 percent vermicompost. Similar findings were also observed by Singh (2002), while working with contaminated/ degraded soils amended with rice husk compost. Composting is the best natural approach to reduce the plant pathogens and increase the crop yield (Mehta et al., 2014).

As evident from the data of direct assay method that comparatively higher percent inhibition of conidia germination in *F. solani* and *A. alternata* was recorded in the soils amended with vermicompost over the vermicompost unamended control soils and fungistastic activity increased with increase in vermicompost concentration. In case of direct assay method, *F. solani* showed greater percent inhibition of conidia germination in comparison to *A. alternata* in the soils amended with vermicompost over the vermicompost unamended control soils. *F. solani* showed minimum percent inhibition (69 percent) of conidia germination in the soils amended with 5 percent vermicompost. However, this value increased with the increase in the doses of vermicompost amendments and attained its maximum level (84%) in the soils amended with 25 percent vermicompost. *A. alternata* showed maximum percent inhibition of conidia germination (78 percent) in the soils amended with 25 percent vermicompost.

Chinn (1967), Kanaujia (1973), Pandey (1976) and Agrawal (1983) observed that fertilizer amendments of the soil resulted in enhancement of fungistasis. The increased level of fungistasis in fertilizer amended soil was attributed to the increased fungal population in such soil. Likewise, in the present study, higher population of fungi, bacteria and actinomycetes were recorded in the soils amended with varying concentration of vermicompost) (Table 16.5).

The fungistatic activity as observed in various assay methods can be attributed in the following order: direct assay method > agar disc method > soil emanation agar method. These results were also supported by the findings of earlier workers (Dixit, 2002; Singh, 2002).

Fungistasis at initial doses of vermicompost amendment was positively correlated with the microbial population. Contrary to the present findings, an annulment of fungistasis has been widely observed by the addition of organic substances or energy rich compounds to the soil (Chin and Ledingham, 1957; Powelson and Patil, 1963; Agnihotri and Vaartaja, 1967; Papavizas and Adams, 1969; Hsu and Lockwood, 1973; Lockwood, 1977; El-Abyad and Ismail, 1984; Singh, 2002 and Dixit, 2002). It has been proved that vermicompost act as an organic fertilizer as well as biological control agent by suppressing many plant-borne pathogens and pests also (Yurdagul, 2011).

4. CONCLUSION

The growing volume of research conducted on vermicompost production and utilization in recent years serves as an indicator of increasing interest in vermicompost's potential inclusion as a soil amendment and as waste management option for agricultural system using earthworms. This study attests the multiple benefits associated with vermicompost application to saline soils, including improved porosity, water holding capacity, nutrients levels with reduced electrical conductivity and exchangeable sodium content.

Further, research is needed to more fully elucidate some of these constraints to indicate situations, when vermicompost production and use might be economically viable. Indeed, closer scrutiny of vermicompost's integral role as a waste management technology and as a soil amendment in suitable and transitional farming systems is advisable prior to its widespread promotion and adaptation.

References

Agarwal AK (1983). Ecological studies on soil fungi with special reference to fungistasis and biological control. Ph.D. Thesis, Jiwaji University, Gwalior, India.

Agnihotri VP, and Vaartaja O (1967). Effect of amendments, soil moisture contents and temperature on germination of *Phythium* sporangia under influence of soil mycostasis. J Phytopathol 57:1116–1120.

Ajit SN, and Hooda IS (2002). Influence of integrated use of organic manures and inorganic fertilizers on wheat yields and soil properties. Res Crops 3(1):11–16.

Albanell E, Plaixats J, and Cabrero T (1988). Chemical changes during vermicomposting (*Eisenia fetida*) of sheep manure mixed with cotton industrial wastes. Biol Fertile Soil 6: 266–269.

Alexander M (1971). Microbial Ecology. John-Willey and Sons, Inc., New York.

Aneja KR (1996). Experiments in microbiology, plant pathology, tissue culture and mushroom cultivation. 2nd edn. Wishwa Prakashan, New Age International (P) Ltd., New Delhi.

Ansari AA (2002). Vermitechnology: A potential technology for reclamation of sodic soil. Ph.D. Thesis, University of Madras, India.

Badale SB, and More SD (2000). Soil organic carbon status as influenced by organic and inorganic nutrient source in vertisol. J. Maharastra Agricul. Univ. 25 (2): 220–222.

Balaji SK (1994). Effect of vermicompost on growth and flower yield of China aster (*Casllistephus chinensis*). M.Sc. (Ag.) Thesis, University of Agricultural Sciences, Dharwad, India.

Bharadwaj A, Agarwal AK, and Singh S (2007). Comparative efficacy of compost and vermicompost on physico-chemical characteristics of salt affected soils. Flora & Fauna 13:335–339.

Bhargava GP, and Abrol IP (1978). Characteristics of some typical salt affected soils of Uttar Pradesh. Report No. 6, CSSRI, Karnal, India.

Chinn SHF, and Ledingham RJ (1957). Application of a new laboratory method for the determination of the survival of *Helminthosporium sativum* spores in soil. Can J Bot 35: 697–701.

Chinn SHF (1967). Differences in fungistasis in some Saskatchewan soils with special reference to *Cochliobolus sativus*. J Phytopathol 57: 224–226.

Cohen C (1949). Fungi as indicators of soil conditions. Nature 164: 408–409.

Coleman DA (1985). Environmental factors influencing the activities of soil fungi. J Soil Sci 2:1–65.

Dargan KS, Singh OP, and Gupta OC (1982). Crop production in salt affected soils. Oxford and IBH Publ. Co., New Delhi, p. 276.

Dixit P (2002). Soil nutrient enrichment with Eucalyptus plant residue and its impact on soil microbes with emphasis on soil fungistasis. Ph.D. Thesis, Dr. B.R. Ambedkar University, Agra, India.

El-Abyad MS, Ismail IK (1984). Influence of adding host and other plant amendments to natural soil on the fungistatic sensitivity of some fungi. Z Microbiologie 139: 43–48.

Elvira C, Sampedro L, Benitez E, and Nogales R (1998). Vermicomposting of sludges from paper mill and dairy industries with *Eisenia andrei*: A pilot-scale study. Biores Technol 63: 205–211.

Garg VK, and Singh B (2003). Macronutrient dynamics and use efficiency in three species of short rotation forestry development on sodic soils in north India. J Trop Forest Sci 15(2): 289–302.

Garrett SD (1956). Biology of root infecting fungi. Cambridge University Press, London.

Garrett SD (1963). Soil fungi and soil fertility. Pergamon Press, Oxford.

Gautam M (2002**).** Studies on the role of organic amendments in reclamation and management of saline/alkali soils of Mathura. Ph.D. Thesis, Dr. Bhim Rao Ambekar University, Agra, India.

Giri B, Giang PH, Kumari R, Prasad R, and Varma A (2005). Microbial diversity in soils. In: Buscot F, Varma A (eds) Microorganisms in soils: Roles in Genesis and Functions. Springer-Verlag Berlin Heidelberg 19–58.

Goel A (2007). Physico-chemical properties of sodic soils of Lucknow and their sustainable utilization through organic amendments. Ph.D.Thesis, Dr. Bhim Rao Ambekar University, Agra, India.

Gunadi B, Blount C and Edwards CA (1999). The growth and fecundity of *Eisenia fetida* in cattle solids precomposted for different periods. Pedobiol 46 (1): 15–33.

Gupta BR, and Bajpai PD (1974). Microbiological and chemical alterations in salt-affected soil under the influence of some inorganic ameliorants. J Indian Soc Soil Sci 17 (2): 66–68.

Gupta RN, and Singh GN (2004). Sodic land reclamation: Overview of Uttar Pradesh. In: Proceedings International Conference on Sustainable Management of Sodic Lands, U.P. Council of Agriculture Research, Lucknow, pp. 6–11.

Hora TS, and Baker R (1970). Volatile factor in soil fungistasis. Nature (London) 225: 1071–1072.

Hsu SC, and Lockwood JL (1973). Soil fungistasis behavior of nutrient independent spores and sclerotia in a model system. J Phytopathol 63: 334–337.

Iswaran V. and Marwaha S. (1980). A modified rapid Kjeldahl method for determination of total nitrogen in agricultural and biological materials, Geobios, 7(6): 281-282.

Jackson ML (1973). Soil chemical analysis. Prentice-Hall India Pvt. Ltd. New Delhi, India.

Jackson RM (1958). An investigation on fungistasis in Nigerian soils. J Gen Microbiol 18: 248–258.

James N (1959). Plate counts of bacteria and fungi in a saline soil. Can J Microbiol 5: 431–436.

Johnson LE, and Curl EA (1972). Methods on research on ecology of soil born plant pathogens. Burges Publishing Co., Minneapolis, Minnesota.

Joshi RC, Haokip DD, and Singh KN (1990). Effect of green manuring on the improvement of physical conditions of soil under rice-wheat cropping system. Proceedings International Symposium on National Resource Management for Sustainable Agriculture, New Delhi, pp. 61–62.

Kanaujia RS (1973). Studies on succession of fungi on root regions of *Pennisetum typhoides (Burm f.)* stapf and Hubb. Ph.D. Thesis, University of Gorakhpur, Gorakhpur, India.

Kannan P, Saravanan A, Krishnakumar S, and Natarajan SK (2005). Biological properties of soil as influenced by different organic manures. J Agricul Biol Sci 1(2):181–183.

Katyal JC (1993). Integrated nitrogen management and supply: An overview. Proc Indian Natl Sci Acad Part- B 59: 161–172.

Kaviraj, and Sharma S (2003). Municipal solid waste management through vermicomposting employing exotic and local species of earthworms. Biores Technol 90: 169–173.

Lewis JA, and Papavizas GC (1975). Survival and multiplication of soil-borne plant pathogens as affected by plant tissue amendments. Pages 84-89 In: Bruchl GW (ed) *Biology and Control of Soil Borne Plant Pathogens*. Am Phytopathol Soc, St. Paul. Minn, p. 216.

Lingappa BT, and Lockwood JL (1963). Direct assay of soils for fungistasis. J Phythopathol 53: 529–531.

Lockwood JL (1977). Fungistasis in soils. Biol Rev 52: 1–43.

Mac-Millian A (1956). The entry of ammonia into fungal cells. J Exp Bot 7: 113–126.

Maheswarappa HP, Nanjappa HV, and Hegde MR (1999). Influence of organic manures on yield of arrowroot, soil physico-chemical properties when grown as intercrop in coconut garden. Ann Agricul Res 20(3): 318–323.

Mahmood K, Malik KA, Lodhi MAK, and Sheikh KH (1969). Seed germination and salinity tolerance in plant species growing on saline waste lands. Biologia Plant 38 (2): 309–315.

Mallik MAB, and Rice EI (1966). Relation between soil fungi and seed plant in three successional forests communities in Oklahoma. Bot Gaz 127: 120–127.

Martin JP (1950). Use of acid, rose bengal and streptomycin in the plate method for estimating soil fungi. J Soil Sci 69: 215–232.

Masciandaro G, Ceccanti B, and Garcia C (2000b). "*In situ*" vermicomposting of biological sludges and impacts on soil quality. Soil Biol Biochem 32: 1015–1024.

Mehta CM, Palni U, Franke-Whittlec IH, and Sharma AK (2014). Compost: Its role, mechanism and impact on reducing soil borne plant diseases. Waste Management 34 (3) 607–622.

Ohi ME and Ebo PO (1995). The effects of organic amendments on soil physical properties and maize production in a severely degraded sandy soil in Southern Nigeria. Biores Technol 51: 117–123.

Pandey KK (1976). Investigation of soil fungistatis and effects of fertilizer amendments on soil fungistatis. Fertil Technol 13: 306–308.

Papavizas GC, and Adams PB (1969). Survival of root infecting fungi in soil. X. Sensitivity of propagules of Thielaviopsis basicola to soil fungistasis in natural and alfalfa amended soil. J Phytopathol 59: 135.

Peyronel B (1956). Cosiderazioni sulle micro-cenosi esui metodi per studiarle. Allionia, 3: 85–109.

Piper CS (1966). Chemical analysis saline soil. Soil and Plants Analysis. Hans Publication, Bombay, India.

Powelson RL and Patil SS (1963). Influence of oat and alfalfa residues on soil respiration, fungistasis and survival of *Verticillium albo-atrum*. J Phytopathol 53: 1141.

Ram H and Prasad R (1982). Effect of pyrites on properties of alkali soil. In: Proceedings of FAI-PPCL-GADA Seminar on Management of Salt Affected Calcarious Soils. IARI, Pusa, New Delhi, pp. 91–95.

Ranganathan LS, and Parthasarathi K (2000). Enhanced phosphatase activity in earthworm casts is more of microbial origin. Curr Sci 79: 1158–1159.

Ranganathan LS, and Vinotha SP (1998). Influence of pressmud on the variations in the different reproductive stages of *Eudrilus eugeniae.* Curr Sci 74: 634–635.

Robbins CW (1986). Sodic caicarious soil reclamation as affected by different amendments and crops. J Agron 78: 916–920.

Saksena RK, Krishna Nand, and Sarbhoy AK (1966). Ecology of the soil fungi of Uttar Pradesh IV. Bundelkhand and Gangetic tract soils. Proc Natl Inst Sci India. 33 B: 5–6.

Saksena SB (1955). Ecological factors governing distribution of soil microfungi in some forest soils of Sagar. J Ind Bot Soc 24: 262–298.

Sharma BD, Kumar R, Arora H, and Sethi M (2004). Micronutrient distribution in salt affected southern western district of Punjab and their correlation with soil properties. In: Proceedings International Conference Sustainable Management of Sodic Lands. UP Council of Agricultural Research, Lucknow, pp. 207–210.

Shukla AK, and Mishra PN (1993). Improvements of sodic soils under tree cover. Indian Fores 119: 43–52.

Singh B, Garg VK, Singh PK, and Tripathi KP (2004). Diversity and productivity effect on amelioration of afforested sodic soils. Indian Fores 130: 14–25.

Singh A (2002). Studies on the microbial ecology of contaminated and degraded soils with emphasis on bioremediation. Ph.D. Thesis, Dr. Bhim Rao Ambekar University, Agra, India.

Singh B, Goel VL, Srivastava N, Tripathi KP, and Behl HM (2000). Development of social forestry plantations on sodic wastelands. In: Proceedings International Conference on Managing Natural Resources for Sustainable Agricultural Production in the 21st Century. IARI, New Delhi, (India), pp. 1005–1007.

Singh RN (1950). Reclamation of 'Usar' lands in India through blue-green algae. Nature (London). 165: 325–326.

Singh R (2006). Impact of compost enrichment on physico-chemical and microbiological characteristics of degraded soils with emphasis on remediation. Ph.D. Thesis, Dr. Bhim Rao Ambedkar University, Agra, India.

Somani LL (1981). Reclamation of a saline alkali soil using organic materials and inorganic amendments. Agoken Talajt 30: 333–350.

Subba Rao NS (1986). Soil Microbiology. 4th edn.: Soil microorganisms and plant growth. Oxford and IBH Publishing Company Pvt. Ltd., New Delhi.

Subler S, Edward CA, Metzger J (1998). Comparing vermicompost and compost. BioCycle 39: 63–66.

Tansley AG (1946). Introduction to plant ecology. George Allen and Unwin Ltd., London.

Tresner HD, Backus MP, and Curtis JT (1954). Soil microfungi in relation to hard wood forest continnum in southern Wisconsin. Mycologia 46: 314–333.

Venkatesh S (1995). Effect of vermiculture on soil composition, growth yield and quality of Thompson seedless grapes (*Virtis Vinifera*). M.Sc. (Ag.) Thesis, Univ Agricul Sci Dharwad.

Waksman SA (1917). Is there a fungus flora of the soil. J Soil Sci 3: 565–589.

Warcup JH (1950). The ecology of soil fungi. Trans. Brit Mycol Soc 34: 376–399.

Warman PR, and AngLopez MJ (2002). The chemical properties of vermicompost derived from different feed stocks. In: Proceedings International Composting and Compost Science Symposium, Columbus, Ohio, 6-8 June.

Weaver JE, and Clements FE (1938). Plant Ecology. McGraw-Hill Book Co., London and New York.

Yurdagul SE (2011). The use of vermicompost products to control plant diseases and pests. Soil Biology 24: 191–213.

Index

N

O

P